APPLIED THERMODYNAMICS

Combustion: An International Series

Norman Chigier, *Editor*

Bayvel and Orzechowski, Liquid Atomization
Chigier, Combustion Measurements
Kuznetsov and Sabel'nikov, Turbulence and Combustion
Lefebvre, Atomization and Sprays
Li, Applied Thermodynamics: Availability Method and Energy Conversion

Forthcoming Titles
Chen and Jaw, Fundamentals of Turbulence Modeling
Libby, Introduction to Turbulence

APPLIED THERMODYNAMICS: Availability Method And Energy Conversion

Kam W. Li
North Dakota State University
Fargo, North Dakota

Taylor & Francis
Publishers since 1798

USA	Publishing Office:	Taylor & Francis 1101 Vermont Avenue, N.W., Suite 200 Washington, DC 20005-3521 Tel: (202) 289-2174 Fax: (202) 289-3665
	Distribution Center:	Taylor & Francis 1900 Frost Road, Suite 101 Bristol, PA 19007-1598 Tel: (215) 785-5800 Fax: (215) 785-5515
UK		Taylor & Francis Ltd. 4 John St. London WC1N 2ET Tel: 071 405 2237 Fax: 071 831 2035

APPLIED THERMODYNAMICS: Availability Method and Energy Conversion

1 2 3 4 5 6 7 8 9 0 EBEB 9 8 7 6 5

This book was set in Times Roman by Technical Typesetting Inc. The editor was Seth A. Maislin. The pre-press supervisor was Bonny Gaston. Cover design by Michelle Fleitz. Printing and binding by Edwards Brothers, Inc.

A CIP catalog record for this book is available from the British Library.

∞ The paper in this publication meets the requirements of the ANSI Standard Z39.48-1984 (Permanence of Paper)

Library of Congress Cataloging-in-Publication Data

Li, Kam W.
Applied thermodynamics: availability method and energy conversion
/Kam W. Li
p. cm.
Includes index.

1. Thermodynamics. I. Title. II. Series: Combustion (New York, N.Y.: 1989)
TJ265.L47 1995
621.402′1—dc20 95-23915
CIP

ISBN 1-56032-349-3
ISSN 1040-2756

CONTENTS

PREFACE

The textbook is the outgrowth of the author's consulting work and teaching in electric power generation and applied thermodynamics. It was written to meet the needs of engineering students and practicing engineers. There are many research papers and reports in the area of availability (exergy) analysis and yet few textbooks are available to introduce this subject. It is the author's observation that engineers are still very much uncomfortable with the availability (exergy) method and do not apply it frequently to their designs and analyses.

The purpose of the textbook is to present the availability method and its application in general to power plant system designs and energy conversions. The author's approach is to keep the text easy-to-follow. It starts with concepts engineers feel comfortable with, such as the conservation of energy, and reviews some basic thermodynamics. Also reviewed are reversibility and irreversibility of processes and the works produced under these conditions. Of the different types of work, the available work is the most critical and important in the development of the availability method. The available work is the work from the process that is internally and externally reversible. It represents the maximum amount of work that would be produced by a device between two given states. Once the concept of available work is formulated, it will be used as a basis to develop the concept of availability, which may be either thermomechanical or of the diffusion and chemical types.

While the methodology is adequately presented in the text, emphasis will be placed on the applications of the availability method. It is the author's intention that for every conversion process such as turbine, pump, and heat exchanger, there will be at least one illustrative example. The availability method is also applied to various conversion systems, including the coal-fired power plant, the combustion engine, and the cogeneration plant. Again, examples are prepared for each energy conversion system. It is believed that by this approach both practicing engineers and students would have a better understanding of the availability method.

The scope of the textbook includes (1) the development of working equations needed for an availability analysis and (2) its applications to the energy conversion processes as well as systems. Basically, the materials in the text are organized into two parts. The first part covers the development and formulation of the availability method and consists essentially of Chapters 2 and 3. The second part is the application of the availability method. Chapters 4, 5, and 6 present an availability analysis of energy conversion processes. These processes included expansion and compression, cooling and heating, compressible and incompressible fluid flow, reactive and nonreactive, and so on. Some of the materials in Chapters 5 and 6 may be treated as further formulation of availability concepts and methods. The applications of the availability method to various energy conversion systems are in the last five chapters. These include systems for electric-power generation, heat pump, refrigeration, and cogeneration. The materials in the application areas are presented in such a way that omitting some would not in any way disturb the continuity of the text.

The textbook is intended for use at the undergraduate and beginning graduate levels. It should provide sufficient materials, including homework, for one 3-semester-credit course in universities and colleges. The prerequisites are the first course of thermodynamics, heat transfer, and fluid mechanics. This book is also suitable as a reference for engineers in consulting engineering firms and in utility as well as manufacturing companies.

During the preparation of this text, readers were foremost in the author's mind. His objective was to develop in readers the awareness of the availability method and its usefulness in engineering analysis and design. Efforts were made to demonstrate by examples the application of the availability method. No claim is made for complete originality of the text. The text has been tested twice at North Dakota State University. The helpful input from students and faculty over the two-year period is greatly appreciated.

The author would like to take this opportunity to thank the North Dakota State University for the creative and challenging environment in which he has been fortunate to work for many years. In particular, thanks are given to Patricia Kallis and Donna Alby for their patience and skill in the preparation of this manuscript. Finally, the author is very grateful to his wife, Margaret, and children, Christopher and Charles, without whose love and support this book would not have been written.

NOMENCLATURE

LIST OF SYMBOLS

a, A	specific availability, availability
a_f	flow availability
A/F	air-fuel ratio
c	unit cost
cop	coefficient of performance
c_p	specific heat at constant pressure
c_v	specific heat at constant volume
e, E	energy per unit of mass, energy
F	fuel consumption
g	acceleration of gravity
g, G	Gibbs function per unit of mass, Gibbs function
g_c	constant that relates force, mass, length, and time
h, H	enthalpy per unit of mass, enthalpy
$\bar{h}_f^\circ$	enthalpy of formation per mole at standard state
HHV	higher heating value
I, $\dot{I}$	irreversibility, irreversibility rate
k	specific heat ratio: c_p/c_v
k_l	temperature ratio
ke, KE	kinetic energy per unit of mass, kinetic energy
LHV	lower heating value
lw, LW	lost work per unit mass, lost work
m	mass
$\dot{m}$	mass flow rate
M	molecular weight

mf	mass fraction
n	number of moles and polytropic exponent
p	pressure
p_i	pressure associated with mixture component i, partial pressure of i
p_r	reduced pressure: p/p_c
pe, PE	potential energy per unit of mass, potential energy
q, Q	heat transfer per unit mass, heat transfer
$\dot{Q}$	heat transfer rate
r_v	compression ratio
R	gas constant: $\overline{R}/M$
$\overline{R}$	universal gas constant
s, S	entropy per unit of mass, entropy
s°, s'	entropy function as used in gas tables, absolute entropy at the standard reference state
T	temperature
T_r	reduced temperature: T/T_c
u, U	internal energy per unit of mass, internal energy
v, V	specific volume, volume
V	Velocity
v_r	relative volume as used in gas tables
w, W	work per unit mass, work
$\dot{W}$	rate of work, or power
x	quality
z	elevation
Z	compressibility factor and firing ratio

GREEK LETTERS

α	cutoff ratio and heat to electric energy ratio
β	the second law efficiency and availability ratio
Δ	change: final minus initial
ε	second law efficiency or effectiveness
ϕ	relative humidity
η	efficiency
ρ	density
ρ_p	insentropic temperature ratio
σ, $\dot{\sigma}$	entropy production, rate of entropy production
Σ	summation
τ	time
ω	humidity ratio

SUBSCRIPTS

a	dry air
act	actual
ad	adiabatic
ava	available
b	burner and boiler
c	property at the critical point, compressor, and Carnot cycle
cg	cogeneration
cy	cycle
cv	control volume
d	diffuser and dead state
db	dry bulb
e	state of a substance exiting a control volume
f	fuel, property of saturated liquid and final value
fg	difference in property for saturated vapor and saturated liquid
g	gas and property of saturated vapor
gt	gas turbine
h	high temperature reservoir
i	state of a substance entering a control volume, mixture component
i	initial value
int	internal
j	portion of the boundary, number of components present in a mixture
l	low-temperature reservoir
n	net change and nozzle
p	pump and products
q	heat transfer
r	reservoir and reactants
ref	reference value or state and refrigerator
reg	regenerator
res	reservoir
rev	internally reversible
s	isentropic and steam
sat	saturated
st	steam, steam turbine
sur	surroundings
t	turbine
tot	total
o	property at the environmental state and overall system
v	vapor
w	water and work
wb	wet bulb
ext	external
1, 2, 3	different states of a system, different locations in space

SUPERSCRIPTS

ch	chemical contribution to availability
di	diffusion contribution to availability
th	thermomechanical contribution to availability
–	bar over symbol denotes property on a molar basis and normalized values
·	dot over symbol denotes time rate
∘	property at standard state
*	ideal gas

CHAPTER
ONE
INTRODUCTION

Ever since the dawn of civilization, energy consumption has been increasing. It was especially evident in the last several decades. Every indication shows that the energy consumption will continue its upward trend in the long run. The reason for this trend is quite simple: As the world population increases, energy consumption will increase. More important, the people in every country have a natural desire to improve their living standard, which is closely related to the per capita energy consumption. Although per capita consumption has been on the increase everywhere, there are great differences in present consumption levels from country to country. For instance, the United States per capita consumption is approximately six times the world average. To bring the whole world up to the level of the United States would require a tremendous increase in the energy consumption.

There is a close relationship between energy consumption and disposable income. (Disposable income may be defined as the gross national product minus the sum of all taxes, undistributed corporate profit, transfer payments, etc.) As the disposable income increases, the energy consumption will increase. This relationship is especially evident in the United States. Because of this, as a country becomes more prosperous, the demand for energy will certainly grow whether or not the population is increased.

For industrial countries such as the United States, several policy options are available to meet the future energy demand. These options include the increased production of coal, the exploration for more oil and gas, the further development of nuclear power, the use of renewable resources of energy (such as solar and wind), and the general conservation of energy. There are many ways to approach energy conservation, including improvement in the efficiency of energy conversion systems, reduction of energy loss by insulation, and the use of waste products as fuel. In general, the best approach to energy conservation essentially is the efficient use of energy resources.

1-1 ENERGY CONSERVATION

The general objective of energy conservation is an efficient use of energy resources. Efficient energy use means low energy consumption for a given product or operation. It also means a low per capita energy consumption for the same standard of living. The importance of energy conservation is evident when consideration is given to the cost and future availability of the fuel. Any reduction in energy demand also has some positive impact on the environment. Not only is pollution reduced, but also more energy resources are conserved for future use.

The potential for a substantial reduction in energy consumption exists almost everywhere. Take the electric-power generating station as an example. In central power stations only about one-third of the energy in coal or oil appears as electric power, and two-thirds of the energy is thrown away in the form of lukewarm water to a cooling tower, to rivers, or to the sea. It is estimated that approximately 40 percent of the world energy consumption is devoted to the production of electricity. Because of this, any improvement in the electric-power generating station will result in a large fuel saving. Table 1-1 presents the fuel utilization factors for various processes.

Potential fuel savings can be realized through a rational system design. For an electric-power generating station, one may utilize the high-pressure and high-temperature steam. Use of a regenerator such as a feedwater heater also will result in an improvement of station efficiency. One convenient approach in power station system design is to utilize the topping-cycle or bottoming-cycle heat engines. When the gas turbine (Brayton cycle) is treated with a topping-cycle heat engine and combined with a steam turbine (Rankine cycle), a higher efficiency will generally result. Figure 1-1 presents a schematic diagram for a

Table 1-1 Fuel utilization factors for various processes

Processes	Fuel utilization factor (%)
Electric power generation	33.0
Residential heater	9.0
Domestic water heater	2.0–3.0
High-pressure steam boiler	50.0
Tobacco dryer	4.0
Coal gasification	46.0
Petroleum refining	10.0
Blast furnace	46.0

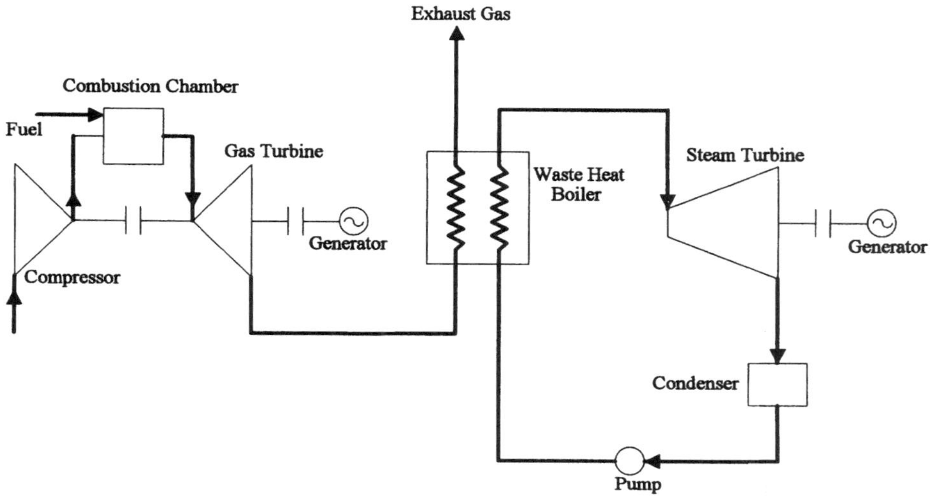

Figure 1-1 Schematic diagram for a combined cycle.

combined cycle. It is seen that the hot exhaust gas from the gas turbine is utilized to generate the steam, which is in turn used to drive the steam turbine. In this combined system, combustion of the fuel is effected only at one point in the cycle, namely in the combustion chamber of the gas turbine, and the cycle work is produced at two different places.

Many industrial processes currently reject low-temperature (300–700°F) waste heat. Occasionally, the heat is fed back into the process by means of recuperators or used to generate steam. However, more often the waste heat is not utilized at all because of marginal economics. An attractive method for converting waste heat to electricity or mechanical power is afforded by the organic Rankine engine used in a bottoming cycle. In this system, an organic vapor such as Freon-12 is generated in a boiler that is fueled by the low-grade energy. The vapor then enters a turbine for work production and condenses in a condenser. Eventually the substance will be pumped back into the boiler in liquid phase to complete its Rankine cycle. Figure 1-2 presents a binary cycle coupling an organic Rankine cycle with a steam Rankine cycle. It is seen that the organic Rankine cycle will serve as the bottoming cycle. In this combined system, the device coupling these two Rankine cycles is a heat exchanger in which steam is condensed and the organic substance is changed from liquid to vapor. To increase the efficiency of the organic side, organic vapor is usually superheated. The superheater frequently is located in the steam generator. Because the organic Rankine cycle is fueled by the waste heat from the steam Rankine cycle, the combined system (Fig. 1-2) is expected to have a high efficiency.

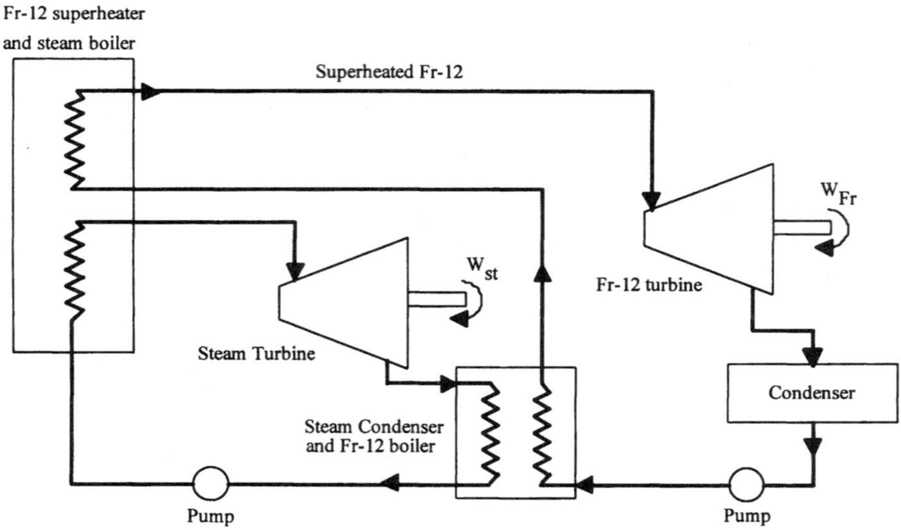

Figure 1-2 Schematic diagram for a binary cycle.

Steam raising is an important process in various industries. It is estimated that the bulk of industrial fuel—approximately 45 percent—is consumed in producing steam. The process usually is very ineffective, and the fuel utilization factor is around 10 percent. If there is any improvement in the steam raising process, there will be a substantial amount of fuel savings. One approach is to design and utilize a cogeneration system to simultaneously produce electricity and process steam. Figure 1-3 shows two basic cogeneration methods. In Fig. 1-3(a), steam is produced at a higher pressure and temperature than required by the steam users and is brought down to the needed conditions through a back-pressure steam turbine. This cogeneration system usually is designed to meet the demand of the thermal energy and to produce electricity as a by-product. In Fig. 1-3(b), the cogeneration system is similar to the conventional power plant. Some of the steam is extracted from the turbine for the steam users, thus eliminating or reducing the use of a separate boiler. In this cogeneration system, the steam is a by-product. The cogeneration system generally has a fuel utilization factor around 80 percent.

There also is great potential for saving fuel in various manufacturing areas. Gyftopoulos, Lazaridis, and Widmer, in their study, "Potential Fuel Effectiveness in Industry" [A-5], indicated the specific fuel consumption for five industrial products as shown in Table 1-2. They suggested that the specific fuel consumption will be greatly reduced if some recently demonstrated technologies are adopted. In their study they calculated the theoretical minimum fuel consumption by the thermodynamic availability method. Compared with the current values, the theoretical minimum fuel consumption will provide the upper limit for fuel savings.

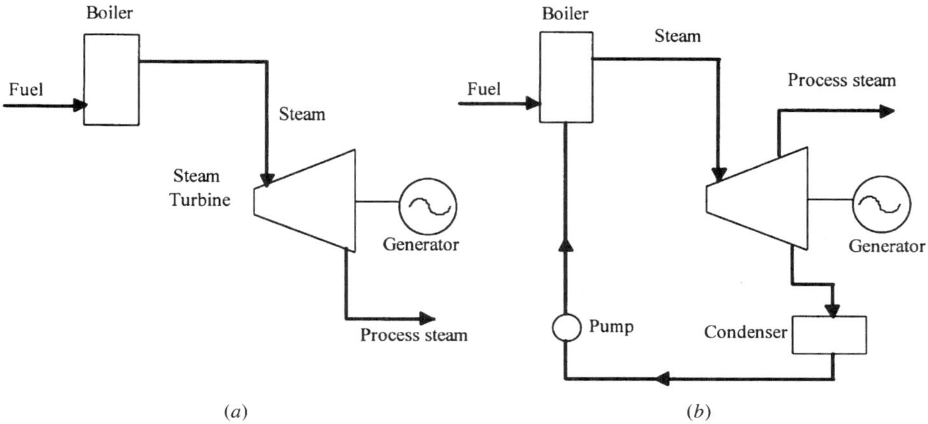

Figure 1-3 Schematic diagram for two basic cogeneration methods: (*a*) electricity as by-product, (*b*) steam as by-product.

1-2 ENERGY CONVERSION AND THE AVAILABILITY METHOD

Energy conversion is one of the main subjects in the science of thermodynamics. Traditionally, it covers such diverse areas as electric-power generating systems, refrigerations and heat pumps, turbines and compressors, nozzles and diffusers, and so on. Some of the conversion processes also may involve chemical reactions and the combustion of hydrocarbons and other industrial fuels. A good example in this area is the coal gasification plant. Recently, unconventional energy conversion systems have become more visible and popular. These include

Table 1-2 Comparison of specific fuel consumption of known processes with theoretical minimum for selected U.S. industries [A-5]

Processes	1968 specific fuel consumption (Btu/ton)	Potential specific fuel consumption using technology existing in 1973 (Btu/ton)	Theoretical minimum specific fuel consumption based upon thermodynamic availability analysis (Btu/ton)
Iron and steel	26.5×10^6	17.2×10^6	6.0×10^6
Petroleum refining	4.4×10^6	3.3×10^6	0.4×10^6
Paper	39.0×10^6 [a]	23.8×10^6 [a]	Greater than -0.2×10^6 [c] Smaller than 0.1×10^6
Primary aluminum production[b]	190.0×10^6	152.0×10^6	25.2×10^6
Cement	7.9×10^6	4.7×10^6	0.8×10^6

[a] Includes 14.5×10^6 Btu/ton of paper produced from waste products consumed as fuel by paper industry.
[b] Does not include effect of scrap recycling.
[c] Negative value means that no fuel is required.

systems that utilize solar, wind, tidal, ocean, and geothermal power. To improve the energy conversion efficiency, engineers frequently combine two different systems and develop a new, more complicated conversion system. The combined system generally results in high fuel utilization and other benefits.

The general objective in the design and analysis of energy conversion systems is to produce a reliable facility capable of operating at high efficiency. By efficiency we generally mean the ratio of the desired output to the input required to achieve that output. For instance, in an electric-power generating system, the desired output is the produced electricity made available to the power users. The required input is simply the energy of the fuel supplied and consumed in the system. In equation form, the efficiency of an electric-power generating system is

$$\eta_0 = \frac{\dot{W}}{\dot{m}_f(\text{LHV})}$$

where $\dot{W}$ is the electric power output, $\dot{m}_f$ the fuel consumption, and LHV the fuel low heating value. Table 1-3 presents the typical conversion efficiencies of various electric-power generating systems. It is seen that the efficiencies vary over a wide range, with the fossil fuel steam plant being the most efficient and the simple gas turbine system the least efficient. In general, these system efficiencies are relatively low and thus result in low fuel utilization.

Capital cost also is an important factor in the design and analysis of an energy conversion system. While the system efficiency affects the annual fuel cost, the capital cost affects the annual fixed cost. Especially at times when the interest rate is relatively high, the capital cost of the energy conversion system may play a more important role than that of the fuel cost. Figure 1-4 indicates a general relationship of the cost components for several alternative designs. It is seen that capital investment increases with efficiency. The benefit of such increase will be realized in lower fuel costs. The operation and maintenance generally remain unchanged or at least insensitive to the change of capital investment. The selection of alternative designs usually is made by comparing the total annual cost. The optimal design frequently is not at a maximum efficiency but at a minimum annual cost.

Table 1-3 Typical conversion efficiency of various electric-power generating systems

Generation type	Unit size (mW)	Thermal efficiency (%)
Steam (oil)	200–800	32–40
Steam (coal)	—	30–38
Steam (BWR and PWR)	500–1100	31–34
Gas turbine (open cycle)	50–100	22–28
Combined gas/steam turbine	300–400	36–40
Diesel engine	10–30	27–30

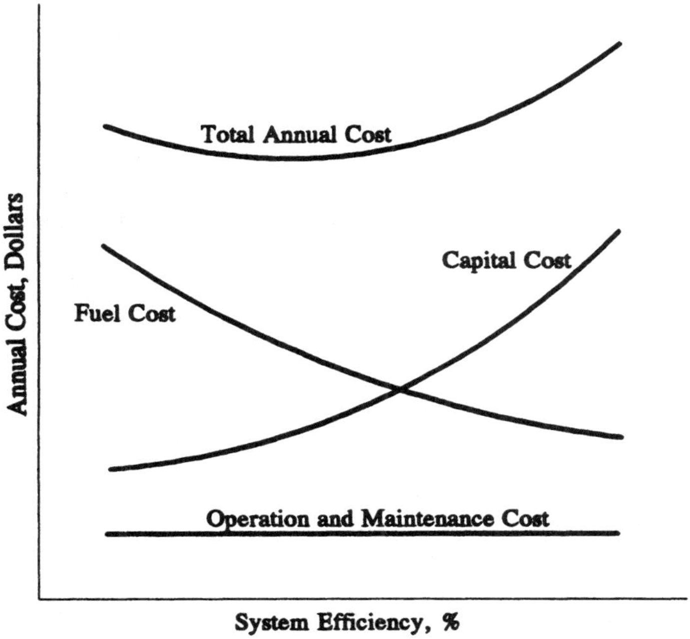

Figure 1-4 General relationship between the annual cost and system efficiency.

Impact of energy conversion systems on the environment is of great concern to all engineers. The magnitude of waste heat is significant because of low system efficiency. Take a coal-fired power generating system as an example. For every kilowatt of electricity generated, there are approximately two kilowatts of waste heat. The availability of cooling water frequently is considered to be an important factor for plant site selection. The plant cooling system affects not only the plant performance but also the plant environment. Air pollution associated with coal-fired power plant operation is a serious concern. Of all air pollutants emitted from coal-fired power plant, sulfur dioxide (SO_2) is the most hazardous. Illness from SO_2 may be acute (short-lived) or chronic (usually permanent and irreversible). Chronic bronchitis may develop in persons exposed to even moderate levels of SO_2 over many years. In past years various SO_2 control devices have been proposed and made available in the market. In the system design and analysis, consideration must be given to the impact of the energy conversion system on the environment.

Energy conversion can take place in a simple process as well as in a complicated system comprised of several processes. However, all energy conversion processes are governed by the first and second laws of thermodynamics. The first law involves the conservation of energy. It indicates that there will be no energy created nor destroyed in any conversion process. In other words, all energy can be accounted for during the energy conversion process. The first law of thermodynamics is widely used in engineering and is the basis of the energy-balance method frequently used in the analysis and design of energy systems.

The second law of thermodynamics has many ramifications with respect to energy conversion processes. The second law enables the engineer to determine whether or not the desired conversion is possible. The classic example in this area is heat flow from a cup of hot coffee to the surroundings. Everyone knows that this heat transfer process is possible and that the reversed process is not, even though the first law of thermodynamics would be fully satisfied. Another example is a coal-fired power plant. In this complicated system, coal energy is converted to electrical energy, with waste heat as a by-product. However, the reversed process—from electrical energy plus waste heat to coal—is impossible. The second law of thermodynamics also enables the engineer to determine the maximum efficiency of the conversion process and system. This upper limit of performance is important information to the engineer. For an existing system, an engineer can estimate the potential for further improvement. As the energy conversion system is being designed, the engineer can apply the second law to determine the deviation of the actual system from an ideal one.

The second law of thermodynamics plays an important role in the design and analysis of energy conversion systems. While the first law deals with the quantity of energy, the second law deals with the quality of energy. Quality of energy is defined by its utility, or its ability to cause a change. It is said in the second law that not all energy has the same quality. In a conversion process, energy is conserved but energy quality deteriorates. To design a conversion process or system, attention must be paid to the energy quantity as well as the quality.

The concept of availability is embodied in the second law of thermodynamics. Availability represents the work-producing potential of the substance. Availability can take many forms, which can be directly related to each other. Consider, for example, steam from a boiler and water behind a dam on a river. Both the steam and the water have substantial energy, although the first has thermal energy and the second has potential energy. Since not all energy has the same ability to cause a change, one form of energy is not equivalent to another form of energy, even though they may be numerically the same. However, if both steam and water are evaluated in terms of their availabilities or their work-producing potentials, this ambiguity can be easily avoided.

The availability method is developed on the basis of the work potential concept. The method is a systematic approach to apply the first and second laws of thermodynamics to conversion processes or conversion systems. In a process the energy usually changes from one form to another, while at the same time a certain amount of availability is consumed (i.e., either lost or destroyed). The availability output of the process is always less than the availability input. The ratio of these two is the availability utilization factor, frequently referred to as the second law efficiency. The ratio also represents the efficiency in the use of the energy resources. In engineering practice, the first law efficiency still is frequently used and formulated with the concept of energy. The first law efficiency is defined as the ratio of the energy in the product to the energy supplied. Because no attention is given to the quality of energy, the first law efficiency is somewhat arbitrary and does not always represent an efficiency of energy utilization.

In energy conversion system design, due regard should be given to both energy and availability. The system must be evaluated not only in terms of energy balance but also in terms of the availability balance, including availability loss and destruction. Measures should be taken to minimize unnecessary availability loss and destruction so that the energy resources will be better utilized.

1-3 OUTLINE OF THIS TEXT

The objective of this text is to present the availability method and its application to energy conversion. The author's approach is to keep the text easy to follow. It starts with concepts engineers feel comfortable with, such as the conservation of energy, and reviews some basic thermodynamics. Also reviewed are reversibility and irreversibility of processes and the work produced under these conditions. Of the different types of work, the available work is the most critical and important in the development of the availability method. The available work is the work from a process that is internally and externally reversible. It represents the maximum amount of work that could be produced by a device between any two given states. Figure 1-5 presents a general relationship between four basic types of work: Reversible, Actual, Available, and Carnot. In this text the reversible work, or, more precisely, internally reversible work, is the work produced in an internally reversible process. As shown in Fig. 1-5, the difference between the reversible work and the actual work is the lost work, and the difference between the available work and the actual work is the unavailable energy (sometimes referred to as unavailable work). The Carnot work is one of the basic types of work in the text. A combination of Carnot work and lost work would be numerically equal to the unavailable energy. All these types of work will be discussed in Chapter 2.

To connect the materials in this text and those in the elementary text of thermodynamics, Chapter 2 also presents a relationship between the unavailable energy and the net change of entropy in the process. Figure 1-6 shows that the unavailable energy is equal to the net change of entropy times the environmental temperature (T_0). Also, the unavailable energy in the process can be interpreted as the combined internal and external irreversibility.

The available work is one of the important concepts presented in this text. Once the concept of available work is formulated, it will be used as a basis to develop the thermochemical, diffusion, and chemical types of availability. Figure 1-7 presents a flow chart for the development of the text materials. For convenience, Chapter 3 covers both the concepts of thermomechanical availabil-

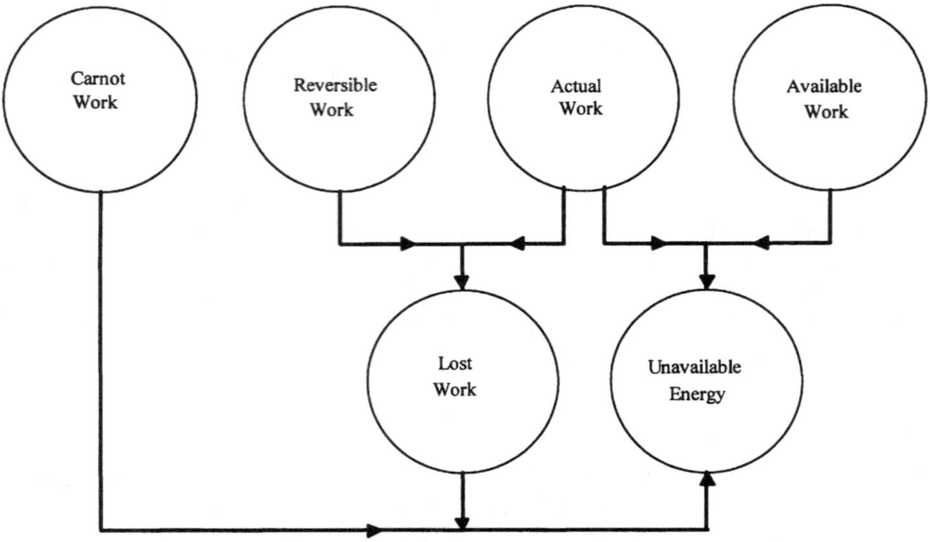

Figure 1-5 A relationship for four basic types of work.

ity and basic availability equations for open as well as closed systems. Chapters 5 and 6 are devoted respectively to diffusion and chemical availability.

Applications of availability method to the engineering processes essentially are presented in Chapter 4. For each process the second law efficiency is defined and demonstrated in use. Also, discussion will be made on the process irreversibility, which is also the availability loss and destruction. The engineering processes to be considered in this text are those associated with compressors, turbines, nozzles, heat exchangers and others. The availability analyses of HVAC and combustion processes are in Chapters 5 and 6.

Applications of availability method to the energy conversion systems are presented in Chapters 7 through 10. These systems are those used for electric-power generation, heat pumps, refrigeration, and cogeneration. For each conversion system, attention will be focused to the availability balance, which generally includes the availability input, output, loss, and destruction. The availability loss and destruction for the components of the energy conversion system also are included in the discussion. The calculation of these terms will enable engineers to identify the critical component in the system for further improvement.

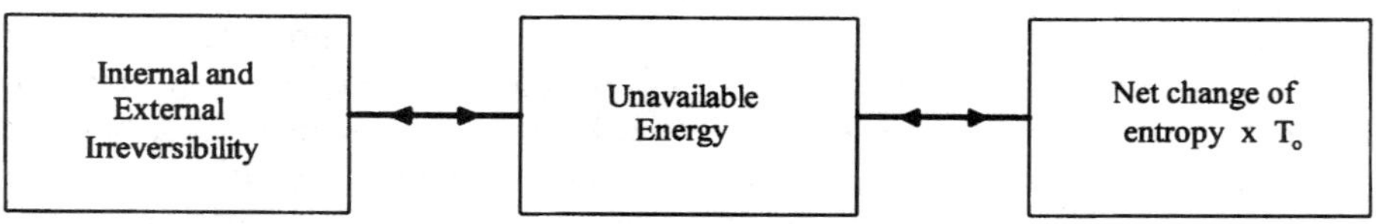

Figure 1-6 A relationship of unavailable energy, net entropy change, and process irreversibility.

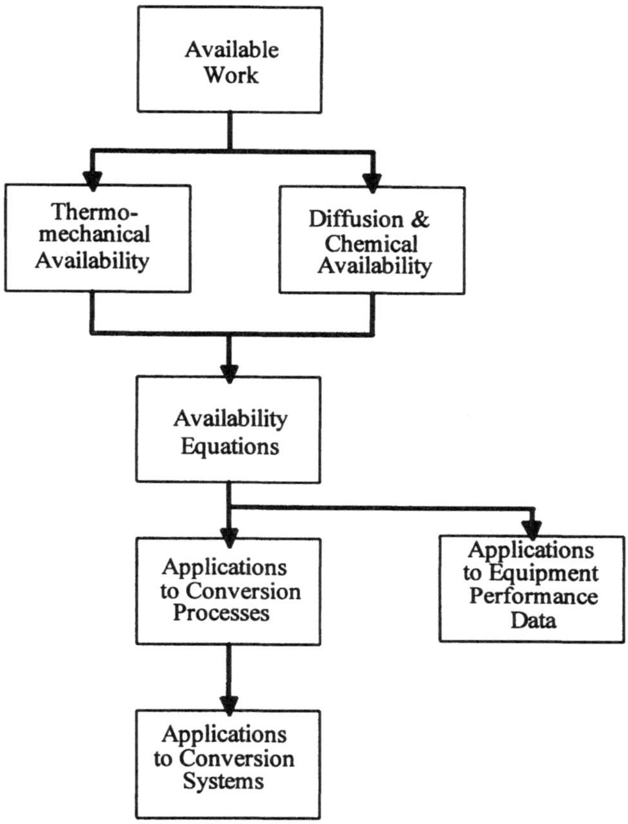

Figure 1-7 Flow chart for the development of the availability method and its applications.

The last chapter is an introduction of availability analysis of equipment performance data. The objective is to determine the second law efficiency and availability loss and destruction for the equipment under various operating conditions. The material in this chapter may be of some interest to those working in industrial and electric-power generating plants.

1-4 A BRIEF REVIEW OF THERMODYNAMICS

Thermodynamics is the science that deals with the transformation of energy from one form into another. In engineering applications the energy may include:

Kinetic energy (KE)
Potential energy (PE)
Internal energy (IE)
Flow work (FW)

Mechanical work (W)
Heat (Q)
Chemical energy (ChE)

The first law of thermodynamics is the principle of energy conservation. It indicates that, when energy is converted from one form into another, no energy will be created nor destroyed. In equation form the first law is expressed as

$$\frac{dE_{cv}}{d\tau} = \dot{Q}_{cv} - \dot{W}_{cv} + \sum_i \dot{m}_i\left(h_i + \frac{V_i^2}{2} + gz_i\right) - \sum_e \dot{m}_e\left(h_e + \frac{V_e^2}{2} + gz_e\right) \quad (1\text{-}1)$$

where the subscript cv denotes the control volume, i the inlet, and e the exit. The summation sign indicates that there may be more than one inlet and/or exit to the control volume. In Eq. (1-1), the energy term E usually represents the sum of internal kinetic and potential energy. Similarly, the mass balance equation for a control volume can be written as

$$\frac{dm_{cv}}{d\tau} = \sum_i \dot{m}_i - \sum_e \dot{m}_e \quad (1\text{-}2)$$

For a steady-state, steady-flow process, both equations are further simplified by the fact that there is no energy and mass accumulation in the control volume. In equation form, Eqs. (1-1) and (1-2) become

$$\dot{Q}_{cv} - \dot{W}_{cv} + \sum_i \dot{m}_i\left(h_i + \frac{V_i^2}{2} + gz_i\right) - \sum_e \dot{m}_e\left(h_e + \frac{V_e^2}{2} + gz_e\right) = 0 \quad (1\text{-}3)$$

and

$$\sum_i \dot{m}_i - \sum_e \dot{m}_e = 0 \quad (1\text{-}4)$$

Eq. (1-1) applies equally well to a closed system where the working substance does not cross the system boundary. The first law equation for a closed system is

$$\frac{dE}{d\tau} = \dot{Q} - \dot{W} \quad (1\text{-}5)$$

The second law of thermodynamics has been expressed by many different statements and equations over the years. One of the popular equations is based on the concept of entropy production and often given by

$$\frac{dS_{cv}}{d\tau} = \sum_i \dot{m}_i s_i - \sum_e \dot{m}_e s_e + \sum_j \frac{\dot{Q}_j}{T_j} + \dot{\sigma} \quad (1\text{-}6)$$

In words, the above equation says that the entropy change inside the control volume must be equal to the net entropy flow in and out of the control volume plus the entropy changes induced by the heat transfer and the internal irreversibility. $\dot{\sigma}$ is the rate of entropy production caused by the presence of internal irreversibility within the control volume. It is always positive in all processes. The heat transfer between the control volume and its surroundings also results in a change of entropy. The entropy change can be positive or negative, depending on the direction of heat transfer. Since the control volume surface may not have constant and uniform temperature, the term $\Sigma_j \dot{Q}_j/T_j$ represents the rates of entropy transfer at various locations on the control volume surface.

For a steady-state, steady-flow process, such as those found in a turbine or a compressor, Eq. (1-6) is simplified by setting the term $(dS_{\text{cv}}/d\tau)$ to zero. That is,

$$\sum_i \dot{m}_i s_i - \sum \dot{m}_e s_e + \sum_j \frac{\dot{Q}_j}{T_j} + \dot{\sigma} = 0 \tag{1-7}$$

Eq. (1-6) is also applicable to a closed system. In this case the terms associated with mass convection would vanish. Therefore, the second law equation for a closed system is

$$\frac{dS}{d\tau} = \frac{\dot{Q}}{T} + \dot{\sigma} \tag{1-8}$$

This equation is simplified further if the closed system undergoes a reversible process. In a reversible process the internally induced entropy becomes zero, and the change of entropy in the system is directly related to the heat transfer. That is,

$$\frac{dS}{d\tau} = \frac{\dot{Q}}{T} \tag{1-9}$$

Integrating this equation will give

$$S_2 - S_1 = \int_1^2 \frac{\delta Q}{T} \tag{1-10}$$

The substance frequently encountered in thermodynamics may be in form of gases (or vapors) as well as liquids. A substance is defined as one that has a homogeneous and constant chemical composition. It may exist in more than one phase. The state of the substance is determined by two independent thermodynamic properties. To review the calculation of property change, the substances are presently organized into two groups: ideal gases and real gases.

An ideal gas is a simple compressible substance which has the equation of state

$$Pv = RT \tag{1-11}$$

where R is the gas constant and is related to the universal gas constant $\overline{R}$

through the molecular weight M. The relationship is

$$R = \frac{\overline{R}}{M} \tag{1-12}$$

It is well known that for an ideal gas, the enthalpy and internal energy are functions of temperature alone. The relation between them is given by

$$h(T) = u(T) + RT \tag{1-13}$$

Differentiating with respect to temperature gives

$$\frac{dh(T)}{dT} = \frac{du(T)}{dT} + R$$

or

$$c_p = c_v + R \tag{1-14}$$

where c_p and c_v are constant-pressure and constant-volume specific heat, respectively. For an ideal gas these two specific heats depend only on temperature. Defining the ratio of specific heats as

$$k = \frac{c_p}{c_v} \tag{1-15}$$

we have the following relationships for an ideal gas:

$$c_p = \frac{kR}{k-1} \tag{1-16}$$

and

$$c_v = \frac{R}{k-1} \tag{1-17}$$

To derive the equations for the change of enthalpy and internal energy, we start with the definitions of the specific heats

$$c_p = \frac{dh(T)}{dT} \tag{1-18}$$

$$c_v = \frac{du(T)}{dT} \tag{1-19}$$

and then integrate them from a reference state to a given state. That is,

$$h(T) - h(T_{\text{ref}}) = \int_{T_{\text{ref}}}^{T} c_p \, dT \tag{1-20}$$

and

$$u(T) - u(T_{\text{ref}}) = \int_{T_{\text{ref}}}^{T} c_v \, dT \tag{1-21}$$

where T_{ref} is the temperature of the reference state. To derive the equation for the change of entropy for an ideal gas, we start with the equation

$$T\,ds = dh - v\,dp \tag{1-22}$$

For an ideal gas, Eq. (1-22) is further reduced to

$$ds = c_p \frac{dT}{T} - R\frac{dP}{P}$$

Integrating this equation between states 1 and 2, we have the entropy change as

$$s_2(T_2, P_2) - s_1(T_1, P_1) = \int_{T_1}^{T_2} c_p \frac{dT}{T} - R \ln \frac{P_2}{P_1}$$

or

$$s_2(T_2, P_2) - s_1(T_1, P_1) = \int_{T_{ref}}^{T_2} c_p \frac{dT}{T} - \int_{T_{ref}}^{T_1} c_p \frac{dT}{T} - R \ln \frac{P_2}{P_1} \tag{1-23}$$

Let

$$s'(T) = \int_{T_{ref}}^{T} c_p \frac{dT}{T} \tag{1-24}$$

Eq. (1-23) becomes

$$s_2(T_2, P_2) - s_1(T_1, P_1) = s'(T_2) - s'(T_1) - R \ln \frac{P_2}{P_1} \tag{1-25}$$

For convenience, the values of the integrals in Eqs. (1-20), (1-21), and (1-24) frequently are determined using the reference temperature (T_{ref}) at a zero-degree Rankine. With the assumption $u(T_{ref}) = h(T_{ref}) = 0$, one can easily see that the internal energy u, enthalpy h, and entropy functions s' are functions of temperature alone. In Appendix C, Table C-5 provides the thermodynamic properties for various ideal gases.

In engineering applications there are many processes in which the temperature change is not substantial and so the specific heats are treated as constants. For these cases, the change of enthalpy between states 1 and 2 is simply

$$h_2 - h_1 = c_p(T_2 - T_1) \tag{1-26}$$

and the change of internal energy is

$$u_2 - u_1 = c_v(T_2 - T_1) \tag{1-27}$$

Similarly, the entropy change for an ideal gas is

$$s_2 - s_1 = c_p \ln \frac{T_2}{T_1} - R \ln \frac{P_2}{P_1} \tag{1-28}$$

The equation of state for an ideal gas is satisfactory in modeling many gases and vapors under certain conditions, such as the temperature more than twice the critical temperature or the pressure less than 1 atm. However, this equation of state definitely is not universally valid. When the behavior of a gas cannot be modeled by the ideal gas equation, the gas is generally referred to as a real gas. The P-v-T relationships of real gases may be presented in several ways. A simple method involves a use of the compressibility factor Z. That is,

$$Pv = ZRT \tag{1-29}$$

For an ideal gas, the compressibility factor Z is obviously unity under all conditions. For a real gas, the value of Z will be different from unity. Thus, the compressibility factor Z gives the deviation from ideal gas behavior. In Appendix C, Fig. C-1 presents a generalized compressibility factor chart. As indicated, the compressibility factor is

$$Z = f(T_r, P_r) \tag{1-30}$$

where T_r and P_r are reduced temperature and reduced pressure, respectively. They are defined as

$$T_r = \frac{T}{T_c} \tag{1-31}$$

$$P_r = \frac{P}{P_c} \tag{1-32}$$

Similarly, Figs. C-2 and C-3 present the generalized enthalpy and entropy departure charts for real gases. In both charts the enthalpy and entropy departures at a given state versus the reduced pressure (P_r) are given for various values of the reduced temperature (T_r). These charts are utilized in the calculation of property change. For instance, the enthalpy change between states 1 and 2 for a real gas is given as

$$h_2 - h_1 = (h_2 - h_2^*) + (h_2^* - h_1^*) + (h_1^* - h_1) \tag{1-33}$$

$$\begin{pmatrix}\text{Enthalpy}\\ \text{change}\\ \text{for real}\\ \text{gas}\end{pmatrix} = \begin{pmatrix}\text{Enthalpy}\\ \text{departure}\\ \text{at state 2}\end{pmatrix} + \begin{pmatrix}\text{Enthalpy}\\ \text{change}\\ \text{from ideal}\\ \text{gas}\\ \text{equation}\end{pmatrix} + \begin{pmatrix}\text{Enthalpy}\\ \text{departure}\\ \text{at state 1}\end{pmatrix}$$

Similarly, the entropy change for a real gas is

$$s_2 - s_1 = (s_2 - s_2^*) + (s_2^* - s_1^*) + (s_1^* - s_1) \tag{1-34}$$

$$\begin{pmatrix}\text{Entropy}\\ \text{change}\\ \text{for real}\\ \text{gas}\end{pmatrix} = \begin{pmatrix}\text{Entropy}\\ \text{departure}\\ \text{at state 2}\end{pmatrix} + \begin{pmatrix}\text{Entropy}\\ \text{change}\\ \text{from ideal}\\ \text{gas equation}\end{pmatrix} + \begin{pmatrix}\text{Entropy}\\ \text{departure}\\ \text{at state 1}\end{pmatrix}$$

The generalized charts are quite easy to use, but the accuracy is not always satisfactory. For this reason, the thermodynamic properties for some important substances frequently are presented in tabular form. In Appendix C, Tables C-2 through C-4 provide the thermodynamic properties for water, Freon-12, and ammonia. The thermodynamic properties in these tables include specific volume, internal energy, enthalpy, and entropy.

The quality of vapor is a thermodynamic property frequently used when a substance is in a saturated state. The quality is defined as the ratio of the mass of vapor to the total mass of vapor plus liquid. That is,

$$x = \frac{m_g}{m_g + m_f} \tag{1-35}$$

In a saturation state, the substance has the specific volume given by

$$v = (1 - x)v_f + xv_g \tag{1-36}$$

Similarly, its enthalpy and entropy are respectively

$$h = (1 - x)h_f + xh_g \tag{1-37}$$

and

$$s = (1 - x)s_f + xs_g \tag{1-38}$$

When the temperature of substance is below the saturation point, the substance usually is referred to as a subcooled or compressed liquid. For a substance in the subcooled region, it is sometimes inconvenient to determine the thermodynamic property values from the tables. As an approximation we frequently treat the subcooled liquid (or liquid in general) as an incompressible substance. Thus we can write the constant-volume specific heat for liquids as

$$c_v = \frac{du}{dT}$$

Rearranging gives

$$du = c_v\, dT$$

Integrating the above equation from the saturated state to the subcooled state, the internal energy of the subcooled liquid is

$$u(T) = u(T_{\text{sat}}) + c_v(T - T_{\text{sat}}) \tag{1-39}$$

Since

$$h = u + Pv$$

the enthalpy of subcooled liquid can be written as

$$h(T) = h(T_{\text{sat}}) + c_v(T - T_{\text{sat}}) + v_f(P - P_{\text{sat}}) \tag{1-40}$$

To derive the equation for the entropy of subcooled liquid, we start with the equation

$$T\,ds = du + P\,dv$$

Because the liquid is treated as an incompressible substance, the term $P\,dv$ will disappear, and so the equation is reduced to

$$T\,ds = du$$

or

$$ds = c_v \frac{dT}{T}$$

Integrating the equation from the saturated state to the subcooled state, the entropy of the subcooled liquid is

$$s(T) = s(T_{\text{sat}}) + c_v \ln \frac{T}{T_{\text{sat}}} \tag{1-41}$$

For liquids, the constant-volume specific heat is numerically equal to the constant-pressure specific heat. Also, the pressure has no influence on the values of internal energy and entropy.

1-5 PROBLEMS

1-1 Nitrogen is heated from 100°F to 1500°F in a constant-pressure process. Calculate the enthalpy and entropy changes per unit mass by assuming constant specific heat (0.248 Btu/lb-R) and using the values from Table C-5. Discuss the accuracy of these results.

1-2 Methane (CH_4) is compressed from 15 atm at 250°F to 55 atm in an isentropic process. Calculate the enthalpy change per unit mass by using both the ideal and real gas models. Also discuss the accuracy of the results.

1-3 Determine the following thermodynamics properties for compressed water at 600 psia and 80°F: (a) internal energy, (b) enthalpy, and (c) entropy.

1-4 Steam enters a high-pressure turbine at 3515 psia and 1000°F and expands to 750 psia and 600°F in an adiabatic process. The steam flow rate is approximately 5×10^6 pounds per hour. Estimate the turbine output in kilowatts.

1-5 Air enters a compressor at 14.7 psia and 80°F and leaves at 150 psia and 650°F in an adiabatic process. The air flow is 1000 cfm at the compressor inlet. Determine the power required to drive the compressor.

1-6 Saturated ammonia liquid at 153 psia enters an expansion valve and leaves at 38.5 psia. Calculate the change in the following thermodynamic properties: (a) specific volume, (b) internal energy, and (c) entropy.

1-7 Steam at 100 psia and 500°F enters a nozzle with a velocity of 100 ft/s and leaves at a pressure of 20 psia and a velocity of 1000 ft/s. The heat loss from the nozzle is negligible. Determine the conditions of the steam leaving the nozzle.

1-8 Freon-12 flows through the condenser in a refrigeration system and experiences a change from 90 psia and 140°F to 85 psia and 60°F. Calculate the heat transfer per unit mass of Freon-12.

CHAPTER

TWO

REVERSIBILITY AND WORK

This chapter begins with the concept of reversibility in a process. As part of the presentation, equations for various types of work in a process are introduced. Several examples are included to illustrate the relationship between the process reversibility and the type of work.

2-1 REVERSIBLE WORK

Figure 2-1 presents a gas in a cylinder with a movable piston. If the gas is defined as a closed system and has a pressure infinitesimally greater than the external pressure, the gas will expand quasi-statically. If the gas pressure is maintained infinitesimally less than the external pressure, the gas will be compressed quasi-statically. In the limit these two processes follow the same path in opposite directions. In thermodynamics, a quasi-static process is an ideal process. It is approximately realized by making the system change very slowly. The quasi-static process is sometimes referred to as a reversible process or, to be more accurate, as an internally reversible process. The reversible work w_{rev} for a closed system is given by

$$w_{\text{rev}} = \int_1^2 P\,dv \tag{2-1}$$

Figure 2-1 also indicates that the reversible work produced in a given process will depend not only on the initial and final states but also on the nature of the process. Different processes between the same two states will result in different

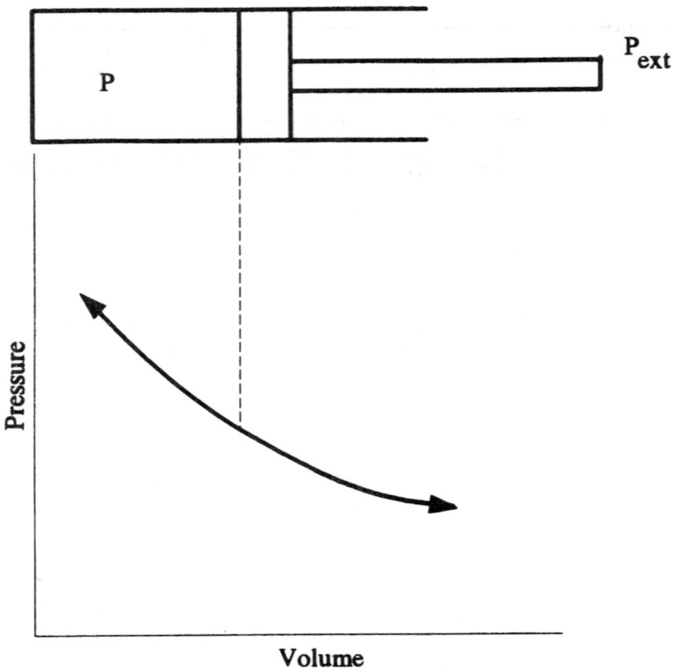

Figure 2-1 Quasi-static expansion and compression of a gas in a cylinder.

amounts of work. For example, if Pv = constant, substituting this relationship into Eq. (2-1) and integrating the equation from state 1 to state 2 would give

$$w_{\text{rev}} = C\int_1^2 \frac{dv}{v} = P_1 v_1 \ln\frac{v_2}{v_1} \tag{2-2}$$

Similarly, the polytropic process with Pv^n = constant, the reversible work would become

$$w_{\text{rev}} = C\int_1^2 \frac{dv}{v^n} = \frac{P_2 v_2 - P_1 v_1}{1-n} \tag{2-3}$$

In engineering applications open systems such as pumps and compressors are frequently encountered. The reversible work for an open system is given by

$$w_{\text{rev}} = -\int_1^2 v\,dP + \frac{V_1^2 - V_2^2}{2} + (z_1 - z_2)g \tag{2-4}$$

For the processes in which the changes in kinetic and potential energy are negligible, Eq. (2-4) is further reduced to

$$w_{\text{rev}} = -\int_1^2 v\,dP$$

Table 2-1 Ideal gas equations for reversible work

Processes	Equations for reversible work Closed system	Open system
Constant volume (v = constant)	0	$-v(P_2 - P_1)$
Constant pressure (P = constant)	$P(v_2 - v_1)$	0
Constant temperature (T = constant)	$P_1 v_1 \ln \frac{v_2}{v_1}$	$-P_1 v_1 \ln \frac{P_2}{P_1}$
Constant entropy (S = constant)	$\frac{P_2 v_2 - P_1 v_1}{1 - k}$	$\frac{k(P_2 v_2 - P_1 v_1)}{1 - k}$
Polytropic (Pv^n = constant)	$\frac{P_2 v_2 - P_1 v_1}{1 - n}$	$\frac{n(P_2 v_2 - P_1 v_1)}{1 - n}$

Table 2-1 presents the list of equations for reversible work in the various ideal gas processes. The reversible work can be also calculated by using the first law equation. The following examples are included for illustrations.

Example 2-1. Air at 14.7 psia and 70°F enters a compressor at a velocity of 10 ft/s and discharges at 150 psia and 300 ft/s. Assuming that the air flow rate is 2000 lbm/min and that the process is reversible and adiabatic, determine the power required for the compressor.

Solution. Consider the adiabatic compressor as a control volume. It is evident from the available data that the process must be steady-state and steady-flow with negligible change in potential energy. Thus, the first law equation for an adiabatic process becomes

$$\dot{m}_i\left(h_i + \frac{V_i^2}{2}\right) = \dot{W}_{\text{rev}} + \dot{m}_e\left(h_e + \frac{V_e^2}{2}\right)$$

Because the air is treated as an ideal gas, the above equation can be further simplified and reduced to

$$\dot{W}_{\text{rev}} = \dot{m}c_p(T_i - T_e) + \frac{\dot{m}\left(V_i^2 - V_e^2\right)}{2} \qquad \text{(2-a)}$$

Since the process is adiabatic and reversible, the exit air temperature T_e is given by the equation

$$\frac{T_e}{T_i} = \left(\frac{P_e}{P_i}\right)^{k-1/k} \tag{2-b}$$

Substituting the numeric values into Eq. (2-b) would give

$$T_e = 530\left(\frac{150}{14.7}\right)^{1.4-1/1.4} = 1029 \text{ R}$$

With the exit air temperature T_e and other numeric values, Eq. (2-a) yields the reversible work

$$\dot{W}_{\text{rev}} = (2000)(0.24)(530 - 1029) + (2000)\frac{(10)^2 - (300)^2}{2 \times 32.2 \times 778}$$

$$\dot{W}_{\text{rev}} = -243{,}184 \text{ Btu/min}$$

Thus, the power required for the compressor is 5733 hp. ■

Example 2-2. Saturated steam at 300 psia enters the turbine and expands isentropically to 1 psia. Calculate the turbine work per pound of steam.

Solution. It is always convenient to use the Mollier diagram for the calculation of steam turbine work. First, the turbine inlet conditions (300 psia, saturated vapor) are used to define state 1, as shown in Fig. 2-2. Since the turbine process is reversible and adiabatic, the process must be represented by a vertical and straight line in the $h = S$ diagram. The interception between this line and the constant pressure curve (1 psia) will locate state 2, which represents the state of steam at the turbine exit. From the Mollier diagram we have $h_1 = 1204$ Btu/lbm, $h_2 = 844$ Btu/lbm, and $s_1 = s_2 = 1.5105$ Btu/lbm-R. Then by the first law equation the turbine reversible work is $w_{\text{rev}} = h_1 - h_2 = 1204 - 844 =$ 360 Btu/lbm. ■

2-2 AVAILABLE WORK

The quasi-static process defined in the last section is internally reversible. Whether the process is externally reversible will depend on the conditions under which the heat transfer between the system and its surroundings takes place. If the heat transfer process is realized through a finite temperature difference, the heat transfer process is not reversible (or irreversible). In reality, all heat transfer processes are irreversible. An interesting question arises now. What, then, is a reversible heat transfer process? Generally, a reversible heat transfer process is defined as the process in which the heat is transferred through an infinitesimal temperature difference. Because of the infinitesimal temperature difference, the process will require an infinite amount of time or infinite heat

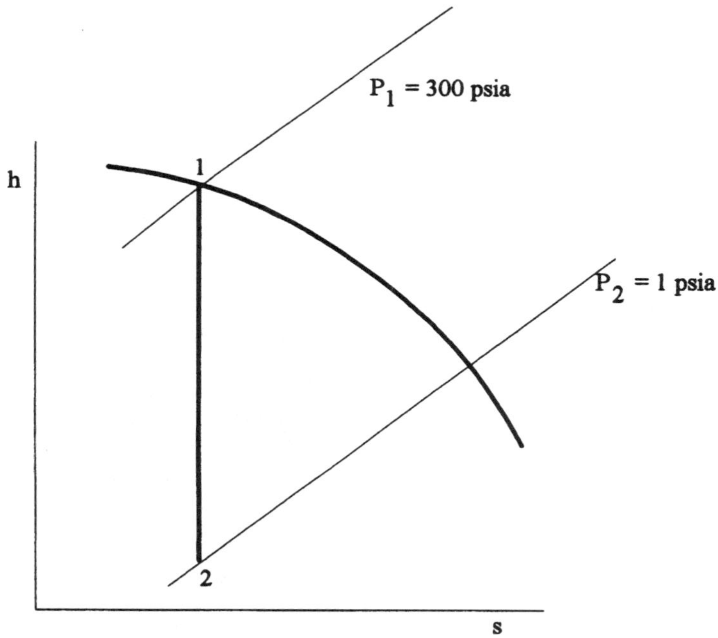

Figure 2-2 Mollier diagram for a steam turbine process.

transfer area. Figure 2-3 indicates the difference between the reversible and irreversible heat transfer processes from body A to body B. While the actual heat transfer process is characterized by a finite temperature difference, the reversible heat transfer process is characterized by an infinite number of fictitious bodies between the actual bodies A and B, each fictitious body performing heat source and sink functions and reducing the temperature by an infinitesimal amount in the direction of heat flow. Like the quasi-static process, the reversible heat transfer process is an ideal process.

Figure 2-4 shows a hot body being cooled in its environment. This actual heat transfer process is not reversible. To make the process entirely reversible, one can install a Carnot engine between the hot body and the environment. Because the Carnot engine is a reversible device, all processes in the Carnot system must be reversible. In other words, the hot body loses heat to its environment by δQ_h in a reversible process and, as a result, produces reversible work of δW_c (i.e., Carnot work). If the hot body has a finite capacity and its temperature is decreased in the heat transfer process, one can make the process reversible by using an infinite number of Carnot engines between the hot body and its environment, one engine operating at a time and having the hot-body temperature reduced by an infinitesimal amount. The sum of all these Carnot engine outputs will be the reversible work in this process.

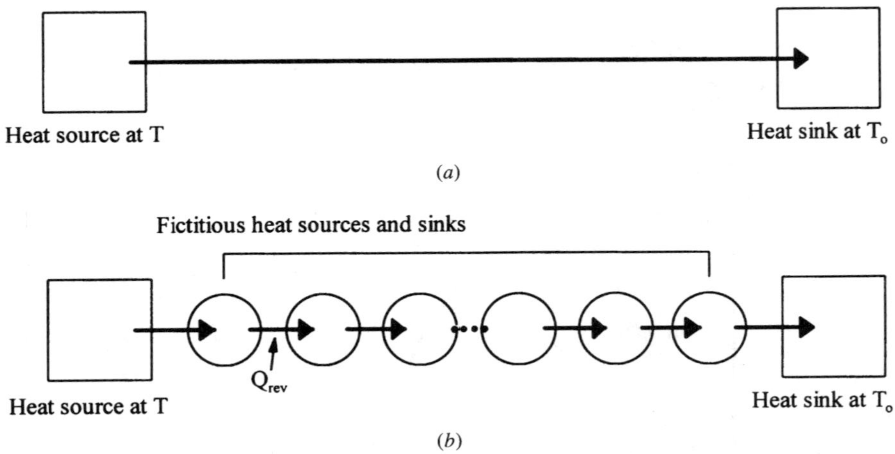

Figure 2-3 Reversible and irreversible heat transfer processes: (*a*) irreversible heat transfer, (*b*) reversible heat transfer process.

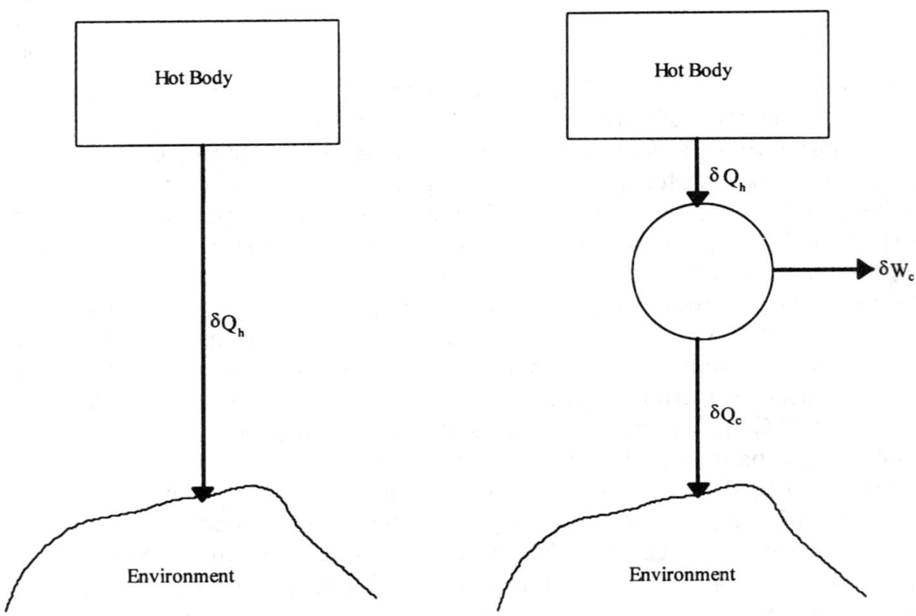

Figure 2-4 Reversible and irreversible heat transfer process between a hot body and its environment.

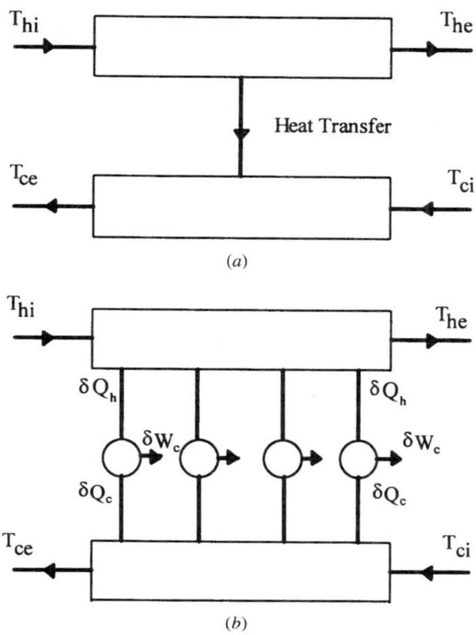

Figure 2-5 Reversible and irreversible heat exchangers: (*a*) irreversible heat exchanger, (*b*) reversible heat exchanger.

Figure 2-5(a) presents a typical heat exchanger in which heat is transferred from a high temperature stream to a low temperature stream. The heat transfer process can be made entirely reversible by placing an infinite number of Carnot engines between the streams. As shown in Fig. 2-5(b), each Carnot engine operates between the streams of different temperatures, and the sum of all Carnot engine outputs will be the reversible work of this heat transfer process. It must be emphasized that the Carnot engine does not exist and that the reversible work represents the amount of mechanical work which ideally could have been obtained but actually was not produced in the process.

Now we would like to develop an energy equation for an open system that operates reversibly in every respect. Consider the two control volumes X and Y in Fig. 2-6. Control volume X undergoes an internally reversible process. Since control volume X's surface temperature is different from the temperature of the environment, the process is not externally reversible. Because of this, the work by control volume X is not the maximum quantity of work that could be produced or the minimum quantity of work that must be supplied. To make the process externally reversible, one or more reversible heat engines (such as Carnot engines) must be attached to control volume X, with each engine operating between a given surface temperature T and the environmental temperature T_0. As indicated in Fig. 2-6, control volume X and the Carnot engine are combined to form a new control volume Y, which undergoes an internally

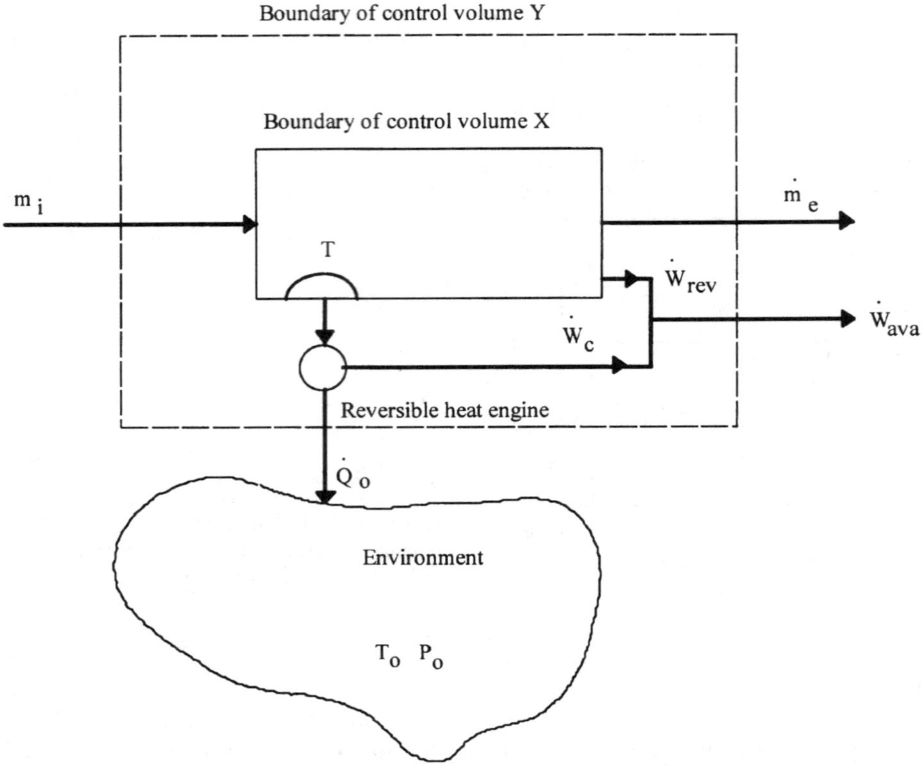

Figure 2-6 An internally and externally reversible process for the control volume Y.

and externally reversible process. Thus, the work produced by control volume Y must be the sum of the reversible work by control volume X and the work by the Carnot engine. This is the work one would have expected from an ideal process between any two given states. In this text, the work from the externally and internally reversible process is defined as the *available work*, w_{ava}. In the following, an equation will be derived for the available work of an open system under steady-state and steady-flow conditions.

Applying the first law equation to control volume Y in Fig. 2-6 gives

$$\dot{W}_{\text{ava}} = \dot{m}_i\left(h_i + \frac{V_i^2}{2} + gz_i\right) - \dot{m}_e\left(h_e + \frac{V_e^2}{2} + gz_e\right) - \dot{Q}_0 \qquad (2\text{-}5)$$

Since the process is internally reversible, there will be no internally produced entropy. The second law equation for the control volume Y will become

$$\dot{m}_i s_i - \dot{m}_e s_e - \frac{\dot{Q}_0}{T_0} = 0 \qquad (2\text{-}6)$$

Combining these two equations yields the expression for the available work per unit mass of working substance:

$$w_{\text{ava}} = \frac{\dot{W}_{\text{ava}}}{\dot{m}_i} = (h_i - h_e) - T_0(s_i - s_e) + \left(\frac{V_i^2}{2} - \frac{V_e^2}{2}\right) + g(z_i - z_e) \quad (2\text{-}7)$$

For processes in which the changes in kinetic and potential energies are negligible, Eq. (2-7) is further reduced to

$$w_{\text{ava}} = (h_i - h_e) - T_0(s_i - s_e) \quad (2\text{-}8)$$

It should be pointed out that the available work, unlike the reversible work, is not a function of the process path. It depends only on the end states of the process for a given environmental temperature T_0. The available work frequently is referred to as the available energy of the process. The available work is the maximum amount of work that could be produced by a device between any given two states. If the device is work-absorbing, such as a compressor or pump, the available work of the process is the minimum amount of work that must be supplied.

The available work by a closed system can be derived in a similar manner. That is, the first law and second law equations first are established for a closed system that would undergo an internally and externally reversible process. These two equations then are combined to yield the available work as

$$w_{\text{ava}} = (u_1 - u_2) - T_0(s_1 - s_2) \quad (2\text{-}9)$$

There is another approach in developing Eq. (2-9). We start with the fact that the available work is equal to the reversible work produced directly by the closed system plus the work produced by the Carnot engine, as shown in Fig. 2-7. The equation in differential form is

$$\delta w_{\text{ava}} = \delta w_{\text{rev}} + \delta w_c \quad (2\text{-}10)$$

Also, the work from Carnot engine is given as

$$\delta w_c = -\delta q_{\text{rev}}\left(1 - \frac{T_0}{T}\right) \quad (2\text{-}11)$$

where the minus sign for δq_{rev} denotes the direction of heat transfer from the closed system to the Carnot engine. Combining these two equations yields

$$\begin{aligned}
\delta w_{\text{ava}} &= \delta w_{\text{rev}} - \delta q_{\text{rev}}\left(1 - \frac{T_0}{T}\right) \\
&= -du_{\text{rev}} + T_0\frac{\delta q_{\text{rev}}}{T} \\
&= -du_{\text{rev}} + T_0\, ds
\end{aligned}$$

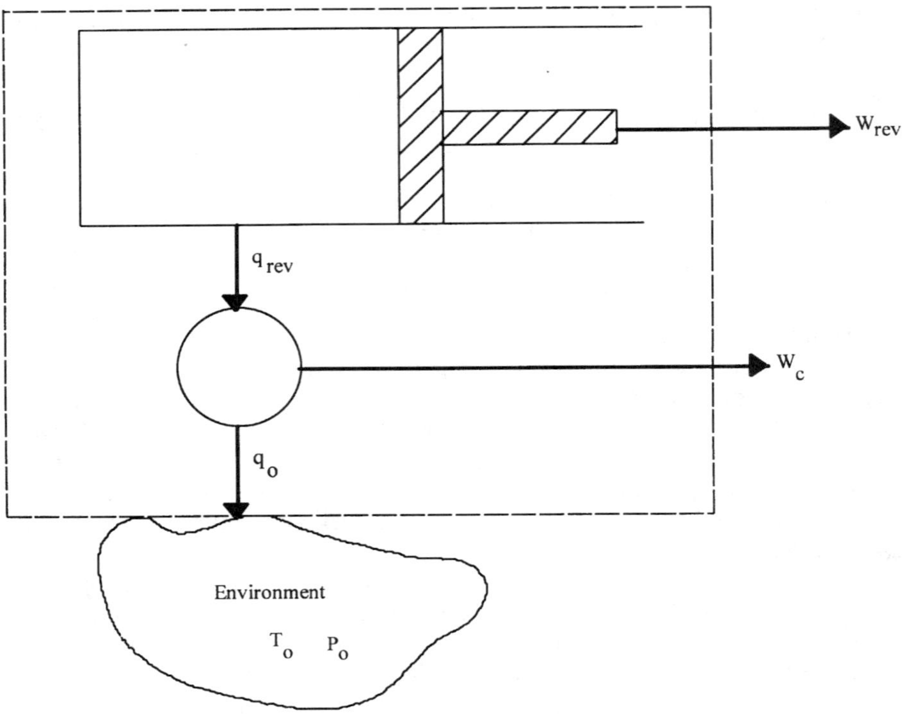

Figure 2-7 Available work produced by a closed system.

Integrating the above equation between 1 and 2 will lead to

$$w_{\text{ava}} = (u_1 - u_2) - T_0(s_1 - s_2)$$

The above equation is identical to Eq. (2-9) and is used to determine the available work by a closed system. As in the case of the open system, the available work depends on the initial and final states of the process for a given environmental temperature T_0. In other words, the process path between the end states will not affect the value of the available work.

A more general expression for the available work is given by

$$W_{\text{ava}} = \sum \Delta m_i(h_i - s_i) - \sum \Delta m_e(h_e - T_0 s_e) + [m_1(u_1 - T_0 s_1) - m_2(u_2 - T_0 s_2)]_{\text{cv}} \tag{2-12}$$

For a steady-state, steady-flow process, we recognize

$$[m_1(u_1 - T_0 s_1) - m_2(u_2 - T_0 s_2)]_{\text{cv}} = 0$$

Thus, the available work for a steady-state, steady-flow process would become

$$W_{ava} = \sum \Delta m_i (h_i - T_0 s_i) - \sum \Delta m_e (h_e - T_0 s_e)$$

or

$$W_{ava} = (H_i - H_e) - T_0 (S_i - S_e)$$

The above equation is identical to Eq. (2-8). For a closed system there is no flow across the boundary of the system, and therefore the first two terms in Eq. (2-12) will disappear. Thus for a closed system, the available work is given by

$$W_{ava} = [m_1(u_1 - T_0 s_1) - m_2(u_2 - T_0 s_2)]_{cv} = (U_1 - U_2) - T_0(S_1 - S_2)$$

The above equation is identical to Eq. (2-9). The following are two illustrative examples. The environmental temperature is assumed to be 77°F unless otherwise indicated.

Example 2-3. Air enters a turbine at 150 psia and 2000°F and expands to 14.7 psia and 1000°F. The heat loss from the turbine is approximately 2 percent of the turbine power output. Calculate the actual work and the available work for this turbine process.

Solution. The turbine process in *T-s* diagram is shown in Fig. 2-8. The

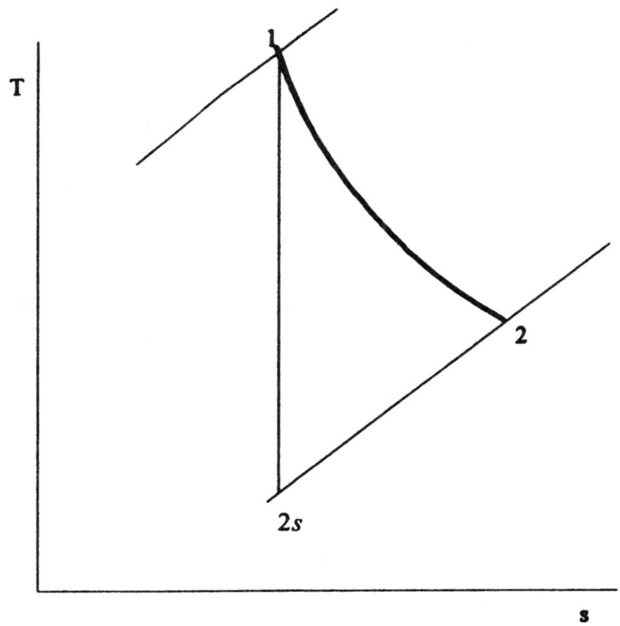

Figure 2-8 A turbine process 1-2 in a *T-s* diagram.

turbine actual work is given by the first-law equation

$$q + h_1 = w_{act} + h_2$$

When the heat loss from the turbine is taken into account, the above equation will become

$$-0.02w_{act} + h_1 = w_{act} + h_2$$

or

$$1.02w_{act} = c_p(T_1 - T_2)$$

Substituting the numeric values into the equation, we have the turbine actual work

$$w_{act} = \frac{0.24(2460 - 1460)}{1.02} = 235.3 \text{ Btu/lbm}$$

The turbine available work is given by the equation

$$w_{ava} = (h_1 - h_2) - T_0(s_1 - s_2)$$

Since air is assumed to be an ideal gas, the equation is further reduced to

$$w_{ava} = c_p(T_1 - T_2) - T_0\left(c_p \ln\frac{T_1}{T_2} - R\ln\frac{P_1}{P_2}\right)$$

With the numeric values, we have the turbine available work

$$w_{ava} = 0.24(2460 - 1460) - 537\left(0.24\ln\frac{2460}{1460} - \frac{53.3}{778}\ln\frac{150}{14.7}\right)$$
$$= 258.2 \text{ Btu/lbm}$$

As expected, the turbine available work is greater than the actual work from the turbine. The difference should represent a potential for further turbine improvement. ■

Example 2-4. Air is compressed adiabatically in a cylinder-piston arrangement. For initial conditions of 20 psia and 100°F and final conditions of 50 psia and 297.5°F, calculate the actual work and the available work in the process.

Solution. The actual work is given by the first law equation

$$q - w_{act} = u_2 - u_1$$

Since the process is adiabatic and air is treated as an ideal gas, the above equation is reduced to

$$w_{act} = c_v(T_1 - T_2)$$

Substituting the numeric values into the equation, we have the actual work

$$w_{act} = 0.17(560 - 757.5) = -33.58 \text{ Btu/lbm}$$

The available work for the same process is

$$w_{\text{ava}} = (u_1 - u_2) - T_0(s_1 - s_2)$$
$$= c_v(T_1 - T_2) - T_0\left(c_p \ln\frac{T_1}{T_2} - R\ln\frac{P_1}{P_2}\right)$$

With the numeric values we have

$$w_{\text{ava}} = 0.17(560 - 757.5) - 537\left(0.24\ln\frac{560}{757.5} - \frac{53.3}{778}\ln\frac{20}{50}\right) = -28.37 \text{ Btu/lbm}$$

In words, the closed system requires the mechanical work of 33.58 Btu/lbm for the compression process. If the system could be operated in an externally and internally reversible manner, the required work would be only 28.37 Btu/lbm. ■

2-3 UNAVAILABLE ENERGY

The work produced by a system in the real world is not equal to either the reversible work or the available work. The difference between the reversible work and the actual work is defined as the *lost work*. Lost work is the work which could have been obtained but was not actually produced because of internal irreversibility. In an equation form, the lost work (LW) is

$$\delta(\text{LW}) = \delta W_{\text{rev}} - \delta W_{\text{act}} \tag{2-13}$$

For a closed system, δW_{rev} is expressed as

$$\delta W_{\text{rev}} = \delta Q_{\text{rev}} - dU_{\text{rev}} \tag{2-14}$$

and δW_{act} is given by

$$\delta W_{\text{act}} = \delta Q_{\text{act}} - dU_{\text{act}} \tag{2-15}$$

Substituting Eqs. (2-14) and (2-15) into Eq. (2-13) and taking into consideration $dU_{\text{rev}} = dU_{\text{act}}$, we have

$$\delta(\text{LW}) = \delta Q_{\text{rev}} - \delta Q_{\text{act}} \tag{2-16}$$

Also, for a reversible process we have

$$\delta Q_{\text{rev}} = T\,dS \tag{2-17}$$

Combining Eqs. (2-16) and (2-17) yields

$$dS = \frac{\delta Q_{\text{act}}}{T} + \frac{\delta(\text{LW})}{T} \tag{2-18}$$

Eq. (2-18) is an important equation. It shows that the entropy change is caused by the heat transfer between the system and its surroundings and by the lost work within the system. The heat transfer can increase or decrease the entropy in the system, depending on the direction of heat flow. However, the lost work is always positive and therefore will always result in an increase of entropy. The entropy induced by the lost work frequently is referred to as the *entropy production*. When the system undergoes an internally reversible process, there will be no lost work, and Eq. (2-18) will become

$$dS = \left(\frac{\delta Q}{T}\right)_{\text{rev}}$$

Eq. (2-18) can be also written in a rate form and expressed as

$$\frac{dS}{d\tau} = \frac{\dot{Q}_{\text{act}}}{T} + \frac{\delta\left(\dfrac{\text{LW}}{T}\right)}{d\tau} \tag{2-19}$$

By comparing Eq. (2-19) with Eq. (1-8), one would expect the relationship between the lost work and the internally generated entropy in a process to be

$$\sigma = \frac{\text{LW}}{T} \tag{2-20}$$

While the lost work is caused by internal irreversibility, the Carnot work discussed in the last section is the work lost due to external irreversibility. In the real world the Carnot work is the work one would have obtained if Carnot engines were available. Since there is no Carnot engine, the work expected from it is imaginary and does not exist at all. In a sense, the Carnot work is also lost, and the lost work due to an external irreversibility is simply $\text{LW}_c = W_c$. Combining this equation with Eq. (2-11), we have the equation for the Carnot work:

$$\text{LW}_c = \int \delta W_c = -\int\left(1 - \frac{T_0}{T}\right) dQ_{\text{rev}} \tag{2-21}$$

It should be pointed out that the values of lost work and Carnot work are process dependent. For given initial and final states the lost work and Carnot work will be determined by the nature of the process; however, the sum of these two losses will remain constant when the initial and final states are specified. In this text, the combination of lost work and Carnot work ($\text{LW} + \text{LW}_c$) is referred to as the *unavailable work* or *unavailable energy*. The following is a brief discussion of unavailable energy and its use in engineering applications.

We start with the definition of available work:

$$\delta W_{\text{ava}} = \delta W_{\text{rev}} + \delta \text{LW}_c$$

Making use of Eq. (2-13), we have

$$\delta W_{\text{ava}} = \delta W_{\text{act}} + \delta \text{LW} + \delta \text{LW}_c$$

or

$$\delta(\text{LW} + \text{LW}_c) = \delta \text{W}_{\text{ava}} - \delta \text{W}_{\text{act}} \tag{2-22}$$

In words, Eq. (2-22) says that the unavailable energy in the process is the difference between the available work and the actual work. It is the amount of energy that is not available for work conversion. In engineering applications we should minimize the unavailable energy. With Eq. (2-22), we can develop the second law equation in a convenient form. For a closed system, Eq. (2-22) is reduced to

$$\begin{aligned}\delta(\mathrm{LW} + \mathrm{LW}_c) &= -dU_{\mathrm{rev}} + T_0\, dS + dU_{\mathrm{act}} - \delta Q_{\mathrm{act}} \\ &= T_0\, dS - \delta Q_{\mathrm{act}} \qquad (2\text{-}23)\end{aligned}$$

The result may be rearranged and expressed as

$$dS = \frac{\delta(\mathrm{LW} + \mathrm{LW}_c)}{T_0} + \frac{\delta Q_{\mathrm{act}}}{T_0} \qquad (2\text{-}24)$$

Eq. (2-24) can be also expressed in a rate form as

$$\frac{dS}{d\tau} = \frac{1}{T_0}\left(\frac{\delta(\mathrm{LW} + \mathrm{LW}_c)}{\delta\tau} + \dot{Q}_{\mathrm{act}}\right) \qquad (2\text{-}25)$$

When the entropy change due to the mass flow in and out of a control volume is taken into consideration, the second law equation for an open system can be written as

$$\frac{dS_{\mathrm{cv}}}{d\tau} = \sum_i \dot{m}_i s_i - \sum_e \dot{m}_e s_e + \frac{1}{T_0}\sum_j \left(\mathrm{L\dot{W}} + \mathrm{L\dot{W}}_c\right)_j + \frac{1}{T_0}\dot{Q}_{\mathrm{cv}} \qquad (2\text{-}26)$$

Eqs. (2-25) and (2-26) represent another form of the second law equation. These two equations can be used to determine the unavailable energy in the process. As previously stated, the unavailable energy has a unique value that is independent of the process path used in its evaluation. More discussion about these two equations will be presented in the next section.

Example 2-5. For the turbine operation specified in Example 2-3, it is assumed that the air first expands isentropically to 14.7 psia and then changes to the final state by a reversible and isobaric process. Calculate for this entire turbine process (a) the reversible work, (b) the Carnot work, (c) the lost work, and (d) the unavailable energy.

Solution. The turbine process *T-s* diagram is shown in Fig. 2-9. The process 1-x is isentropic, and the process x-2 is reversible and isobaric. The air is treated as an ideal gas with a constant-pressure specific heat of 0.24 Btu/lbm-R.

(a) The reversible work for the entire process is given by

$$w_{\mathrm{rev}} = w_{\mathrm{rev}_{1\text{-}\mathrm{x}}} + w_{\mathrm{rev}_{\mathrm{x}\text{-}2}} = c_p(T_1 - T_{\mathrm{x}}) + 0 \qquad (2\text{-c})$$

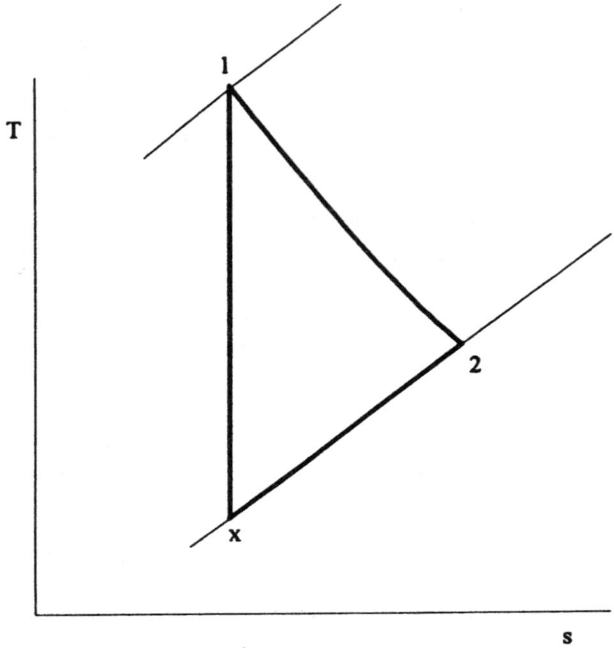

Figure 2-9 A turbine process 1-x-2 in a *T-s* diagram.

The temperature T_x is calculated by using the isentropic relationship

$$\frac{T_x}{T_1} = \left(\frac{P_x}{P_1}\right)^{k-1/k}$$

Substituting the numeric values into the equation gives

$$T_x = 2460\left(\frac{14.7}{150}\right)^{1.4-1.0/1.4} = 1265.9 \text{ R}$$

Then we find the reversible work by using Eq. (2-c)

$$w_{\text{rev}} = 0.24(2460 - 1265.9) = 286.6 \text{ Btu/lbm}$$

(b) The Carnot work for the entire process must be the sum of the Carnot work for the processes 1-x and x-2. That is,

$$w_c = w_{c_{1\text{-x}}} + w_{c_{\text{x-2}}} = 0 - \int_{\text{x}}^{2}\left(1 - \frac{T_0}{T}\right)\delta q \qquad \text{(2-d)}$$

Since the air is an ideal gas, the expression for δq is given by

$$\delta q = c_p \, dT \tag{2-e}$$

First combining Eqs. (2-d) and (2-e) and then integrating along the process x-2, we have the Carnot work for the entire turbine process:

$$w_c = -c_p(T_2 - T_x) + T_0 c_p \ln \frac{T_2}{T_x}$$

$$= -0.24(1460 - 1265.9) + 537(0.24)\ln \frac{1460}{1265.9} = -28.2 \text{ Btu/lbm}$$

The Carnot work is also equal to the available work minus the reversible work. For checking, we make use of calculations in Example 2-3 and we have $w_c = w_{ava} - w_{rev} = 258.2 - 286.6 = -28.4$ Btu/lbm. The results of both methods are in good agreement.

(c) The lost work due to the internal irreversibility is the difference between the reversible work and the actual work. For the turbine process 1-x-2, the lost work is $\text{lw} = w_{rev} - w_{act} = 286.6 - 235.3 = 51.3$ Btu/lbm.

(d) The unavailable energy for the turbine process 1-x-2 is the sum of the Carnot work and the lost work. Numerically it is given by $\text{lw} + \text{lw}_c = 51.3 - 28.4 = 22.9$ Btu/lbm.

The unavailable energy also equals the available work minus the actual work. For checking, we make use of numeric calculations in Example 2-3 and have $\text{lw} + \text{lw}_c = w_{ava} - w_{act} = 258.2 - 235.3 = 22.9$ Btu/lbm. The results by both methods are in good agreement. ■

At this point it must be emphasized that the exact process between the turbine inlet and exit is always unknown. It is very difficult, if not impossible, to determine the path the working substance would follow. The assumptions about the process, like those in the previous examples, are very much needed for calculations and generally are selected for convenience. Because of this, the numeric values of lost work and Carnot work are not unique and depend on the choice of process assumptions. However, this situation does not present any difficulty in application, because the unavailable energy—the sum of lost work and Carnot work—is our main interest. The unavailable energy will remain unchanged when the initial and final states are specified. The following is an example for a closed system.

Example 2-6. For the compression specified in Example 2-4, it is assumed that the air is first compressed isentropically and then expanded reversibly and isothermally to the final state. For this entire process calculate (a) the reversible work, (b) the lost work, (c) the Carnot work, and (d) the unavailable energy.

Solution. The process 1-x-2 in *T-s* diagram is shown in Fig. 2-10.

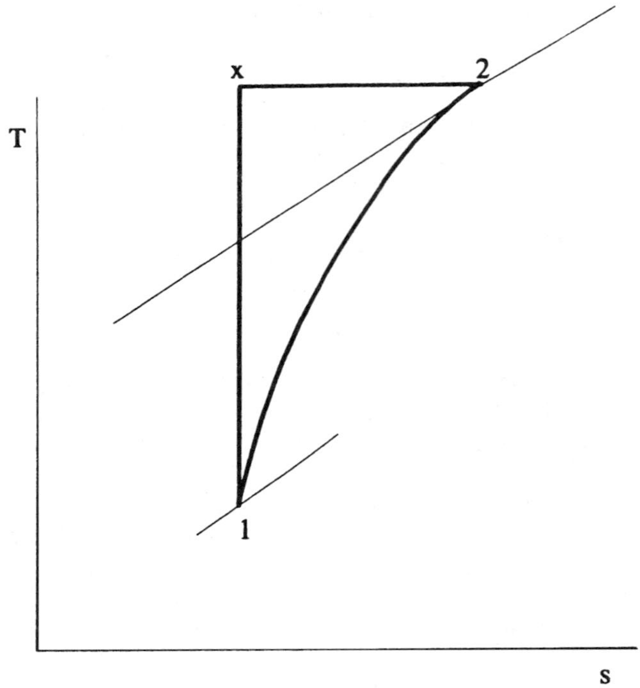

Figure 2-10 A compression process 1-x-2 in a *T-s* diagram.

(a) The reversible work is given by

$$w_{\text{rev}} = w_{\text{rev}_{1\text{-x}}} + w_{\text{rev}_{\text{x-2}}} \tag{2-f}$$

By applying the first-law equation, the reversible work for the process 1-x is

$$w_{\text{rev}_{1\text{-x}}} = c_v(T_x - T_1) = 0.17(757.5 - 560) = -33.58 \text{ Btu/lbm} \tag{2-g}$$

Similarly, the reversible work for the process x-2 is given by the equation

$$w_{\text{rev}_{\text{x-2}}} = T_2(s_2 - s_x) \tag{2-h}$$

Since $s_x = s_1$ in an isentropic process, Eq. (2-h) is further reduced to

$$w_{\text{rev}_{\text{x-2}}} = T_2\left(c_p \ln\frac{T_2}{T_1} - R\ln\frac{P_2}{P_1}\right)$$

Substituting the numeric values into the above equation, we have

$$w_{\text{rev}_{\text{x-2}}} = 757.5\left(0.24\ln\frac{757.5}{560} - \frac{53.3}{778}\ln\frac{50}{20}\right) = 7.35 \text{ Btu/lbm}$$

Thus, using Eq. (2-f), the reversible work for the process 1-x-2 is $w_{\text{rev}} = -33.58 + 7.35 = -26.23$ Btu/lbm.

(b) The lost work is the reversible work minus the actual work. That is, $\text{lw} = w_{\text{rev}} - w_{\text{act}}$. With numeric values, we have $\text{lw} = -26.23 - (-33.58) = 7.35$ Btu/lbm.

(c) The Carnot work for the entire process is given by

$$w_c = w_{c_{1\text{-x}}} + w_{c_{\text{x-2}}} = 0 - \int_{\text{x}}^{2} \left(1 - \frac{T_0}{T_2}\right) \delta q \tag{2-i}$$

Since $\delta q = T_2\, ds$ in the reversible and isothermal process, Eq. (2-i) is further reduced to

$$w_c = -(T_2 - T_0)(s_2 - s_{\text{x}})$$

or

$$w_c = -(T_2 - T_0)\left(c_p \ln \frac{T_2}{T_1} - R \ln \frac{P_2}{P_1}\right)$$

With the numeric values the Carnot work is

$$w_c = -(757.5 - 537)\left(0.24 \ln \frac{757.5}{560} - \frac{53.3}{778} \ln \frac{50}{20}\right) = -2.14 \text{ Btu/lbm}$$

(d) The unavailable energy in the entire process 1-x-2 is the lost work plus the Carnot work. That is, $\text{lw} + \text{lw}_c = 7.35 - 2.14 = 5.21$ Btu/lbm.

The unavailable energy also equals the available work minus the actual work in the process. For checking, we make use of the results in Example 2-4 and have $\text{lw} + \text{lw}_c = w_{\text{ava}} - w_{\text{act}} = -28.37 - (33.58) = 5.21$ Btu/lbm. The results by these two methods are in good agreement. ■

2-4 TOTAL ENTROPY CHANGE

In this section attention will be focused on the total change in the entropy of a system and its surroundings as the system undergoes a process. This consideration will lead to the establishment of another statement for the second law of thermodynamics. In addition, it provides a relationship between the unavailable energy and the total entropy change.

For an open system the second law equation has been developed and expressed as Eq. (2-26). For convenience, the equation is reproduced as follows:

$$\frac{dS_{\text{cv}}}{d\tau} = \sum_i \dot{m}_i \dot{s}_i - \sum_e \dot{m}_e \dot{s}_e + \frac{1}{T_0} \sum_j \left(\dot{\text{LW}} + \dot{\text{LW}}_c\right)_j + \frac{1}{T_0} \dot{Q}_{\text{cv}} \tag{2-27}$$

When this equation is applied to the surroundings of the open system, the second law equation becomes

$$\frac{dS_{sur}}{d\tau} = \sum_e \dot{m}_e s_e - \sum_i \dot{m}_i \dot{s}_i - \frac{1}{T_0}\dot{Q}_{cv} \tag{2-28}$$

where all irreversibilities in the surroundings are assumed to be zero and the heat rejected by the system is exactly equal to that received by the surroundings. Combining Eqs. (2-27) and (2-28) will produce

$$\frac{dS_{cv}}{d\tau} + \frac{dS_{sur}}{d\tau} = \frac{1}{T_0}\sum_j \left(\dot{LW} + \dot{LW}_c\right)_j \tag{2-29}$$

or

$$\frac{dS_{tot}}{d\tau} = \frac{1}{T_0}\sum_j \left(\dot{LW} + \dot{LW}_c\right)_j \tag{2-30}$$

In words, the above equation says that the rate of total entropy change in the system and its surroundings is numerically equal to the rate of change in the unavailable energy divided by the surrounding temperature (T_0). Since the rate of change in unavailable energy is always positive and becomes zero only for the system that undergoes an internally and externally reversible process, Eq. (2-30) can be simply expressed as

$$\frac{dS_{tot}}{d\tau} \geq 0 \tag{2-31}$$

where the equality holds for reversible processes and the inequality for irreversible processes. This equation frequently is referred to as the *principle of entropy production.* The rate of entropy production is used to indicate the quality of a design or the performance of a system operation. When the rate of entropy production is reduced, an improvement in design or system performance has been achieved. At the limit, when the system undergoes an ideal process, the rate of entropy production is expected to be zero.

The concept of unavailable energy is similar to that of entropy production in the system and its surroundings. For a real process, the rate of change in unavailable energy is always positive and would become zero at the limit. It should be pointed out that while the unavailable energy ($lw + lw_c$) must be positive, the Carnot work (lw_c) can be either positive or negative. A similar observation can be made for the net entropy production. That is, the rate of entropy production for the system and its surroundings is always positive, but the rate for the individual system or its surroundings may be positive or negative. There is another similarity between the unavailable energy and the total entropy change between any two given states. Both terms are not process dependent. For a given condition in the surroundings, the unavailable energy and the total change of entropy can be determined by the conditions at the end states.

While Eq. (2-31) has been developed for an open system, the equation is also valid for a closed system. For the closed system the total entropy change refers to the entropy change in the closed system plus the entropy change in its surroundings. Since the thermodynamic properties for a closed system are uniform and well defined at a given moment, Eq. (2-30) can be further simplified to

$$\frac{dS_{tot}}{d\tau} = \frac{\dot{LW} + \dot{LW}_c}{T_0} \tag{2-32}$$

The following are two illustrative examples. The surrounding temperature T_0 is assumed to be 77°F unless otherwise indicated.

Example 2-7. Calculate the total entropy change per pound of air for the turbine process described in Example 2-3.

Solution. The total change of entropy in the turbine process is

$$\Delta s_{tot} = \Delta s_{cv} + \Delta s_{sur} \tag{2-j}$$

Because of the steady-state and steady-flow condition, the term Δs_{cv} must vanish. Then Eq. (2-j) becomes

$$\begin{aligned} \Delta s_{tot} &= \Delta s_{sur} \\ &= s_2 - s_1 + \frac{q_0}{T_0} \\ &= c_p \ln\frac{T_2}{T_1} - R \ln\frac{P_2}{P_1} + \frac{0.02 w_{act}}{T_0} \end{aligned}$$

Substituting the numeric values into the equation, we have the total entropy change as

$$\Delta s_{tot} = 0.24 \ln\frac{1460}{2460} - \frac{53.3}{778}\ln\frac{14.7}{150} + \frac{(0.02)(235.3)}{537} = 0.0428 \text{ Btu/lbm-R}$$

For checking, we multiply the total change of entropy by the surrounding temperature T_0 and have $T_0 \Delta s_{tot} = (537)(0.0428) = 22.9$ Btm/lbm. The result is equal to the value of the unavailable energy (lw + lw_c) as indicated in Example 2-5. ■

Example 2-8. Superheated steam at 2400 psia and 1000°F is throttled to 1800 psia by a valve. Calculate the net change of entropy and the unavailable energy in the process.

Solution. When the valve is treated as a control volume, the net change of entropy for the process is

$$\Delta s_n = \Delta s_{cv} + \Delta s_{sur}$$

Because the process is adiabatic, steady-state, and steady-flow, the equation is further reduced to

$$\Delta s_n = s_e - s_i \tag{2-k}$$

The unavailable energy for this process is $\text{lw} + \text{lw}_c = w_{\text{ava}} - w_{\text{act}}$. Because there is no work produced and the steam enthalpy remains constant in the process, we have

$$\text{lw} + \text{lw}_c = T_0(s_e - s_i) \tag{2-l}$$

Combining Eqs. (2-k) and (2-l) will yield

$$\text{lw} + \text{lw}_c = T_0 \Delta s_n \tag{2-m}$$

It is seen in Eq. (2-m) that the unavailable energy in the process is indeed equal to the net change of entropy times the temperature of the environment. Using the numeric values $h_i = 1460.6$ Btu/lbm, $s_i = 1.5326$ Btu/lbm-R, and $s_e = 1.5642$ Btu/lbm-R, we have the net change of entropy in the process $\Delta s_n = 1.5642 - 1.5326 = 0.0316$ Btu/lbm-R and the unavailable energy in the process $\text{lw} + \text{lw}_c = 537(0.0316) = 16.97$ Btu/lbm. ■

It is of interest to know that Eq. (2-31) can be developed also by using the second law equation, Eq. (1-6). This equation involves the use of lost work or the internally generated entropy. For convenience, Eq. (1-6) is reproduced and expressed as

$$\frac{dS_{\text{cv}}}{d\tau} = \sum_i \dot{m}_i s_i - \sum_e \dot{m}_e s_e + \sum_j \frac{\dot{Q}_i}{T_j} + \dot{\sigma} \tag{2-33}$$

When the second law equation is applied to the surroundings of the system, it gives

$$\frac{dS_{\text{sur}}}{d\tau} = \sum_e \dot{m}_e s_e - \sum_i \dot{m}_i s_i - \sum_j \frac{\dot{Q}_i}{T_0} \tag{2-34}$$

Eq. (2-34) is based on the assumption that there is no internally generated entropy in the surroundings. The assumption is reasonable because engineers are concerned only with the system and not with other activities in the surroundings. When Eqs. (2-33) and (2-34) are added together, it gives the rate of total change in entropy as

$$\frac{dS_{\text{cv}}}{d\tau} + \frac{dS_{\text{sur}}}{d\tau} = \dot{\sigma} + \sum_j \dot{Q}_j \left(\frac{1}{T_j} - \frac{1}{T_0} \right) \tag{2-35}$$

In Eq. (2-35), the first term $\dot{\sigma}$ is always positive by nature and would become zero only when the process is internally reversible. The second term also is positive because both $\dot{Q}_j$ and the quantity $(1/T_j - 1/T_0)$ are positive or negative at the same time. In case that the process is externally reversible (i.e.,

$T_j = T_0$), the second term would vanish, and thus Eq. (2-35) can be reduced to Eq. (2-31).

Eq. (2-35) frequently is rearranged in terms of internal and external irreversibility. For this purpose we modify Eq. (2-35) and express it as follows:

$$T_0\left(\frac{dS_{cv}}{d\tau} + \frac{dS_{sur}}{d\tau}\right) = T_0\dot{\sigma} - \sum_j \dot{Q}_j\left(1 - \frac{T_0}{T_j}\right) \tag{2-36}$$

Let the rate of internal irreversibility be

$$\dot{I}_{int} = T_0\dot{\sigma} \tag{2-37}$$

and the rate of external irreversibility be

$$\dot{I}_{ext} = -\sum_j \dot{Q}_j\left(1 - \frac{T_0}{T_j}\right) \tag{2-38}$$

Then Eq. (2-36) becomes

$$T_0\left(\frac{dS_{cv}}{d\tau} + \frac{dS_{sur}}{d\tau}\right) = \dot{I}_{int} + \dot{I}_{ext} \tag{2-39}$$

Eq. (2-39) indicates that the rate of process irreversibility is linearly related to the rate of the total entropy change in the system and its surroundings. It is also seen that the rate of internal irreversibility is numerically equal to the internally induced entropy times the temperature of the surroundings. Physically, it is related to the lost work in the process. In Eq. (2-38), the rate of external irreversibility is the rate at which the Carnot work is produced. Since a Carnot engine is not available, the Carnot work represents the lost work (lw_c) in the process.

2-5 ADDITIONAL REMARKS ON AVAILABLE WORK

In the preceding sections we have considered closed and open systems that exchange heat solely with the environment. This approach is quite realistic. There are many engineering processes that can be approximated and treated in this fashion. However, in more general circumstances there are some systems that exchange heat with the environment as well as with other bodies. For instance, the water and steam in a boiler receive heat from the products of combustion. Obviously the temperature of combustion products is substantially higher than that of the environment. Another example is a simple heat exchanger. On one side the hot substance loses the heat, while on another side the cold substance receives heat. Both substances have temperatures quite different from that of the environment.

In this section we apply the available work equation to this new situation. Let us consider the heat reservoir to which the system is connected. The heat reservoir is a body of infinite capacity. When the heat reservoir undergoes a process, it has a constant temperature. The heat reservoir produces no work and exchanges heat only with the surroundings. On this basis we start with Eq. (2-9) for the available work by a heat reservoir:

$$dW_{\text{ava}} = -dU + T_0\, dS \tag{2-40}$$

Since a heat reservoir undergoes only an internally reversible process and produces no work, we have $dU = \delta Q_r$ and $dS = \delta Q_r/T_r$. Substituting these two terms into Eq. (2-40), we have the expression for the available work by a heat reservoir as

$$\delta W_{\text{ava}} = -\left(1 - \frac{T_0}{T_r}\right)\delta Q_r$$

Integrating the above equation gives

$$W_{\text{ava}} = -\left(1 - \frac{T_0}{T_r}\right)Q_r \tag{2-41}$$

where T_r is the temperature of the reservoir and Q_r is the heat transferred from the reservoir. Physically, the available work represents the maximum amount of work that could be produced by the heat reservoir when it releases the heat. Equation (2-41) is equally valid for a heat reservoir when it is connected with an open system.

When the system exchanges heat with its environment as well as with the heat reservoir, the available work for the process simply is the sum of the available work for the system and the available work for the heat reservoir. In equation form, for a closed system we have

$$W_{\text{ava}} = (U_1 - U_2) - T_0(S_1 - S_2) - \left(1 - \frac{T_0}{T_r}\right)Q_r \tag{2-42}$$

and for an open system we have

$$W_{\text{ava}} = (H_1 - H_2) - T_0(S_1 - S_2) - \left(1 - \frac{T_0}{T_r}\right)Q_r \tag{2-43}$$

Example 2-9. Compressed air is stored in a vessel with a volume of 20 ft^3. Because of heat transfer from a heat reservoir at 1500 R, the compressed air is changed from 25 psia and 100°F to 660°F. Calculate the available work for the process.

Solution. Compressed air is treated as an ideal gas. The mass of air in the vessel is given by

$$m_1 = \frac{P_1V_1}{RT_1} = \frac{(25)(144)(20)}{(53.3)(560)} = 2.41 \text{ lbm}$$

Since the reservoir has a temperature substantially greater than the compressed air, the heat transfer between them is calculated by the first law equation. That is, $Q - W = m_1(u_2 - u_1)$. Since $W = 0$, we have $Q = m_1c_v(T_2 - T_1) = 2.41 \times 0.171 \times (1120 - 560) = 231$ Btu.

The available work for the process is the available work for the compressed air plus the available work for the heat reservoir. First, we calculate the available work for the compressed air as

$$\begin{aligned}
W_{\text{ava, air}} &= m_1[(u_1 - u_2) - T_0(s_1 - s_2)] \\
&= m_1\left[c_v(T_1 - T_2) - T_0\left(c_v \ln\frac{T_1}{T_2}\right)\right] \\
&= 2.41\left[0.171(560 - 1120) - 537 \times 0171 \ln\frac{560}{1120}\right] \\
&= -77.39 \text{ Btu}
\end{aligned}$$

Second, the available work by the heat reservoir is calculated as

$$W_{\text{ava, res}} = -Q_r\left(1 - \frac{T_0}{T_r}\right) = 231\left(1 - \frac{537}{1500}\right) = 148.3 \text{ Btu}$$

Thus, the available work for the process is $W_{\text{ava}} = W_{\text{ava, air}} + W_{\text{ava, res}} = -77.39 + 148.3 = 70.91$ Btu. ■

For the above example, there is no work produced and, therefore, the unavailable energy is numerically equal to the available work. Physically, it represents the amount of energy not available for work conversion because of irreversibility between the compressed air (i.e., the system) and the heat reservoir.

Now we extend the above considerations to a heat exchanger. When the heat exchanger is well insulated and the two streams (hot and cold) are treated as two individual systems, we will see that one system exchanges heat with another system and that the environment is not involved. Following the procedure discussed, we first calculate the available works for the individual hot and cold streams as

$$W_{\text{ava}_h} = m_h(h_i - h_e)_h - m_hT_0(s_i - s_e)_h \tag{2-44}$$

and

$$W_{\text{ava}_c} = m_c(h_i - h_e)_c - m_cT_0(s_i - s_e)_c \tag{2-45}$$

Then, the available work for the heat exchanger process is simply

$$W_{\text{ava}} = W_{\text{ava}_h} + W_{\text{ava}_c} \tag{2-46}$$

The numerical calculation is left as an exercise in the text.

2-6 PROBLEMS

2-1 Air enters a turbine at 140 psia and 2000°F and expands in an adiabatic process to 15 psia and 1200°F. Calculate the actual work and available work per unit mass, assuming a constant specific heat of 0.24 Btu/lbm-R.

2-2 For Problem 2-1, calculate the reversible work, lost work, and the Carnot work per unit mass using the following two sets of conditions: (a) The air first undergoes an isentropic process and then an isothermal process to reach the final state; (b) The air first undergoes an isothermal process and then an isentropic process to reach the final state.

2-3 A rigid vessel of 10 ft^3 contains steam at the conditions of 300°F and 80% quality. The vessel is then cooled to 80°F. Calculate the available work in the process. What is the physical meaning of this available work?

2-4 Consider the throttling process across the expansion valve in a vapor-compression refrigeration system. Estimate (a) actual work, (b) available work, and (c) reversible work for the conditions below:

Substance:	Ammonia
Inlet conditions:	200 psia, 90°F
Exit conditions:	40 psia

For the calculation of reversible work, ammonia is assumed to undergo first a reversible and isothermal process and then an isentropic process.

2-5 Saturated water enters a boiler feedwater pump at 160 psia and leaves at 3000 psia. Assuming the pump has an efficiency of 70%, calculate the actual work and the available work per pound of water. What is the physical meaning of this available work?

2-6 Steam enters an insulated nozzle at 100 psia and 440°F at low velocity and exits at 30 psia. The nozzle has an efficiency of 95%. Calculate the available work for this nozzle process.

2-7 Air enters a compressor at 15 psia and 80°F and leaves at 120 psia. Assuming the process is isothermal and internally reversible, calculate (a) the actual work, (b) the available work, and (c) the reversible work.

2-8 Two perfectly insulated vessels (A and B) are connected by a valve. Vessel A has a volume of 20 ft^3 and contains steam at 20 psia and 400°F. Vessel B has a volume of 10 ft^3 and contains steam at 80 psia and 80% quality. The valve is then opened, and the steam in the two vessels is allowed to come to a uniform state. What is the final pressure? What is the available work in the process?

2-9 Air preheaters frequently are installed in fossil fuel power plants. Consider the air preheater operating under the conditions below:

Hot side	
Flow rate	100,000 lbm/hr
Specific heat	0.26 Btu/lbm-R
Inlet temperature	600°F
Outlet temperature	400°F
Cold side	
Flow rate	96,000 lbm/hr
Specific heat	0.24 Btu/lbm-R
Inlet temperature	90°F

The heat loss from the air preheater is negligible. What is the available work for this heat exchanger process?

2-10 Compressed air enters an underground storage system at the conditions of 300 psia and 100°F. The storage system has a volume 1,000,000 ft^3 and initial conditions of 50 psia and 80°F. The charging process stops at the time when the storage pressure reaches 300 psia. Assuming a negligible heat loss from the storage system, determine the available work and the net change of entropy for the process.

CHAPTER
THREE
AVAILABILITY EQUATIONS

There are several approaches that engineers may use to quantify the irreversible nature of real processes and to perform analyses based on the combined first and second laws of thermodynamics. One approach makes use of the concept of unavailable energy. As shown in Chapter 2, the unavailable energy is the work that could have been produced but actually was not because of the irreversibility in the process. In design, engineers take every step to minimize the unavailable energy for a given process or engineering system. Another approach is the application of the principle of entropy production. According to this principle, the total change of entropy measures the departure of the real process from the ideal process. The total change of entropy will approach zero as the process approaches both internal and external reversibility. Thus the process design is an optimization by minimizing the total change of entropy in the process.

In this chapter we will develop a convenient approach for thermodynamic analysis. First we define availability, and then we make use of the concept of availability to develop the availability equations. As shown later, the availability equations are based on the combination of the first and second laws of thermodynamics. The analysis involving the use of availability equations is referred to as *availability analysis* in this text.

3-1 AVAILABILITY

The available work discussed in Chapter 2 represents the work one would expect from an ideal process. By an ideal process we mean that the process is both internally and externally reversible. Numerically, the available work is the maximum amount of the work that could be extracted from a substance as it undergoes a process. For a substance entering an open system at state 1 and leaving at state 2 under steady-state, steady-flow conditions, the maximum work we can extract is given by

$$w_{ava} = (h_1 - h_2) - T_0(s_1 - s_2) \tag{3-1}$$

provided that the changes in kinetic and potential energy are negligible. Physically, the available work by Eq. (3-1) is the work-producing potential of the substance for the process between the states 1 and 2. When the conditions at the system exit are identical to those in the environment, the available work between state 1 and the environment state would represent the work potential for the substance at state 1. In this text the work potential of the substance at any state is defined as the *availability* of the substance, which is given by

$$a_f = (h - h_0) - T_0(s - s_0) \tag{3-2}$$

The subscript f denotes an open system (or flow system). To be more precise, the availability defined by Eq. (3-2) should be referred to as a flow availability. In Eq. (3-2), T_0, h_0, and s_0 are properties of the substance in temperature and pressure equilibrium with the environment (T_0, P_0).

Equation (3-2) indicates that the flow availability is a composite quantity. The flow availability at a state is a function of the properties at the same state for a specified environmental condition. In other words, the flow availability would act just like any other thermodynamic property for a given set of environmental temperature and pressure (T_0, P_0). Unless otherwise indicated, the environmental temperature and pressure are assumed to be 77°F and 14.7 psia, respectively.

Similarly, we can develop the definition for the nonflow availability of a substance. For a closed system, if the changes in kinetic and potential energy are negligible, the available work between any two states 1 and 2 is given by

$$w_{ava} = (u_1 - u_2) - T_0(s_1 - s_2) \tag{3-3}$$

It should be pointed out that the available work from Eq. (3-3) is not entirely useful. A portion of this work must be used against the surroundings as the system expands from the initial state to the final state. The work done against the surroundings based on a unit of mass of the substance is

$$W_{sur} = P_0(v_2 - v_1) \tag{3-4}$$

Thus the useful available work between any two states 1 and 2 is

$$\begin{aligned} w_{ava,\,useful} &= w_{ava} - w_{sur} \\ &= (u_1 - u_2) + P_0(v_1 - v_2) - T_0(s_1 - s_2) \end{aligned} \tag{3-5}$$

When the final state of a substance is in equilibrium with the environment, the available work by Eq. (3-5) will become the work-producing potential of the substance at state 1. In this text the work potential at a state is defined as the *nonflow availability* of the substance. In equation form, the nonflow availability per unit of mass is given by

$$a = (u - u_0) + P_0(v - v_0) - T_0(s - s_0) \tag{3-6}$$

So far, we have developed the flow and nonflow availabilities. Both represent the work-producing potentials for a substance at a given condition in the environment. The flow availability and nonflow availability are related by the equation

$$a = a_f + v(P_0 - P) \tag{3-7}$$

There is also a relationship between the available work and the change of availability in a process. We have for an open system

$$w_{\text{ava}} = a_{f_1} - a_{f_2} \tag{3-8}$$

and for a closed system

$$w_{\text{ava, useful}} = a_1 - a_2 \tag{3-9}$$

Example 3-1. Consider the steam throttling process through an expansion valve. Steam enters the valve at 1900 psia and 1000°F and leaves at 1800 psia. Calculate the flow availability per unit mass of steam at the valve inlet and exit.

Solution. The flow availability for steam at the valve inlet is $a_{f_i} = (h_i - h_0) - T_0(s_i - s_0)$. The state at the inlet is determined by $P_i = 1900$ psia and $T_i = 1000$°F. From the steam table we have $h_i = 1477.4$ Btu/lbm and $s_i = 1.5672$ Btu/lbm-R. The steam conditions at the environmental state are approximated by the properties of saturated water at $T_0 = 77$°F. That is, $h_0 = 45.09$ Btu/lbm and $s_0 = 0.0877$ Btu/lbm-R. Substituting these numerical values into the above equation, we have the steam flow availability at the valve inlet $a_{f_i} = (1477.4 - 45.09) - 537(1.5672 - 0.0877) = 637.8$ Btu/lbm.

Similarly, the flow availability for steam at the valve exit is $a_{f_e} = (h_e - h_0) - T_0(s_e - s_0) = (1477.4 - 45.09) - 537(1.5725 - 0.0877) = 634.9$ Btu/lbm. The throttling process has resulted in a decrease of flow availability. From the thermodynamic point of view, the throttling process is generally inefficient. ■

Example 3-2. Calculate the nonflow availability for air at two states: 150 psia and 2000°F, and 10 psia and 20°F.

Solution. The nonflow availability is given by the equation

$$a = (u - u_0) + P_0(v - v_0) - T_0(s - s_0) \tag{3-a}$$

For state 1, we have

$$u_1 - u_0 = c_v(T_1 - T_0) = 0.171(2460 - 537) = 328.8 \text{ Btu/lbm}$$

$$\begin{aligned} P_0(v_1 - v_0) &= P_0 R\left(\frac{T_1}{P_1} - \frac{T_0}{P_0}\right) \\ &= \frac{(14.7)(53.3)}{778}\left(\frac{2460}{150} - \frac{537}{14.7}\right) \\ &= -20.27 \text{ Btu/lbm} \end{aligned}$$

$$\begin{aligned} T_0(s_1 - s_0) &= T_0\left(c_p \ln\frac{T_1}{T_0} - R\ln\frac{P_1}{P_0}\right) \\ &= 537\left(0.24\ln\frac{2460}{537} - \frac{53.3}{778}\ln\frac{150}{14.7}\right) \\ &= 110.71 \text{ Btu/lbm} \end{aligned}$$

Substituting these values into Eq. (3-a), we have the nonflow availability at state 1: $a_1 = 328.8 - 20.27 - 110.71 = 197.82$ Btu/lbm. Similarly, the nonflow availability at state 2 is

$$\begin{aligned} a_2 &= 0.171(480 - 537) + \frac{(14.7)(53.3)}{778}\left(\frac{480}{10} - \frac{537}{14.7}\right) \\ &\quad - 537\left(0.24\ln\frac{480}{537} - \frac{53.3}{778}\ln\frac{10}{14.7}\right) \\ &= 2.09 \text{ Btu/lbm} \end{aligned}$$

At state 2, the nonflow availability for air is positive even though the pressure and temperature are lower than those of the environment. In fact, the nonflow availability is always positive for all substances. ■

3-2 FURTHER ASPECTS OF AVAILABILITY

Availability is a measure of the departure for the state of the system from the state of the environment. When the state of the system becomes identical to that of the environment, there will be no availability in the system, and the system will have no work potential relative to the environment. Based on the concept of work potential, the system should have additional availability if the system's kinetic and potential energies are different from those in the environment. Since all kinetic and potential energy theoretically can be converted into useful work, the availability associated with these energies must be numerically equal to those of the energies. Thus, when the kinetic and potential energy are taken into consideration, the flow availability and nonflow availability can be respectively rewritten as

$$a_f = (h - h_0) - T_0(s - s_0) + \frac{V^2}{2} + gz \tag{3-10}$$

and

$$a = (u - u_0) + P_0(v - v_0) - T_0(s - s_0) + \frac{V^2}{2} + gz \tag{3-11}$$

The kinetic and potential energies in the environment are assumed to be zero.

The concept of availability associated with work potential is useful. Consider a system that receives work from the surroundings. For this work interaction the system will have its work potential increased, and therefore its availability will also increase. Since 100 percent of the work received theoretically can be converted back into work by the system, the availability increase in the system must be exactly equal to the work received by the system. In equation form, the availability associated with work transfer is

$$a_w = -w_{\text{act}} \tag{3-12}$$

The negative sign accounts for the fact that the work is positive when it is produced by the system and negative when it is received by the system. In other words, the system will have an availability change when the system has a work interaction with the surroundings. When the system experiences a volume change during a process, the work done against the surroundings also must be considered. In this case, Eq. (3-12) should be rewritten as

$$a_w = -(w_{\text{act}} - w_{\text{sur}}) \tag{3-13}$$

or

$$a_w = -w_{\text{act}} + P_0(v_2 - v_1) \tag{3-14}$$

In this text, a_w is called the *work availability transfer*. The availability change associated with the heat transfer between the system and the surroundings also can be determined by using the concept of work potential. Consider a system that loses heat to the surroundings. Because of the heat loss, the system is expected to have a reduction in work potential. Since not all heat energy can be converted into work, the reduction of work potential must be somewhat less than the amount of heat transferred out from the system. In an ideal process the conversion is given by

$$\delta w_c = -\left(1 - \frac{T_0}{T}\right)\delta q \tag{3-15}$$

where δq and δw_c represent the heat loss from the system and the reduction of work potential, respectively. The right-hand side of Eq. (3-15) is recognized as the work by a Carnot engine or any other reversible cycle device operating between the temperature T and the temperature of environment T_0. Thus the availability change for the system becomes

$$\delta a_q = \left(1 - \frac{T_0}{T}\right)\delta q \tag{3-16}$$

For a process between any two states 1 and 2, therefore, the availability change associated with the heat transfer between the system and the surroundings is given by

$$a_q = \int_1^2 \left(1 - \frac{T_0}{T}\right)\delta q \tag{3-17}$$

In summation form Eq. (3-17) becomes

$$a_q = \sum_j \left(1 - \frac{T_0}{T_j}\right)\delta q_j \tag{3-18}$$

In this text, a_q is called the *heat availability transfer*. Before we conclude this section, let us summarize the availability and the specific availability as below:

Nonflow availability	$A = ma$
Flow availability	$A_f = ma_f$
Work availability transfer	$A_w = ma_w$
Heat availability transfer	$A_q = ma_q$

3-3 AVAILABILITY EQUATION FOR A CLOSED SYSTEM

When a closed system has a unit mass of a substance and has a negligible change in kinetic and potential energies in a process 1-2, the first law equation is given by

$$u_2 - u_1 = \int_1^2 \delta q - w \tag{3-19}$$

and the second law equation is

$$s_2 - s_1 = \int_1^2 \frac{\delta q}{T} + \sigma \tag{3-20}$$

where σ is the entropy production due to internal irreversibility. Multiplying Eq. (3-20) by the temperature of environment T_0 and then subtracting the resultant equation from Eq. (3-19), we have

$$u_2 - u_1 - T_0(s_2 - s_1) = \int_1^2 \left(1 - \frac{T_0}{T}\right)\delta q - w - T_0\sigma \tag{3-21}$$

Since the nonflow availabilities at states 1 and 2 are respectively

$$a_1 = (u_1 - u_0) + P_0(v_1 - v_0) - T_0(s_1 - s_0)$$

and

$$a_2 = (u_2 - u_0) + P_0(v_2 - v_0) - T_0(s_2 - s_0)$$

Eq. (3-21) can be rewritten as

$$a_2 - a_1 = \int_1^2 \left(1 - \frac{T_0}{T}\right) \delta q - w + P_0(v_2 - v_1) - T_0 \sigma \tag{3-22}$$

We recognize the heat availability transfer as

$$a_q = \int_1^2 \left(1 - \frac{T_0}{T}\right) \delta q \tag{3-23}$$

the work availability transfer as

$$a_w = -w + P_0(v_2 - v_1) \tag{3-24}$$

and the internal irreversibility as

$$i = T_0 \sigma \tag{3-25}$$

Substituting these terms into Eq. (3-22), we have

$$a_2 - a_1 = a_q + a_w - i \tag{3-26}$$

Eq. (3-26) generally is referred to as the *availability equation for closed systems*. It says that the availability change in a closed system can be accounted for in terms of heat availability transfer (a_q), work availability transfer (a_w), and internal irreversibility (i). As indicated in the last chapter, the internal irreversibility (i) is related to the lost work (LW) and now in Eq. (3-26) can be regarded as the availability consumed in a process. The term i frequently is referred to also as the *availability destruction*. By nature, the availability destruction is always positive and would approach zero only if the process were internally reversible. The heat availability transfer (a_q), however, can be positive or negative, depending on the direction of the heat flow. When the system transfers heat to its environment, it usually transfers the availability to its environment. This availability change is called the availability loss in the process. In this sense, Eq. (3-26) is the equation of availability balance for a closed system.

In engineering applications the availability equation frequently is needed for a closed system for which the mass is different from unity. For this case, we multiply Eq. (3-26) by the mass of the system and write the equation as

$$A_2 - A_1 = A_q + A_w - I \tag{3-27}$$

where

$$A_1 = ma_1$$

$$A_2 = ma_2$$

$$A_q = ma_q = \int_1^2 \left(1 - \frac{T_0}{T}\right) \delta Q$$

$$A_w = ma_w = -W + P_0(V_2 - V_1)$$

$$I = mi = mT_0 \sigma$$

In a rate form, Eq. (3-27) is written simply as

$$\frac{dA}{d\tau} = \frac{dA_q}{d\tau} + \frac{dA_w}{d\tau} - \frac{dI}{d\tau}$$

or

$$\dot{A} = \dot{A}_q + \dot{A}_w - \dot{I} \tag{3-28}$$

The following examples are used to illustrate the use of the availability equation for closed systems.

Example 3-3. A steam storage tank has a volume of 0.8 ft^3. The tank is filled with dry saturated steam at a pressure of 20 psia, and all valves to the tank are closed. After a period of time, the steam pressure in the tank drops to 10 psia. For the steam in the process, calculate (a) the heat loss, (b) the nonflow availability change, and (c) the availability loss and destruction.

Solution. The steam in the tank is a closed system. The heat transferred from the steam is determined by the first law equation, $\delta Q - \delta W = dU$. Since there is no work in the process, integrating the equation from the initial state to the final state will give

$$Q = m_1(u_2 - u_1) \tag{3-b}$$

At this point we determine the steam properties in both states. The results are summarized in Table 3-1.

Table 3-1 Steam properties for tank in Example 3-3.

State location	v (ft^3/lbm)	x	u (Btu/lbm)	s (Btu/lbm-R)
1	20.09	1.0	1082.0	1.7320
2	20.09	0.523	637.4	1.0698

The mass inside the tank is $m = V/v = 0.8/20.09 = 0.0389$ lbm. Substituting this value and the steam properties into Eq. (3-b), we have the heat loss from the steam $Q = 0.0389(637.4 - 1082) = -17.29$ Btu.

The nonflow availability change for the steam is given by

$$A_2 - A_1 = m[(u_2 - u_1) + P_0(v_2 - v_1) - T_0(S_2 - S_1)]$$

Substituting the numeric values into the equation, we have $A_2 - A_1 = 0.0389[(637.4 - 1082) + 0 - 537(1.0698 - 1.7320)] = -3.46$ Btu.

To determine the availability loss and destruction in the process, we apply the availability equation $I - A_q = A_w + A_1 - A_2$. Since $A_w = 0$, we have the availability loss and destruction equal to the decrease of flow availability, that is, $I - A_q = 3.46$ Btu.

Before the example is concluded, several observations should be made. First, the steam in the tank experiences an availability drop in the process. This availability drop represents the availability loss and destruction for the process. Since the process is internally reversible, there is no availability destruction in the process. Thus, the availability drop is due entirely to the heat transferred from the storage system. In this example the availability transfer associated with the heat loss is an availability loss. Since the heat availability transfer is not utilized, it is the same as the availability destruction. In fact, in some thermodynamic analyses these two terms frequently are grouped into one term. ■

Let us again consider the same example, but this time let us treat the tank (including the steam inside) and its immediate surroundings as a closed system. For this newly defined closed system, there will be, as before, no work. Because of this, the availability drop in the steam will be exactly equal to the availability destruction in this newly defined closed system.

Example 3-4. A rigid, well-insulated tank has a volume of 10 ft^3 and initially is filled with water at 100°F and 20 psia. The water is heated to 200°F by an electrical resistor. Calculate the change of nonflow availability in the water and the availability destruction in the process.

Solution. We take the water inside the tank as a closed system. The mass of the water is $m = V/v_1 = 10/0.01613 = 620$ lbm. The first law equation for the closed system is $Q - W = \Delta U$. Since $Q = 0$, we have the work in the process as $W = -m(u_2 - u_1) = -mc_v(T_2 - T_1) = -620(1)(200 - 100) = -62 \times 10^3$ Btu.

To determine the change of nonflow availability in the water, we apply the equation $A_2 - A_1 = m(a_2 - a_1) = m[(u_2 - u_1) - T_0(s_2 - s_1) + P_0(v_2 - v_1)]$. Since the volume of water remains essentially unchanged, the above equation can be further modified to

$$A_2 - A_1 = m\left[c_v(T_2 - T_1) - T_0 c_v \ln \frac{T_2}{T_1}\right]$$

Substituting the numeric values into the equation, we have the change of nonflow availability as

$$A_2 - A_1 = 620\left[(1)(660 - 560) - 537(1)\ln\frac{660}{560}\right]$$
$$= 7297 \text{ Btu}$$

The availability destruction in the process is determined by the availability equation $I = A_w - (A_2 - A_1)$. Since $A_w = -W = 62 \times 10^3$ Btu, we have the availability destruction as $I = 62{,}000 - 7297 = 54{,}703$ Btu.

Evidently the water-heating process in this example is very inefficient. The system receives 62,000 Btu of electrical energy and increases the availability in the water by only 7297 Btu. The difference is the amount of availability consumed in the process. In this example the availability consumption, or destruction, constitutes approximately 88 percent of the process input. ■

Example 3-5. Air at 600 psia and 600°F expands in a reversible and isothermal process in a cylinder-piston arrangement. The final pressure in the process is 14.7 psia. On the basis of unit mass, calculate (a) the nonflow availability change, (b) the heat availability transfer, and (c) the availability destruction in the process.

Solution. When the air in the cylinder is treated as a closed system, the availability equation for this process is given by Eq. (3-22), that is,

$$a_2 - a_1 = \int_1^2 \left(1 - \frac{T_0}{T}\right)\delta q - w + P_0(v_2 - v_1) - i \qquad \text{(3-c)}$$

For the nonflow availability change, we have

$$a_2 - a_1 = (u_2 - u_1) + P_0(v_2 - v_1) - T_0(s_2 - s_1)$$
$$= c_v(T_2 - T_1) + P_0\left(\frac{RT_2}{P_2} - \frac{RT_1}{P_1}\right) - T_0\left(c_p \ln\frac{T_2}{T_1} - R\ln\frac{P_2}{P_1}\right)$$
$$= 0 + \frac{14.7}{778}\left(\frac{(53.3)(1060)}{14.7} - \frac{(53.3)(1060)}{600}\right) - 537\left(0 - \frac{53.3}{778}\ln\frac{14.7}{600}\right)$$
$$= -65.65 \text{ Btu/lbm}$$

To determine the availability transfer due to the heat transfer in the process, we first calculate the heat transfer by the equation

$$q = T_1(s_2 - s_1) = T_1\left(c_p \ln\frac{T_2}{T_1} - R\ln\frac{P_2}{P_1}\right) = 1060\left(0 - \frac{53.3}{778}\ln\frac{14.7}{600}\right)$$
$$= 269.35 \text{ Btu/lbm}$$

Then, the heat availability transfer is given by

$$a_q = \int_1^2 \left(1 - \frac{T_0}{T}\right) \delta q$$

After integrating this equation, we have the heat availability transfer as

$$a_q = \left(1 - \frac{T_0}{T}\right) q = \left(1 - \frac{537}{1060}\right)(269.35) = 132.90 \text{ Btu/lbm}$$

To determine the availability destruction for the process, we first calculate the useful work produced by the closed system. In this case the useful work is given by

$$\begin{aligned} w_{\text{useful}} &= w - P_0(v_2 - v_1) = q - P_0\left(\frac{RT_2}{P_2} - \frac{RT_1}{P_1}\right) \\ &= 269.35 - \frac{14.7}{778}\left(\frac{53.3 \times 1060}{14.7} - \frac{53.3 \times 1060}{600}\right) \\ &= 198.54 \text{ Btu/lbm} \end{aligned}$$

Substituting the numeric values into Eq. (3-c) gives $-65.64 = 132.90 - 198.54 - i$, and the availability destruction for the process is therefore $i = 0$ Btu/lbm.

The zero availability destruction confirms that the process is indeed internally reversible. In terms of the availability balance, the calculation indicates that the system receives 132.90 Btu of heat availability transfer and produces 198.54 Btu of work availability. The difference must be equal to the change of nonflow availability change. For this example the nonflow availability decreases by 65.64 Btu. ■

Before the conclusion of this section, it is of interest to apply the availability equation to a thermal reservoir. A thermal reservoir is a body of infinite capacity and therefore would not experience any change in temperature and pressure in an interaction with its surroundings. When the thermal reservoir is treated as a closed system, the availability equation would become

$$A_2 - A_1 = A_q + A_w - I \tag{3-29}$$

The term $(A_2 - A_1)$ is the decrease of availability in the thermal reservoir. It is given by

$$A_2 - A_1 = (U_2 - U_1) - T_0(S_2 - S_1)$$

Since $U_2 - U_1 = Q_{1\text{-}2}$ and $S_2 - S_1 = Q_{1\text{-}2}/T_r$, we have

$$A_2 - A_1 = \left(1 - \frac{T_0}{T_r}\right) Q_{1\text{-}2}$$

where $Q_{1\text{-}2}$ is the amount of heat transferred from the thermal reservoir and T_r is the reservoir temperature.

The term A_q is the heat availability transfer from the thermal reservoir. As previously indicated, it is

$$A_q = \left(1 - \frac{T_0}{T_r}\right) Q_{1\text{-}2}$$

The term A_w is the work availability transfer. Since the thermal reservoir does not produce mechanical work, we have $A_w = 0$. Substituting these terms into Eq. (3-29), we have zero internal irreversibility for the thermal reservoir. In other words, the availability decrease in the reservoir is exactly equal to the heat availability transferred from it.

3-4 AVAILABILITY EQUATION FOR AN OPEN SYSTEM

The availability equation for an open system in a uniform-state, uniform-flow process can be developed with the first and second laws of thermodynamics. Neglecting the changes of kinetic and potential energies, we write the first law equation as

$$\sum_i \Delta m_i h_i - \sum_e \Delta m_e h_e + Q_{\text{cv}} - W_{\text{cv}} = (m_2 u_2 - m_1 u_1)_{\text{cv}} \tag{3-30}$$

and the second law equation as

$$\sum_i \Delta m_i s_i - \sum_e \Delta m_e s_e + \sum_j \frac{\delta Q_j}{T_j} + \sigma = (m_2 s_2 - m_1 s_1)_{\text{cv}} \tag{3-31}$$

Secondly, we multiply Eq. (3-31) by the temperature of the environment, T_0, and then subtract the resulting equation from Eq. (3-30). This will lead to

$$\sum_i \Delta m_i (h_i - T_0 s_i) - \sum_e \Delta m_e (h_e - T_0 s_e) + \sum_j \left(1 - \frac{T_0}{T_j}\right) \delta Q_j - W_{\text{cv}} - T_0 \sigma$$
$$= m_2(u_2 - T_0 s_2) - m_1(u_1 - T_0 s_1) \tag{3-32}$$

Next, we make use of the definitions of the flow and nonflow availability and obtain the equation in the following form:

$$\sum_i \Delta m_i a_{f_i} - \sum_e \Delta m_e a_{f_e} + \sum_j \left(1 - \frac{T_0}{T_j}\right) \delta Q_j - W_{\text{cv}} + P_0(V_2 - V_1) - T_0 \sigma$$
$$= m_2 a_2 - m_1 a_1 \tag{3-33}$$

Eq. (3-33) can be further reduced to

$$A_2 - A_1 = A_q + A_w + A_{f_i} - A_{f_e} - I \tag{3-34}$$

where

$$A_1 = m_1 a_1$$

$$A_2 = m_2 a_2$$

$$A_q = \sum_j \left(1 - \frac{T_0}{T_j}\right) \delta Q_j$$

$$A_w = -W_{\text{cv}} + P_0(V_2 - V_1)$$

$$A_{f_i} = \sum \Delta m_i a_{f_i}$$

$$A_{f_e} = \sum \Delta m_e a_{f_e}$$

$$I = T_0 \sigma$$

The term $(A_2 - A_1)$ represents the change of nonflow availability within the control volume. Under steady-state and steady-flow conditions, this term will vanish. The term A_q denotes the availability transfer due to the heat transfer between the control volume and its surroundings. The term A_w is the work availability transfer. Numerically, the term A_w is equal to the negative value of the work produced by the control volume. In most cases the control volume has a constant volume, and therefore A_w can be further simplified. The terms A_{f_i} and A_{f_e} represent the flow availability and numerically depend on the inlet and exit conditions, respectively. Finally, the term I is the availability destruction in the process. Numerically, the availability destruction is the internally induced entropy in the control volume times the temperature of the environment, T_0. When the process is internally reversible, this term will vanish as expected.

Eq. (3-34) generally is referred to as the *availability equation for an open system*. Under steady-state and steady-flow conditions, there is no change of nonflow availability within the control volume, and so Eq. (3-34) can be further reduced to

$$A_q + A_w + A_{f_i} - A_{f_e} = I \tag{3-35}$$

Both Eqs. (3-34) and (3-35) indicate that availability, unlike energy, does not observe the principle of conservation. In a real process, some availability will be consumed. When the availability consumption (or destruction) is taken into account, the availability equation becomes the availability balance. The term availability balance implies that the availability flow in and out of the control volume must be equal to the change of availability plus the availability destruction within the control volume. The examples below are used to illustrate the use of the availability equation.

Example 3-6. Hot water at 100 psia and 300°F enters an insulated flash evaporator at a constant rate. Saturated steam is drawn from the top of the evaporator, and saturated water is drawn from the bottom. The pressure inside the flash evaporator is maintained at 40 psia. Calculate the availability destruction per pound of water entering the evaporator.

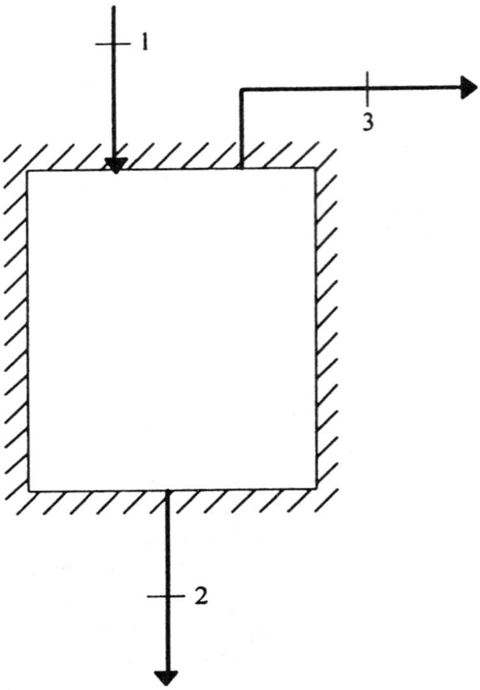

Figure 3-1 An insulated flash evaporator.

Solution. Figure 3-1 shows the control volume with an inlet at location 1 and exits at locations 2 and 3. From the steam table we can determine the steam properties at these three locations and summarize them in Table 3-2.

The flow availabilities are calculated by the equation

$$a_f = (h - h_0) - T_0(s - s_0)$$

The results also are included in Table 3-2. For these calculations, the environmental temperature and pressure are assumed to be 77°F and 14.7 psia, respectively. Accordingly, we have $h_0 = 45.09$ Btu/lbm and $s_0 = 0.0877$ Btu/lbm-R.

Table 3-2 Steam properties for evaporator in Example 3-6.

State location	h (Btu/lbm)	s (Btu/lbm-R)	a_f (Btu/lbm)
1	269.7	0.4370	37.04
2	236.2	0.3921	27.64
3	1170.0	1.6767	271.61

To determine the availability destruction in this process, we first calculate the mass flow at the exit locations from the following two equations:

$$m_1 = m_2 + m_3 \tag{3-d}$$

$$m_1 h_1 = m_2 h_2 + m_3 h_3 \tag{3-e}$$

Setting $m_1 = 1$ lbm and making use of the enthalpies at three locations, we have $m_2 = 0.964$ lbm and $m_3 = 0.036$ lbm.

Because there is no heat loss or work from the control volume, the availability equation can be written as

$$I = m_1 a_{f_1} - m_2 a_{f_2} - m_3 a_{f_3} \tag{3-f}$$

Substituting the numeric values into Eq. (3-f), we have an availability destruction for this process of $I = (1)(37.04) - (0.964)(27.64) - (0.036)(271.6) = 0.64$ Btu/lbm. ■

Example 3-7. Return to Example 2-3 and calculate (a) the flow availability at turbine inlet and exit, (b) the availability transfer due to the heat loss, and (c) the availability destruction within the turbine.

Solution. The flow availability is given by the equation $a_f = (h - h_0) - T_0(s - s_0)$. Since air is treated as an ideal gas, the equation can be further reduced to

$$a_f = c_p(T - T_0) - T_0\left(c_p \ln\frac{T}{T_0} - R\ln\frac{P}{P_0}\right) \tag{3-g}$$

Substituting the numeric values into the equation, we have the flow availability at the turbine inlet

$$\begin{aligned} a_{f_1} &= 0.24(2460 - 537) - 537\left(0.24\ln\frac{2460}{537} - \frac{53.3}{778}\ln\frac{150}{14.7}\right) \\ &= 350.8 \text{ Btu/lbm} \end{aligned}$$

Similarly, the flow availability at turbine exit is $a_{f_2} = 92.6$ Btu/lbm.

The availability transfer due to the heat loss is calculated by

$$a_q = \int_1^2 \left(1 - \frac{T_0}{T_s}\right)\delta q \tag{3-h}$$

The temperature at the boundary surface of the control volume is not constant but varies over a wide range. For this calculation we approximate the surface temperature by the average value of the air temperatures at the turbine inlet and exit. Because of this approximation, Eq. (3-h) is reduced to

$$a_q = \left(1 - \frac{T_0}{T_{s,a}}\right) q$$

Taking $q = -4.71$ Btu/lbm from Example 2-3, we have the heat availability transfer $a_q = [1 - (537/1960)](-4.71) = -3.42$ Btu/lbm.

The availability destruction is determined with the availability equation, $i = a_q + a_w + a_{f_1} - a_{f_2}$. Making use of the results calculated previously, we have the availability destruction $i = -3.42 - 235.3 + 350.8 - 92.6 = 19.48$ Btu/lbm.

The availability balance for the turbine operation is shown in Table 3-3. ■

Before the section is concluded, it should be emphasized that the availability equation shown in Eq. (3-33) is formulated on the basis of the first and second laws of thermodynamics. Similar to the first law equation, Eq. (1-1), and the second law equation, Eq. (1-6), the availability equation is relatively general in nature and valid for closed as well as open systems. It may be of interest to examine the general availability equation in terms of available work and actual work. Let us start with Eq. (3-32) and arrange it as follows:

$$\begin{aligned}\sum_i \Delta m_i(h_i - T_0 s_i) - \sum_e \Delta m_e(h_e - T_0 s_e) - m_2(u_2 - T_0 s_2) \\ + m_1(u_1 - T_0 s_1) - W_{cv} \\ = T_0 \sigma - \sum_j \left(1 - \frac{T_0}{T_j}\right) \delta Q_j \end{aligned} \tag{3-36}$$

Since the available work is given by

$$\begin{aligned} W_{ava} = \sum_i \Delta m_i(h_i - T_0 s_i) - \sum_e \Delta m_e(h_e - T_0 s_e) \\ + [m_1(u_1 - T_0 s_1) - m_2(h_2 - T_0 s_2)]_{cv} \end{aligned} \tag{3-37}$$

Table 3-3 Availability balance for turbine operation in Example 3-7.

Availability type	Availability (Btu/lbm)	
Flow availability decrease ($a_{f_1} - a_{f_2}$)	258.19	
Turbine work (a_w)		−235.29
Loss (a_q)		−3.42
Destruction (i)		−19.48
Total	258.19	−258.19

the availability equation can be further reduced to

$$W_{\text{ava}} - W_{\text{cv}} = T_0\sigma - \sum_j \left(1 - \frac{T_0}{T_j}\right)\delta Q_j \tag{3-38}$$

In words, Eq. (3-38) states that the unavailable energy ($W_{\text{ava}} - W_{\text{cv}}$) is numerically equal to the availability destruction and loss in a process. As indicated in Chapter 2, the unavailable energy is also the net change of entropy multiplied by the environmental temperature, or the combined internal and external irreversibility in a process. In an equation form, they are respectively

$$W_{\text{ava}} - W_{\text{cv}} = T_0 \Delta S_{\text{net}} \tag{3-39}$$

and

$$W_{\text{ava}} - W_{\text{cv}} = I_{\text{int}} + I_{\text{ext}} \tag{3-40}$$

3-5 AVAILABILITY COMPUTATIONS

Flow availability and nonflow availability have been developed in the previous section. They are respectively

$$a_f = (h - h_0) - T_0(s - s_0) \tag{3-41}$$

$$a = (u - u_0) + P_0(v - v_0) - T_0(s - s_0) \tag{3-42}$$

Combining these equations yields the following relationship between the flow and nonflow availability:

$$a = a_f + v(P_0 - P) \tag{3-43}$$

It should be pointed out that for simplicity, the kinetic and potential energies have been omitted from these equations. The objectives of this section are to present numerical calculations for and to determine the effect of temperature and pressure change on availability. The substances may be either ideal gases or real gases. Attention will be focused to the computation of flow availability.

Ideal Gases

In engineering applications the specific heat often is assumed to be constant. Using this assumption we can write the enthalpy and entropy changes for an ideal gas as

$$h - h_0 = c_p(T - T_0)$$

and

$$s - s_0 = c_p \ln\frac{T}{T_0} - R\ln\frac{P}{P_0}$$

Substituting these two expressions into Eq. (3-41), we have the flow availability as

$$a_f = c_p(T - T_0) - T_0\left(c_p \ln\frac{T}{T_0} - R\ln\frac{P}{P_0}\right)$$

To normalize the flow availability a_f, we divide the above equation by the term c_pT_0. Upon arranging them, we have

$$\frac{a_f}{c_pT_0} = \frac{T}{T_0} - 1 - \ln\frac{T}{T_0} + \frac{k-1}{k}\ln\frac{P}{P_0} \tag{3-44}$$

Combining Eqs. (3-43) and (3-44), the normalized nonflow availability is expressed as

$$\frac{a}{c_pT_0} = \frac{a_f}{c_pT_0} + \frac{k-1}{k}\left(\frac{T}{T_0}\right)\left(\frac{P_0}{P} - 1\right) \tag{3-45}$$

Graphical displays of Eqs. (3-44) and (3-45) are shown in Fig. 3-2. It is seen that nonflow availability is always positive in value, whereas flow availability can be positive or negative, depending on the value of the pressure ratio P/P_0. For a given pressure, flow availability will increase if the temperature deviation from the environment increases. When a constant temperature is maintained, an increase in pressure will increase the value of flow availability. The effect of temperature and pressure on the nonflow availability is much more complicated. Figure 3-2(a) shows that for a given temperature, an increase in pressure does not always result in an increase in nonflow availability. When a constant pressure is maintained, the nonflow availability may increase or decrease with an increase in temperature.

Figure 3-3 presents normalized availability contour curves for air. It is again shown that nonflow availability is always positive, whereas flow availability can be positive or negative. The same value of availability can be achieved by the substance at different states.

There are many processes in which the temperature change is substantial and the specific heat cannot be treated as a constant. For these, we write the change of enthalpy and entropy as

$$h - h_0 = \int c_p\, dT$$

and

$$s - s_0 = \int \frac{c_p\, dT}{T} - R\ln\frac{P}{P_0}$$

Substituting these two expressions into Eq. (3-41), we have the flow availability as

$$a_f = \int\left(1 - \frac{T_0}{T}\right)c_p\, dT + RT_0\ln\frac{P}{P_0} \tag{3-46}$$

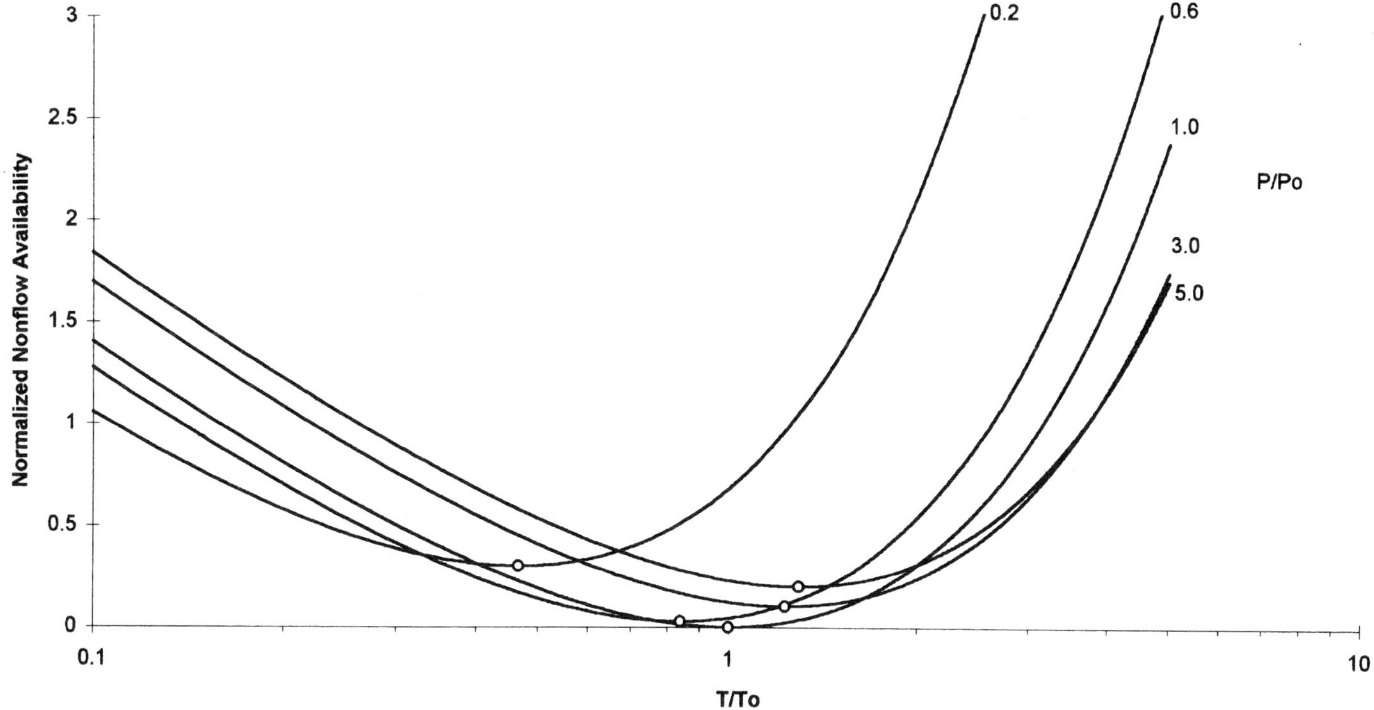

(*a*)

Figure 3-2 (a) Normalized nonflow availability for air. (b) Normalized flow availability for air.

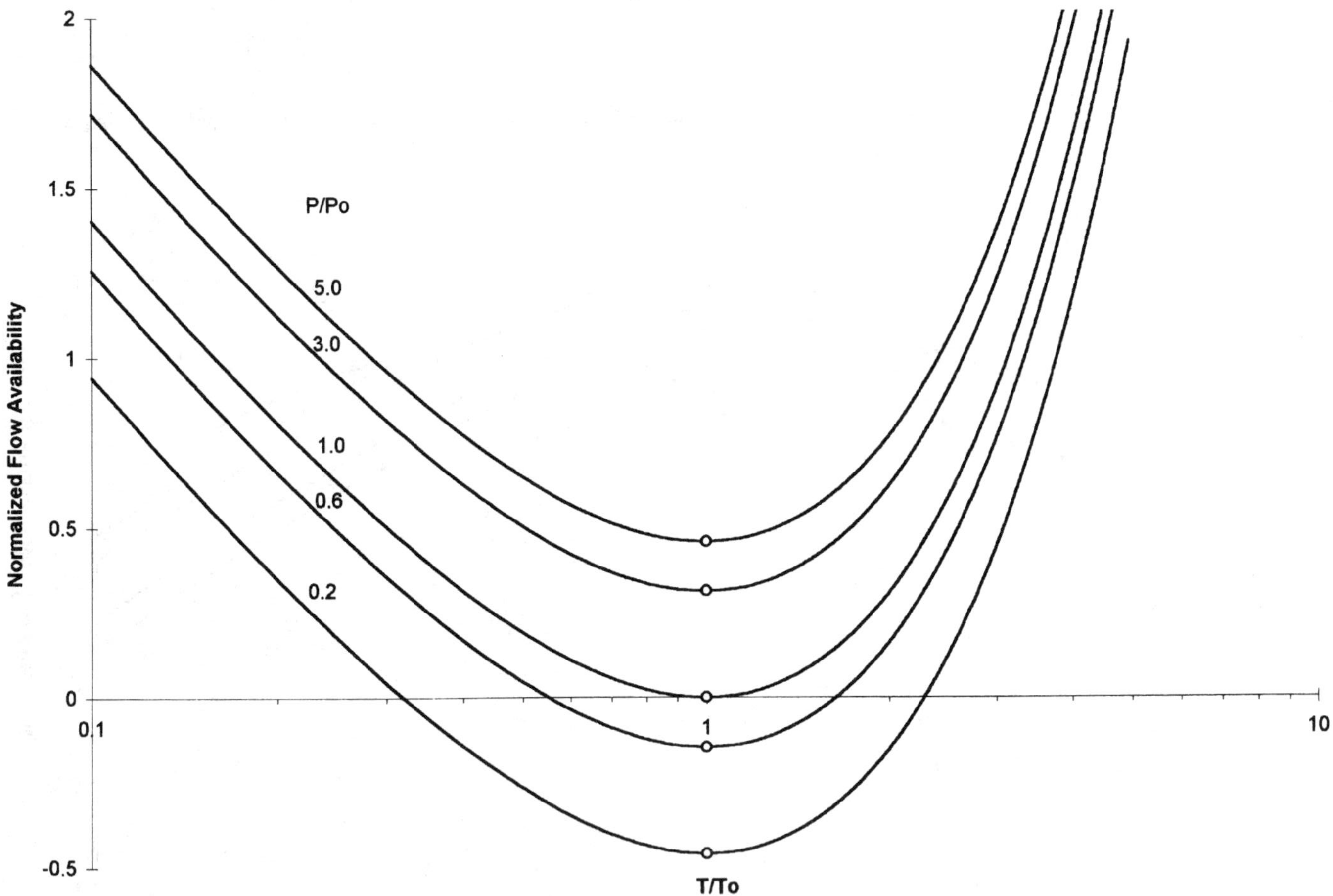

(*b*)

Figure 3-2 Continued.

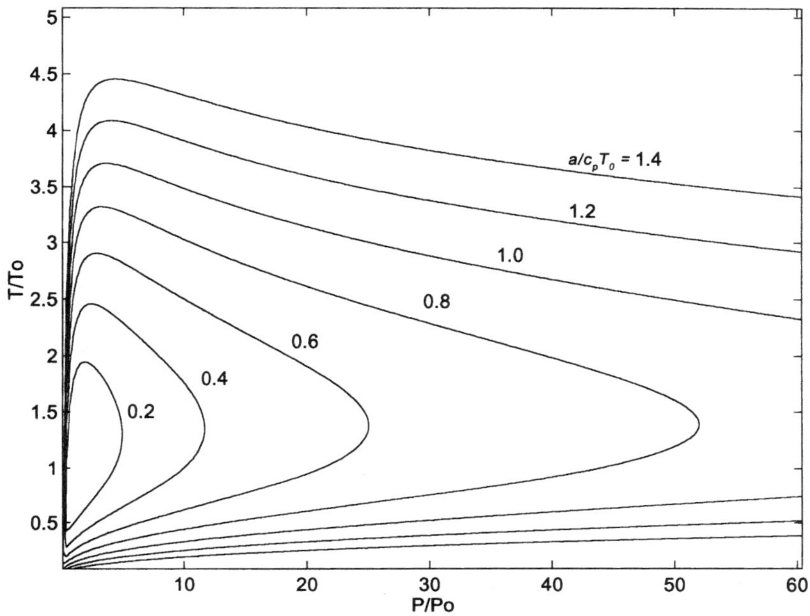

(a)

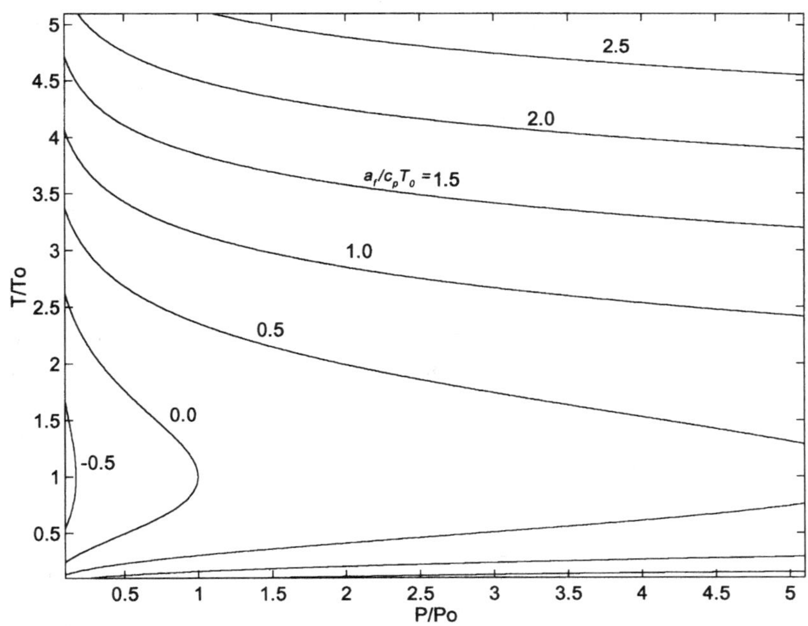

(b)

Figure 3-3 (*a*) Normalized nonflow availability contours for air. (*b*) Normalized flow availability contours for air.

The specific heat c_p for an ideal gas is a function of temperature and can be expressed in polynomial form, such as

$$\bar{c}_p = c_0 + c_1\left(\frac{T}{1000}\right) + c_2\left(\frac{T}{1000}\right)^2 + c_3\left(\frac{T}{1000}\right)^3 + c_4\left(\frac{T}{1000}\right)^4 + c_5\left(\frac{T}{1000}\right)^5 \tag{3-47}$$

where $\bar{c}_p$ is in Btu/lbmole-R and T is in degrees Rankine. The values of c_0, c_1, c_2, c_3, c_4, and c_5 are given in Table 3-4 for various ideal gases. Computation of flow and nonflow availability can be carried out conveniently on a computer. The sample results for nitrogen gas are shown in Table 3-5. A graphical display of flow availability versus entropy for nitrogen gas is shown in Fig. 3-4. The computation of availability is based on the environmental conditions 14.7 psia and 77°F.

Air is a substance of considerable interest to engineers. For the purpose of computing flow and nonflow availability, the composition of air was assumed to be that shown in Table 3-6. Appendix D presents air flow and nonflow availability as a function of temperature and pressure.

The availability for air also can be determined conveniently by using the air table such as that in Table C-5(a). First we rewrite Eq. (3-45) as

$$a_f = h(T) - h(T_0) - T_0(s'(T) - s'(T_0)) + RT_0 \ln\left(\frac{P}{P_0}\right) \tag{3-48}$$

We recognize that the terms $h(T)$, $h(T_0)$, $s'(T)$, and $s'(T_0)$ are functions of temperature. When the temperature is given, these terms can be obtained easily from the air table. The sum of these terms shows the temperature effect on the flow availability. Similarly, the term $RT_0 \ln(P/P_0)$ represents the pressure influence.

Real Gases

The property values for real gases can be determined with the use of the generalized charts. The basic procedure for this calculation has been reviewed

Table 3-4 Values of coefficients for Eq. (3-47).

	Coefficient					
Gases	c_0	c_1	c_2	c_3	c_4	c_5
Oxygen	6.631	3.721	−15.559	28.512	−21.969	6.226
Nitrogen	6.970	−0.278	1.402	−3.312	3.586	−1.235
Carbon dioxide	5.137	8.712	−3.534	0.758	−0.008	0.003
Water vapor	8.146	−1.415	2.770	−1.180	0.247	−0.022
Hydrogen	3.927	11.547	−16.010	10.217	−2.998	0.331
Carbon monoxide	7.102	−0.879	1.196	−0.114	−0.124	0.026

Table 3-5 Flow and nonflow availability for nitrogen.

	14.7 PSIA				50.0 PSIA			
$T(R)$	$\bar{h}$ (Btu/lbmole)	$\bar{s}$ (Btu/lbmole-R)	$\bar{a}$ (Btu/lbmole)	$\bar{a}_f$ (Btu/lbmole)	$\bar{h}$ (Btu/lbmole)	$\bar{s}$ (Btu/lbmole-R)	$\bar{a}$ (Btu/lbmole)	$\bar{a}_f$ (Btu/lbmole)
300	2082.0	41.695	525.77	525.77	2082.0	39.264	1410.64	1831.25
320	2220.9	42.144	423.92	423.92	2220.9	39.713	1280.75	1729.40
340	2359.9	42.565	336.65	336.65	2359.9	40.134	1165.44	1642.13
360	2498.9	42.962	262.34	262.34	2498.9	40.531	1063.09	1567.82
380	2637.9	43.338	199.56	199.56	2637.9	40.907	972.27	1505.04
400	2776.9	43.695	147.12	147.12	2776.9	41.263	891.79	1452.00
420	2915.9	44.034	104.03	104.03	2915.9	41.603	820.66	1409.51
440	3054.9	44.357	69.40	69.40	3054.9	41.926	757.99	1374.88
460	3193.9	44.666	42.51	42.51	3193.9	42.235	703.06	1347.99
480	3333.0	44.962	22.66	22.66	3333.0	42.531	655.17	1328.14
500	3472.0	45.246	9.30	9.30	3472.0	42.815	613.76	1314.78
520	3611.1	45.518	1.92	1.92	3611.1	43.087	578.34	1307.40
540	3750.2	45.781	0.05	0.05	3750.2	43.350	548.44	1305.53
560	3889.3	46.034	3.34	3.34	3880.3	43.603	523.68	1308.82
580	4028.4	46.278	11.38	11.38	4028.4	43.847	503.69	1316.86
600	4167.6	46.514	23.87	23.87	4167.6	44.083	488.13	1329.35
620	4306.9	46.742	40.52	40.52	4306.9	44.311	476.74	1346.00
640	4446.2	46.963	61.07	61.07	4446.2	44.532	469.26	1366.55
660	4585.5	47.178	85.29	85.29	4585.5	44.747	465.44	1390.77
680	4725.0	47.386	112.96	112.96	4725.0	44.955	465.06	1418.44
700	4864.5	47.588	143.89	143.89	4864.5	45.157	467.96	1449.37
720	5004.2	47.785	177.92	177.92	5004.2	45.354	473.94	1483.40
740	5143.9	47.976	214.86	214.86	5143.9	45.545	482.84	1520.34
760	5283.8	48.163	254.57	254.57	5283.8	45.732	494.51	1560.05
780	5423.8	48.345	296.94	296.94	5423.8	45.913	508.84	1602.42
800	5564.0	48.522	341.82	341.82	5564.0	46.091	525.68	1647.30
820	5704.3	48.695	389.12	389.12	5704.3	46.264	544.94	1694.59
840	5844.8	48.865	438.72	438.73	5844.8	46.434	566.50	1766.21
860	5985.6	49.030	490.52	490.52	5985.6	46.599	590.26	1790.00
880	6126.5	49.192	544.47	544.47	6126.5	46.761	616.17	1849.95
900	6267.6	49.351	600.47	600.47	6267.6	46.920	644.13	1905.95
920	6409.0	49.506	658.41	658.41	6409.0	47.075	674.02	1963.89
940	6550.7	49.658	718.27	718.27	6550.7	47.227	705.84	2023.75
960	6692.6	49.808	779.97	779.97	6692.6	47.377	739.50	2085.45
980	6834.8	49.954	843.43	843.43	6834.8	47.523	774.93	2148.91
1000	6977.3	50.098	908.63	908.64	6977.3	47.667	812.09	2214.12

The reference temperature is 537.0 R
The reference pressure is 14.7 psia

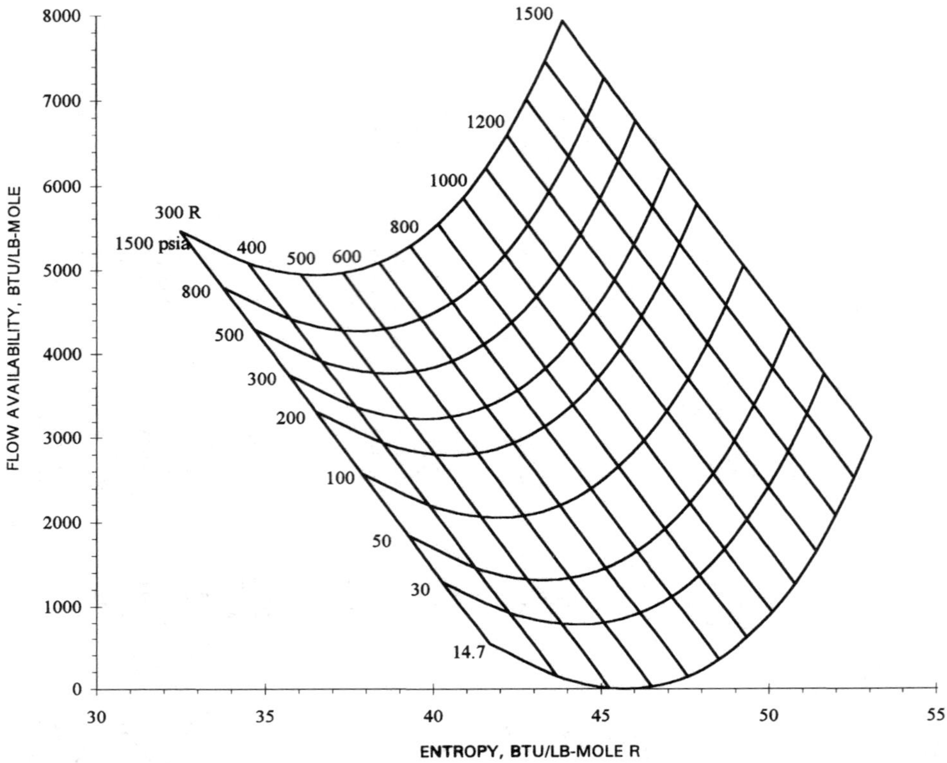

Figure 3-4 An $\bar{a}_f$-s diagram for nitrogen gas.

in Chapter 1. Using this approach, we can write the enthalpy and entropy changes as

$$h - h_0 = (h - h^*) + (h^* - h_0^*) + (h_0^* - h_0) \tag{3-49}$$

and

$$s - s_0 = (s - s^*) + (s^* - s_0^*) + (s_0^* - s_0) \tag{3-50}$$

Substituting Eqs. (3-49) and (3-50) into Eq. (3-41), we have the flow availability as

$$a_f = [(h - h^*) + (h^* - h_0^*) + (h_0^* - h_0)]$$
$$- T_0[(s - s^*) + (s^* - s_0^*) + (s_0^* - s_0)]$$

Table 3-6 Composition of air.

Gas	Molecular weight	Percent by volume
Nitrogen	28.016	78.03
Oxygen	32.000	20.99
Argon	39.950	0.98

Rearranging these terms we have

$$a_f = a_f^* + T_c\left(\frac{h_0^* - h_0}{T_c} - \frac{h^* - h}{T_c}\right) - T_0[(s_0^* - s_0) - (s^* - s)] \quad (3\text{-}51)$$

where

$$a_f^* = (h^* - h_0^*) - T_0(s^* - s_0^*) \quad (3\text{-}52)$$

and T_c is the critical temperature of the substance. When the environmental state is such that the substance would behave as an ideal gas in that state, Eq. (3-51) can be further simplified to

$$a_f = a_f^* - T_c\left(\frac{h^* - h}{T_c} - \frac{T_0}{T_c}(s^* - s)\right) \quad (3\text{-}53)$$

The first term on the right-hand side of Eq. (3-53), a_f^*, is the flow availability for the corresponding ideal gas, and the second term is the departure of the flow availability from the ideal gas value. In the second term it should be recognized that $(h^* - h)/T_c$ and $(s^* - s)$ are the enthalpy and entropy departures at the given state, respectively. As indicated in chapter 1, these values can be determined easily by the generalized charts.

Example 3-8. Calculate the flow availability for nitrogen gas at 300 R and 1000 psia using various methods discussed in this section.

Solution. The flow availability is calculated by the equation

$$a_f = (h - h_0) - T_0(s - s_0) \quad (3\text{-i})$$

Assuming that nitrogen is an ideal gas and has constant specific heat of 0.248

Btu/lbm-R, we have

$$a_f = 0.248(300 - 537) - 537\left(0.248 \ln \frac{300}{537} - \frac{55.15}{778} \ln \frac{1000}{14.7}\right) = 179.4 \text{ Btu/lbm}$$

In the second method, we still assume that nitrogen gas is an ideal gas, but we treat its specific heat as a function of temperature. For this calculation, we apply Eq. (3-48):

$$a_f = (h(T) - h(T_0)) - T_0(s'(T) - s'(T_0)) + RT_0 \ln\left(\frac{p}{p_0}\right) \tag{3-j}$$

Using the values from the table and taking the molecular weight of nitrogen as 28, we have the flow availability as

$$\begin{aligned} a_f &= (74.36 - 133.20) - 537(1.4891 - 1.6337) + \frac{(55.15)(537)}{778} \ln \frac{1000}{14.7} \\ &= 179.5 \text{ Btu/lbm} \end{aligned}$$

In the third method, we treat nitrogen gas as a real gas and calculate the flow availability by using the generalized charts. We start with the equation

$$a_f = a_f^* - T_c\left(\frac{h^* - h}{T_c} - \frac{T_0}{T_c}(s^* - s)\right) \tag{3-k}$$

For nitrogen, $T_c = 227.2$ R and $P_c = 493$ psia. Then the reduced pressure and temperature at the given state respectively are

$$P_r = \frac{P}{P_c} = \frac{1000}{493} = 2.03$$

and

$$T_r = \frac{T}{T_c} = \frac{300}{227.2} = 1.32$$

Using the generalized charts, we obtain

$$\frac{\bar{h}^* - \bar{h}}{T_c} \doteq 3 \text{ Btu/lbmole-R}$$

$$\bar{S}^* - \bar{S} \doteq 1.7 \text{ Btu/lbmole-R}$$

Substituting the numeric values into Eq. (3-k), we have the flow availability

$$a_f = 179.5 - 227.2\left(\frac{3}{28} - \frac{537}{227.2} \times \frac{1.7}{28}\right) = 187.8 \text{ Btu/lbm.} \quad \blacksquare$$

Water

Water is a substance frequently used in energy conversion systems. The flow availability for water can be determined with Eq. (3-41). In this text the conditions in the environmental state are selected as 14.7 psia and 77°F. Under these conditions water is assumed to be a saturated liquid at 77°F. Because of this, we can write the flow availability as

$$a_f(T, p) = h(T, p) - h_f(T_0) - T_0\big(s(T, p) - s_f(T_0)\big) \qquad (3\text{-}54)$$

Since $h_f(T_0) - T_0 s_f(T_0) = h_g(T_0) - T_0 s_g(T_0)$, Eq. (3-53) can be written also as

$$a_f(T, p) = h(T, p) - h_g(T_0) - T_0\big(s(T, p) - s_g(T_0)\big) \qquad (3\text{-}55)$$

In words, when the flow availability for water is calculated, it is immaterial whether water is assumed as saturated liquid or saturated vapor at the temperature T_0. The pressure P_0 has no influence on the flow availability. It should be emphasized that Eqs. (3-54) and (3-55) are valid for water in all states in the superheated, saturated, and compressed regions. For instance, we can write for saturated liquid

$$a_{f,f}(T) = h_f(T) - h_f(T_0) - T_0\big(s_f(T) - s_f(T_0)\big) \qquad (3\text{-}56)$$

and for saturated vapor

$$a_{f,g}(T) = h_g(T) - h_g(T_0) - T_0\big(s_g(T) - s_g(T_0)\big) \qquad (3\text{-}57)$$

If water is in the saturated state, the flow availability can be derived and expressed as

$$a_f(T, x) = (l - x)a_{f,f} + xa_{f,g} \qquad (3\text{-}58)$$

When water is in the compressed liquid region, we make use of the equations $h(T, p) = h_f(T) + v_f(P - P_f(T))$ and $s(T, p) = s_f(T)$ and rewrite the flow availability for compressed water as

$$a_f(T, p) = h_f(T) - h_f(T_0) + v_f\big(P - P_f(T)\big) - T_0\big(s_f(T) - s_f(T_0)\big) \qquad (3\text{-}59)$$

or

$$a_f(T, p) = a_{f,f}(T) + v_f\big(P - P_f(T)\big) \qquad (3\text{-}60)$$

The flow availability for water can be determined easily when the steam properties such as enthalpy and entropy are available. Figure 3-5 shows the graphical relations between the flow availability and temperature for water. Similarly, the relations between flow availability and other properties are presented in Figs. 3-6 and 3-7.

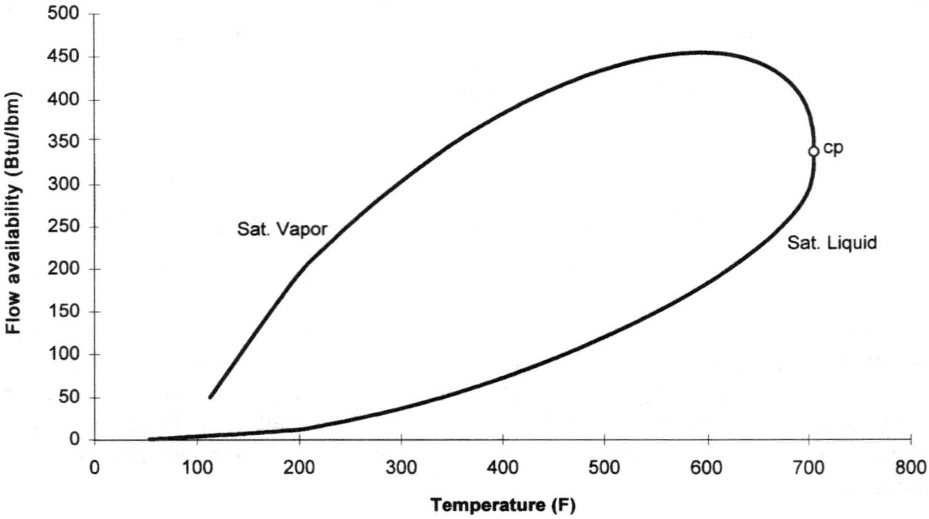

Figure 3-5 Flow availability versus temperature for water.

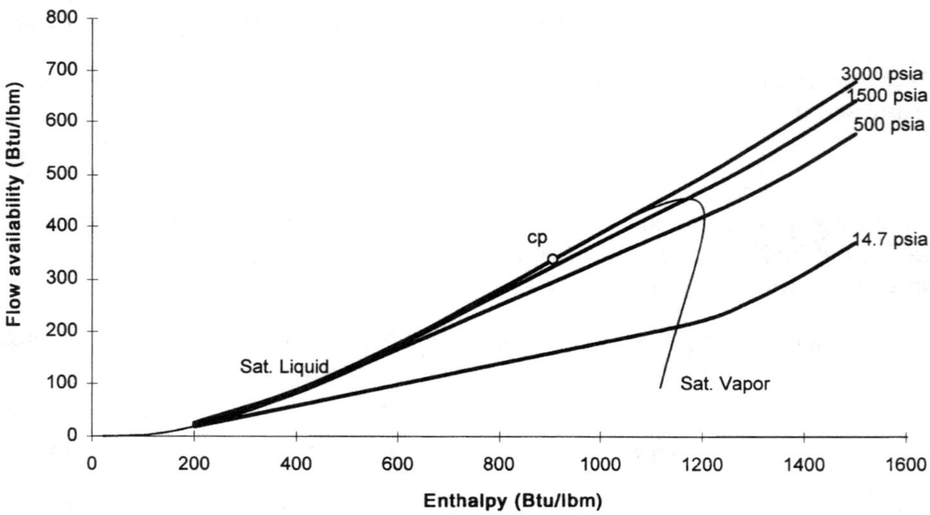

Figure 3-6 Flow availability versus enthalpy for water.

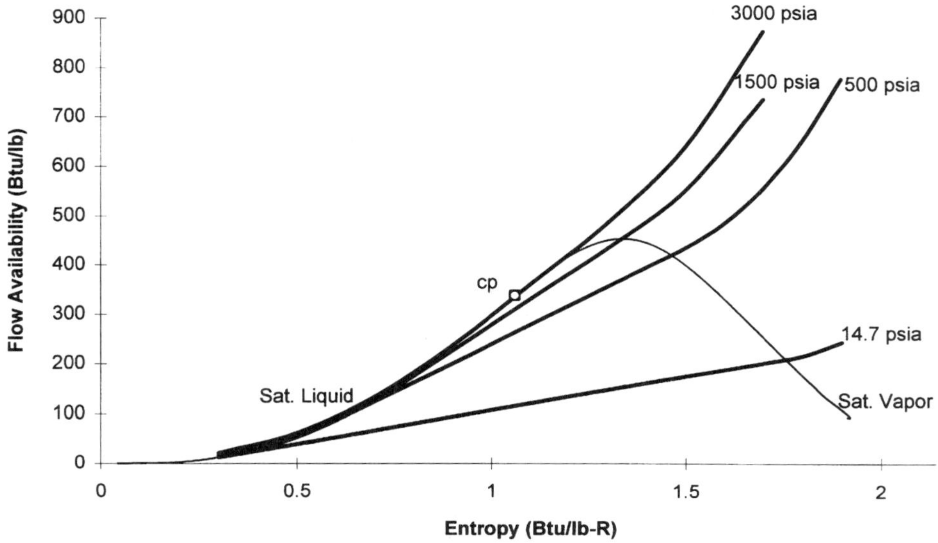

Figure 3-7 Flow availability versus entropy for water.

3-6 MORE ON AVAILABILITY COMPUTATIONS

It has been shown that the change of flow availability is due to the change of temperature and pressure in a system. There are certain availability analyses in which the effect of temperature is separated from that of pressure. In other words, it is sometimes desirable to express the change of availability as a sum of two separate terms, one due to the temperature change and another due to the pressure change, that is,

$$\Delta a_f = \Delta a_{f,\Delta T} + \Delta a_{f,\Delta p} \tag{3-61}$$

where $\Delta a_{f,\Delta T}$ = flow availability change due to temperature, and $\Delta a_{f,\Delta p}$ = flow availability change due to pressure. In the following are the derivations for these two terms.

Ideal Gases

We start with the change of flow availability as

$$\begin{aligned} \Delta a_f &= a_{f_2} - a_{f_1} \\ &= c_p(T_2 - T_1) - T_0\left(c_p \ln\frac{T_2}{T_1} - R\ln\frac{P_2}{P_1}\right) \end{aligned} \tag{3-62}$$

Rearranging the equation gives $\Delta a_f = \Delta a_{f,\Delta T} + \Delta a_{f,\Delta p}$, where

$$\Delta a_{f,\Delta T} = c_p(T_2 - T_1) - T_0 c_p \ln\frac{T_2}{T_1} \tag{3-63}$$

and

$$\Delta a_{f,\Delta p} = T_0 R \ln\frac{P_2}{P_1} \tag{3-64}$$

In terms of temperature change $\Delta T = T_2 - T_1$ and pressure change $\Delta p = P_2 - P_1$, Eqs. (3-63) and (3-64) respectively become

$$\Delta a_{f,\Delta T} = c_p(\Delta T) - T_0 c_p \ln\left(1 + \frac{\Delta T}{T_1}\right) \tag{3-65}$$

and

$$\Delta a_{f,\Delta p} = T_0 R \ln\left(1 + \frac{\Delta p}{P_1}\right) \tag{3-66}$$

At this point it may be of interest to compare the change of flow availability with the change of enthalpy, which traditionally is regarded as a representation of energy. The change of enthalpy for idea gas is given by

$$\Delta h = \Delta h_{\Delta T} + \Delta h_{\Delta p} \tag{3-67}$$

where

$$\Delta h_{\Delta T} = c_p(\Delta T) \tag{3-68}$$

and

$$\Delta h_{\Delta p} = 0 \tag{3-69}$$

In these equations, $\Delta h_{\Delta T}$ and $\Delta h_{\Delta p}$ are the enthalpy changes due to temperature and pressure, respectively. It is seen in Eq. (3-65) that as the initial temperature increases, the flow availability would increase, even for a given ΔT. At a high temperature level the difference between $\Delta a_{f,\Delta T}$ and $\Delta h_{\Delta T}$ will become less evident. Figure 3-8 presents a comparison of enthalpy and flow availability for air.

The effect of pressure change on the flow availability and energy is quite different. The pressure change will not result in any change in energy. However, it is not true for flow availability. Equation (3-64) indicates that as the pressure increases, the flow availability will increase.

Liquids

We assume liquid to be an incompressible substance. The change of flow availability for liquid is $a_{f_2} - a_{f_1} = (h_2 - h_1) - T_0(s_2 - s_1)$. Since $h_2 - h_1 = c_v(T_2 - T_1) + v_1(P_2 - P_1)$ and $s_2 - s_1 = c_v \ln(T_2/T_1)$, we have

$$a_{f_2} - a_{f_1} = c_v(T_2 - T_1) - c_v T_0 \ln\frac{T_2}{T_1} + v_1(P_2 - P_1) \tag{3-70}$$

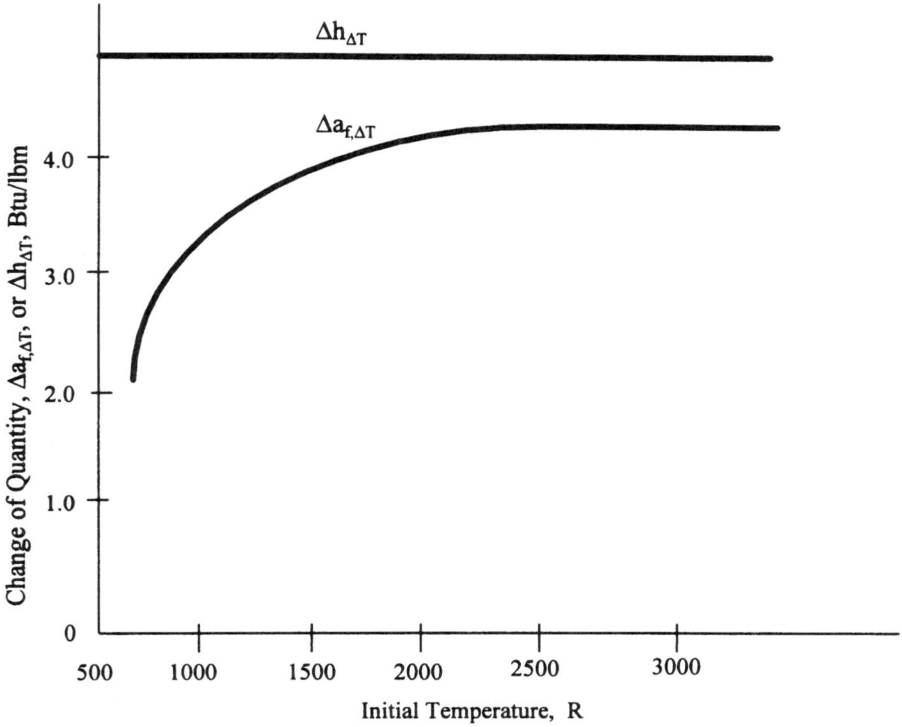

Figure 3-8 Changes of enthalpy and flow availability for air, $\Delta T = 10°\text{F}$.

Comparing Eq. (3-70) with Eq. (3-61), we have the availability change due to temperature as

$$\Delta a_{f,\Delta T} = c_v(\Delta T) - T_0 c_v \ln\left(1 + \frac{\Delta T}{T_1}\right) \tag{3-71}$$

and the availability change due to pressure as

$$\Delta a_{f,\Delta p} = v_1(\Delta p)$$

where ΔT and Δp are the temperature and pressure changes, respectively.

Vapors

For vapors or real gases in general, the change of flow availability is still caused by a change of temperature and pressure. Because the enthalpy and entropy of vapors usually are functions of temperature and pressure, the influence of temperature and pressure on the flow availability cannot be entirely separated. To present the effects of temperature and pressure on the availability change in these situations, the most logical and convenient approach is a graphical method. Figures 3-9 and 3-10 show the flow availability for Freon-12 and ammonia.

There should be a few words about the conditions at the environmental state upon which the availability calculation is based. When the working substance has temperature and pressure equal to those environmental conditions, the working substance loses its capability to do work and therefore has zero availability. Different environmental conditions will result in different numeric calculations. Then the selection of environmental conditions arises. In this text we usually select the reference conditions to be those in the environment surrounding the system. The reasons for this selection are simple. First, the surrounding environment is relatively stable. It has a unique temperature, pressure, and chemical composition. These values remain generally unchanged during the system operation. Second, the use of the surrounding environment as a reference state is a logical and convenient choice. By assigning zero availability to these conditions, the working substance would have positive values of availability in various locations throughout the system. As indicated before, the availability represents a work-producing potential and thus an index of its economic value.

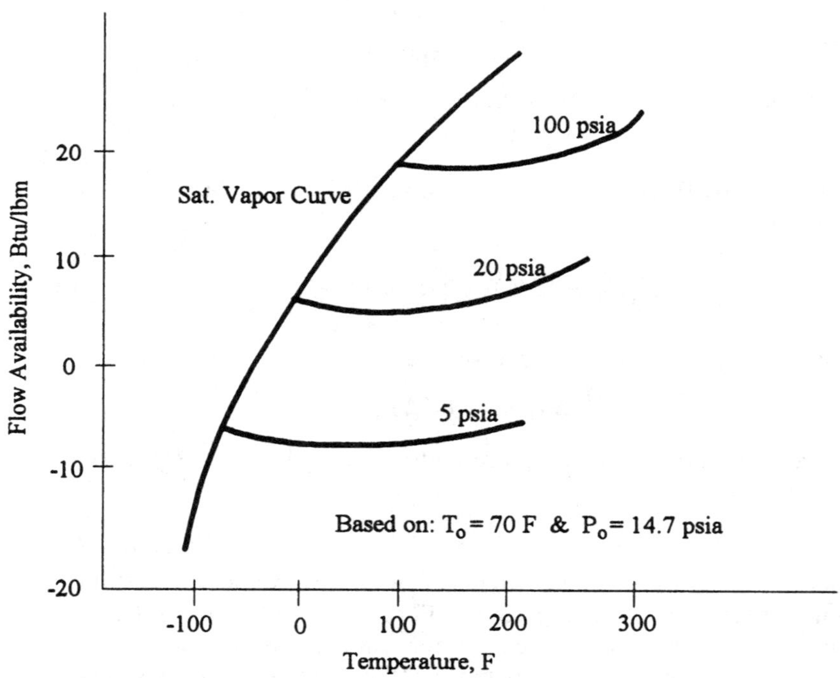

Figure 3-9 Flow availability of Freon-12.

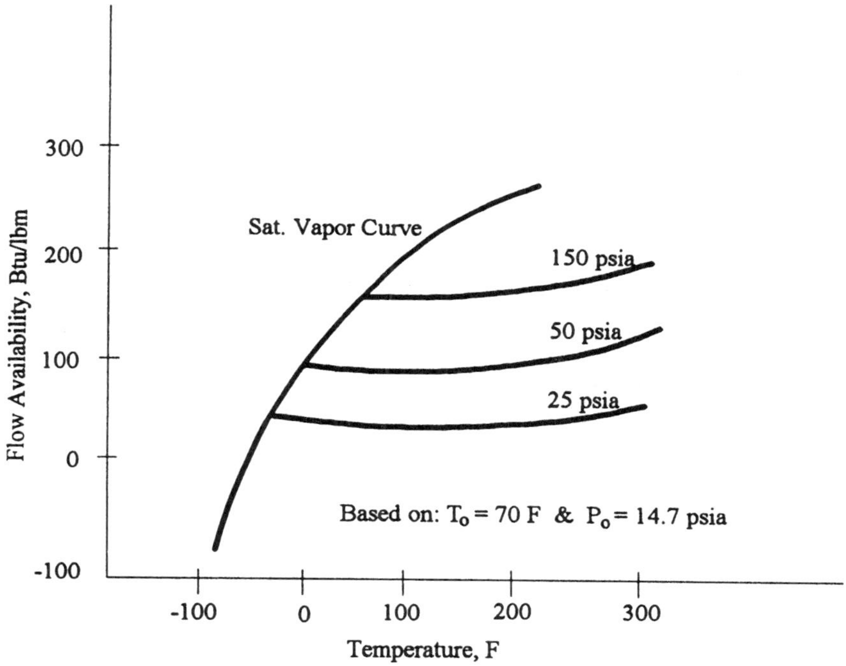

Figure 3-10 Flow availability of ammonia.

In certain availability analyses, other conditions may be selected as the reference conditions. This generally is true for refrigeration systems or the systems involving the refrigerants such as Freon-12 and ammonia. For these cases, the reference conditions generally are those at one location of the physical system. It must be emphasized that the selection of reference conditions is a tool to simplify the availability calculation and to provide the basis for the availability analysis of the energy conversion system.

In summary, there is no one set of reference conditions upon which the availability calculation can be always based. In this text the conditions in the surrounding environment are chosen as the reference conditions. Unless otherwise indicated, the environmental temperature and pressure are assumed to be 77°F and 14.7 psia, respectively.

3-7 PROBLEMS

3-1 Consider steam flow in a horizontal pipe. At one location the steam has a velocity of 200 ft/sec and conditions of 2400 psia and 1000°F. What is the nonflow and flow availability for the steam at this location?

3-2 Water flows in a long pipe. While its temperature and pressure remain approximately constant, the velocity increases from 10 ft/s to 30 ft/s and the elevation decreases by 100 ft. What is the change of flow availability for the water? Would the change of flow availability be equal to that of nonflow availability?

3-3 A 40-ft^3 tank contains saturated ammonia at a pressure of 220 psia. Initially the tank contains 50% liquid and 50% vapor by volume. Because of heat loss, the ammonia undergoes a process and eventually reaches the room temperature of 80°F. Calculate the change of nonflow availability. What is the change of nonflow availability due to the heat loss?

3-4 A rigid and well-insulated vessel contains 10 ft^3 of air. The air, initially at 14.7 psia and 100°F, is heated by an electrical resistor until its temperature is 400°F. Determine the change of nonflow availability in the air and the availability destruction in the process.

3-5 Nitrogen is contained in a cylinder fitted with a piston and expands very slowly according to the relation $Pv^{1.5}$ = constant. The initial conditions of the nitrogen are 100 psia and 500°F. After an expansion, the pressure of the nitrogen drops to 20 psia. Calculate, per pound of nitrogen, (a) the work done, (b) the heat transfer and (c) the availability loss and destruction.

3-6 An initially evacuated tank is to be charged by a charging bottle filled with saturated liquid ammonia at 80°F. The charging bottle is connected to the tank and remains open until the final temperature is 80°F. The tank and bottle have volumes of 50 ft^3 and 0.5 ft^3, respectively. Calculate:

a) the initial and final nonflow availability in the charging bottle,
b) the initial and final nonflow availability in the tank,
c) the availability loss due to the heat transfer, and
d) the availability destruction in the entire process.

3-7 Steam enters a steam turbine at the rate of 5×10^3 lbm/hr. The steam conditions at the turbine inlet and exit are as follows:

	Inlet Conditions	Exit Conditions
Pressure (psia)	3500	750
Temperature (°F)	1000	600
Velocity (ft/s)	200	600
Elevation (ft)	20	10

The heat loss from the steam turbine is negligible. Calculate both the flow availability at the turbine inlet and exit and the availability destruction in the process.

3-8 A steam turbine can be operated at part-load conditions by the use of a control valve at the turbine inlet. Consider the case in which the steam conditions are 2500 psia and 1000°F before throttling and 2400 psia after throttling. Determine the flow availability destruction per pound of steam in the throttling process.

3-9 Prepare a computer program to calculate the nonflow and flow availability for nitrogen, and compare the results with those in Table 3-2.

3-10 Prepare a computer program to calculate the nonflow and flow availability for air, and compare the results with those in Appendix D.

3-11 Air is compressed in a piston-cylinder arrangement from 15 psia to 150 psia. The process is reversible and isothermal at a temperature of 100°F. Calculate (a) the change of nonflow availability, (b) the availability loss due to heat transfer, (c) the availability destruction, and (d) the work done in the process.

3-12 Air enters an insulated nozzle at 100 psia and 600°F at low velocity and exits at 15 psia. The nozzle efficiency is 80%. Determine (a) the exit velocity, (b) the change of flow availability per pound of air, and (c) the availability loss and destruction per pound of air.

3-13 Saturated water vapor at 100 psia is contained in a piston-cylinder assembly. Because of heat transfer, the water vapor undergoes a constant-pressure process and becomes saturated water at the final state. Calculate the change of nonflow availability and the heat availability transfer per unit mass.

3-14 A 10-lbm metal bar initially is in equilibrium at 1500°F in a furnace. It is suddenly removed from the furnace and quenched in a well-stirred water bath at 100°F. After a long period, the bar and water come to an equilibrium temperature of 104°F. Calculate the availability destruction in the process assuming the specific heats for the metal bar and the water are 0.1 and 1.0 Btu/lbm-R, respectively.

3-15 Compressed air is stored in two well-insulated tanks. Each tank has a volume of 30 cubic feet. Initially air is at 100 psia and 600°F in tank A and at 50 psia and 550°F in tank B. A valve in the line between the two tanks is then opened, allowing the contents of both tanks to mix and reach an equilibrium state. Calculate the change of nonflow availability and the destruction of availability in the process.

3-16 For the compressed air storage described in Problem 2-10, determine (a) the initial availability of the air in the storage, (b) the availability of the air in the storage at the end of the charging process, and (c) the availability loss and destruction in the process.

CHAPTER

FOUR

AVAILABILITY ANALYSIS OF ENGINEERING PROCESSES

Engineers encounter many processes in practice. Some processes are simple, and some are relatively complex. In many cases, however, a complex process can be reduced to several simple processes by using assumptions or idealizations. Consequently, it is important for an engineer to acquire the ability to analyze various simple processes. In this chapter we shall apply the availability equations to several common processes, as are organized below:

- Pumps, compressors, and fans
- Turbines
- Nozzles and diffusers
- Heat exchangers
- Fluid flow
- Energy storage

In this chapter we will introduce the second law efficiency, sometimes referred to as *process effectiveness*. It will be seen that the second law efficiency is developed with the concept of availability and is compared with the conventional first law efficiency. It is also the purpose of this chapter to compare the availability analysis with the energy analysis and, in so doing, to present the advantages of the availability method.

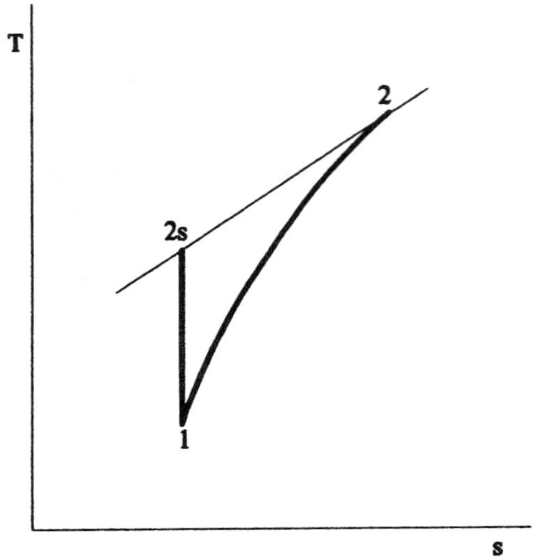

Figure 4-1 T-s diagram for a compression process.

4-1 PUMPS, COMPRESSORS, AND FANS

Pumps, compressors, and fans are used to bring a fluid from a low-pressure state to a high-pressure state. The first law efficiency is usually defined as

$$\eta = \frac{w_{\text{ideal}}}{w_{\text{act}}} \tag{4-1}$$

where w_{ideal} is the isentropic work and w_{act} is the actual work. This comparison is best shown in Fig. 4-1. The definition of the first law efficiency is quite arbitrary and is used to compare the actual process 1-2 with the corresponding ideal process 1-2s.

The second law efficiency is based on the concept of availability. For pumps, compressors, and fans the second law efficiency is defined as

$$\varepsilon = \frac{a_{f_2} - a_{f_1}}{a_w}$$

or

$$\varepsilon = \frac{a_{f_1} - a_{f_2}}{w_{\text{act}}} \tag{4-2}$$

In the above definition the term $(a_{f_1} - a_{f_2})$ represents the minimum work required for the equipment, and the term w_{act} represents the actual work for the same process with identical initial and final states. In other words, the comparison is made on the same process and is very different from the first law efficiency. The second law efficiency as shown by Eq. (4-2) is sometimes referred to as the *process effectiveness*. It shows how effectively the equipment performs its functions. For pumps, compressors, and fans the second law efficiency means the percentage at which the mechanical work is converted into the flow availability (i.e., the work potential of the fluid in the process).

The temperature rise across the pump frequently is required in calculation. For incompressible flow such as water, the equation for the temperature rise is developed simply from the first law equation. Assuming that the process is adiabatic, the first law equation is $h_1 = w_p + h_2$, where w_p is the pump work per unit of mass. For incompressible flow, the pump work is expressed as

$$w_p = -\frac{v_1(P_2 - P_1)}{\eta_p} \tag{4-3}$$

Combining these two equations gives

$$h_1 - h_2 = -\frac{v_1(P_2 - P_1)}{\eta_p}$$

or

$$(u_1 - u_2) + v_1(P_1 - P_2) = -\frac{v_1(P_2 - P_1)}{\eta_p}$$

Rearranging the above equation will yield an expression for the temperature rise across the pump:

$$T_2 - T_1 = \frac{v_1(P_2 - P_1)}{c_v}\left(\frac{1}{\eta_p} - 1\right) \tag{4-4}$$

where c_v is the specific heat of liquid and equals 1 Btu/lbm-R for water. Air flow across the fan frequently is treated as an incompressible flow. Because of this, Eq. (4-4) is equally valid for an estimate of the temperature rise across the fan.

There is an alternative method for the calculation of the air temperature rise. If we assume the process is adiabatic, we have by the first law principle $h_2 - h_1 = -w_f$, or

$$T_2 - T_1 = \frac{-w_f}{c_p} \tag{4-5}$$

Example 4-1. A boiler feedwater pump receives saturated water at a pressure of 130 psia and discharges it as 1900 psia. The pump has a first law efficiency of 80%. Calculate the second law efficiency and the availability destruction per unit mass flowing through the pump.

Solution. The pumping process is assumed to be adiabatic. The actual pump work is calculated by Eq. (4-3). With the numeric inputs we have

$$w_p = -0.01796(1900 - 130)\left(\frac{144}{778}\right)\left(\frac{1}{0.8}\right) = -7.35 \text{ Btu/lbm}$$

Then, we calculate the water temperature at the pump discharge by Eq. (4-4). That is,

$$T_2 = T_1 + \frac{v_1(P_2 - P_1)}{c_v}\left(\frac{1}{\eta_p} - 1\right)$$

Substituting the numeric values into the above equation gives

$$T_2 = 347.4 + 0.01796(1900 - 130)\left(\frac{144}{778}\right)\left(\frac{1}{0.8} - 1\right) = 348.8°\text{F}$$

The flow availability change for water is

$$a_{f_2} - a_{f_1} = (h_2 - h_1) - T_0(s_2 - s_1) \tag{4-a}$$

Since $s_2 - s_1 = c_v \ln(T_2/T_1)$, Eq. (4-a) becomes

$$a_{f_2} - a_{f_1} = (h_2 - h_1) - T_0 c_v \ln\frac{T_2}{T_1}$$

With the numeric values substituted into the above equation, we have the change of flow availability as $a_{f_2} - a_{f_1} = 7.35 - (537)(1)\ln(808.8/807.4) = 6.42$ Btu/lbm. Thus, the second law efficiency is

$$\varepsilon_p = \frac{a_{f_1} - a_{f_2}}{w_p} = \frac{-6.42}{-7.35} = 0.873$$

To determine the availability destruction in the pumping process, we apply the availability equation $i = a_q + a_w + a_{f_1} - a_{f_2}$. Since $a_q = 0$, the availability destruction per unit mass will become $i = a_w + a_{f_1} - a_{f_2} = 7.35 - 6.42 = 0.93$ Btu/lbm.

There will be no energy created or destroyed in any conversion processes, including the above pumping process. This is the essence of the first law of thermodynamics. In terms of flow availability, the conversion is quite different. As indicated in the above calculations, not all mechanical work is converted into the flow availability. The difference is the flow availability consumed in the process. The second law efficiency is the ratio of flow availability increase to the pump work. ■

Example 4-2. A boiler forced-draft fan receives air at 14.7 psia and 90°F. The fan will develop a static pressure head of 30 inches of water at the flow rate of 10^6 cfm. The fan static efficiency is 80%. Calculate the fan power, the second law efficiency, and the availability destruction in the process.

Solution. We first determine the fan developed pressure in pound per square feet as below:

$$P_2 - P_1 = \frac{L\rho g}{g_c} = \frac{(30)(62.4)(32.2)}{(12)(32.2)} = 156 \text{ lbf/ft}^2$$

Next, the mass flow rate in pounds per minute is calculated by the equation

$$\dot{m} = \frac{\dot{V}_1}{v_1} = \dot{V}_1\left(\frac{P_1}{RT_1}\right)$$

Substituting the numeric values into the equation, we have the mass flow rate

$$\dot{m} = 10^6\left(\frac{14.7 \times 144}{53.3 \times 550}\right) = 72{,}209 \text{ lbm/min}$$

To calculate the fan power, we apply the first law equation

$$\dot{W}_f = \frac{-\dot{V}_1(P_2 - P_1)}{\eta_f}$$

With numeric values we have

$$\dot{W}_f = \frac{-10^6(156)}{(778)(0.8)} = -250{,}642 \text{ Btu/min}$$

The air temperature at the fan discharge is given by Eq. (4-5). That is, $T_2 = T_1 - (w_f/c_p)$. Inserting the numeric values into the equation gives the air temperature at the fan discharge:

$$T_2 = 90 + \frac{250{,}642}{72{,}209 \times 0.24} = 104.5°\text{F}$$

The second law efficiency for the fan is

$$\varepsilon_f = \frac{\dot{m}(a_{f_2} - a_{f_1})}{-\dot{W}_f} \tag{4-b}$$

The numerator is the rate of flow availability for air. Since air is treated as an ideal gas, this term is calculated as follows:

$$\dot{m}(a_{f_2} - a_{f_1}) = \dot{m}\left[c_p(T_2 - T_1) - T_0\left(c_p \ln\frac{T_2}{T_1} - R\ln\frac{P_2}{P_1}\right)\right]$$

$$= 72{,}209\left[0.24(564.5 - 550) - 537\left(0.24\ln\frac{564.5}{550} - \frac{53.3}{778}\ln\frac{15.78}{14.7}\right)\right]$$

$$= 197{,}456 \text{ Btu/min}$$

The denominator $\dot{W}_f$ is the actual work in the process. Substituting the values into Eq. (4-b), we have the second law efficiency $\varepsilon_f = (197{,}456/250{,}642) = 0.788$.

To calculate the rate of the availability destruction, we make use of the availability equation: $\dot{I} = \dot{A}_w - \dot{m}(a_{f_2} - a_{f_1}) = 250{,}642 - 197{,}456 = 53{,}186$ Btu/min. ■

Example 4-3. Freon-12 in a refrigeration unit enters the compressor at 20 psia and 20°F and leaves at 200 psia and 220°F. Determine the second law efficiency for the compression process, and present an availability balance for the compressor. Assume that the heat loss from the compressor is 2 Btu per pound of Freon-12.

Solution. We treat the compressor as a control volume. From the Freon-12 table we have enthalpies and entropies at the compressor inlet and exit as below:

$$h_i = 80.4 \text{ Btu/lbm} \qquad h_e = 106.3 \text{ Btu/lbm}$$
$$s_i = 0.1783 \text{ Btu/lbm-R} \qquad s_e = 0.1886 \text{ Btu/lbm-R}$$

The second law efficiency for compressor is

$$\varepsilon_c = \frac{a_{f_i} - a_{f_e}}{w_c}$$

The compressor work is given by $w_c = q + h_i - h_e$. Substituting the numeric values into the above equation, we have $w_c = -2 + 80.4 - 106.3 = -27.9$ Btu/lbm.

The flow availability change for Freon-12 is

$$\begin{aligned} a_{f_i} - a_{f_e} &= (h_i - h_e) - T_0(s_i - s_e) \\ &= (80.4 - 106.3) - 537(0.1783 - 0.1886) \\ &= -20.4 \text{ Btu/lbm} \end{aligned}$$

Thus, the second law efficiency is determined as $\varepsilon_c = (-20.4/-27.9) = 0.731$.

The availability destruction for the compressor is calculated by using the availability equation:

$$i = a_q + a_w + a_{f_i} - a_{f_e} \tag{4-c}$$

The term a_q is the availability transfer due to the heat loss from the compressor and is determined by the equation

$$a_q = \int_1^2 \left(1 - \frac{T_0}{T}\right) \delta q$$

or, if the average surface temperature T_{av} is used,

$$a_q = \left(1 - \frac{T_0}{T_{av}}\right) q$$

In this case, the average temperature simply is the average temperature of 20°F and 220°F. Thus we have

$$a_q = \left(1 - \frac{537}{580}\right)(-2) = -0.148 \text{ Btu/lbm}$$

Table 4-1 Availability balance for compressor in Example 4-3.

Availability type	Availability into the control volume (Btu/lbm)	Availability out from the control volume (Btu/lbm)
Work (a_w)	27.9	
Availability Loss (a_q)		0.15
Availability destruction (i)		7.35
Useful output ($a_{f_e} - a_{f_i}$)		20.40
Total	27.9	27.90

The negative sign shows that the heat availability is transferred from the control volume. Substituting the numeric values into Eq. (4-c), we have the availability destruction per unit mass of Freon-12 flowing through the compressor: $i = -0.148 + 27.9 - 20.4 = 7.35$ Btu/lbm.

The availability balance for the compressor is summarized in Table 4-1.

It must be pointed out that the numeric values of availability destruction and loss would be affected by the choice of the control-volume surface temperature. However, the sum of these two will remain unchanged. Thus, the assumption on the control-volume surface temperature has no serious consequences in engineering practice.

4-2 TURBINES

The turbine is a piece of equipment used to generate mechanical power. In turbines the working fluid is expected to expand and convert its thermal energy into mechanical work. The first law efficiency is defined as

$$\eta_t = \frac{w_{\text{act}}}{w_{\text{ideal}}} \tag{4-6}$$

where w_{act} is the actual turbine work for the process 1-2 in Fig. 4-2, and w_{ideal} is the work of the corresponding isentropic process, such as the process 1-2s.

The second law efficiency is based on the concept of availability. It is defined as the ratio of the turbine actual work to the turbine available work in the same process 1-2. In equation form the second law efficiency is

$$\varepsilon_t = \frac{w_{\text{act}}}{w_{\text{ava}}}$$

or

$$\varepsilon_t = \frac{w_{\text{act}}}{a_{f_1} - a_{f_2}} \tag{4-7}$$

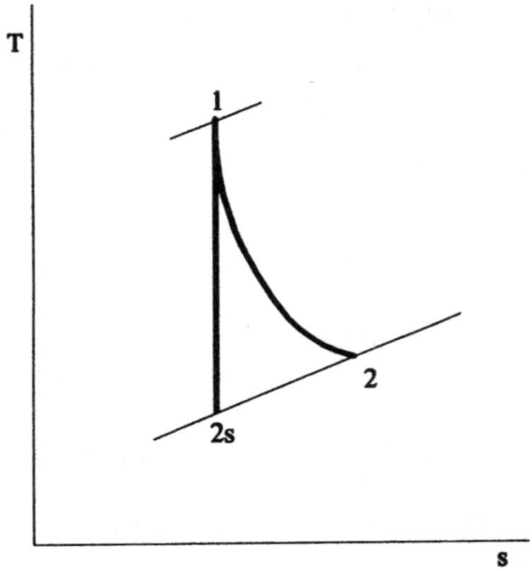

Figure 4-2 T-s diagram for an expansion process.

Since the turbine performs the function opposite to that of a compressor, the turbine expansion process is the reverse of the compression process. From a thermodynamic viewpoint, both processes are the same but are opposite in direction. Thus, the equations for compressors are equally applicable to turbines.

Example 4-4. Superheated steam at 1800 psia and 950°F enters a high-pressure turbine and leaves at 475 psia and 620°F. The steam flow rate is 3×10^6 pounds per hour. Assume that the turbine is perfectly insulated. Calculate the first law and second law efficiencies and the availability destruction in the turbine process.

Solution. From the steam table we can determine the steam conditions at the turbine inlet and exit:

$P_1 = 1800$ psia	$P_2 = 475$ psia
$T_1 = 950°\text{F}$	$T_2 = 620°\text{F}$
$h_1 = 1448.8$ Btu/lbm	$h_2 = 1312.3$ Btu/lbm
$s_1 = 1.5528$ Btu/lbm-R	$s_2 = 1.5769$ Btu/lbm-R

For the isentropic process 1-2s, the steam conditions at the final state are $h_{2s} = 1286.6$ Btu/lbm and $s_{2s} = s_1 = 1.5528$ Btu/lbm-R.

To calculate the first law efficiency, we use the equation

$$\eta_t = \frac{w_t}{w_{\text{ideal}}} \tag{4-d}$$

Since $w_t = h_1 - h_2 = (1448.8 - 1312.3) = 136.5$ Btu/lbm and $w_{\text{ideal}} = h_1 - h_{2s} = (1448.8 - 1286.6) = 162.2$ Btu/lbm, we have the first law efficiency as

$$\eta_t = 136.5/162.2 = 0.842$$

The second law efficiency is given by the equation

$$\varepsilon_t = \frac{w_t}{a_{f_1} - a_{f_2}} \tag{4-e}$$

Since

$$\begin{aligned} a_{f_1} - a_{f_2} &= (h_1 - h_2) - T_0(s_1 - s_2) \\ &= (1448.8 - 1312.3) - 537(1.5528 - 1.5769) \\ &= 149.4 \text{ Btu/lbm} \end{aligned}$$

we have the second law efficiency as $\varepsilon_t = 136.5/149.4 = 0.914$.

The availability destruction for the turbine process is determined with the availability equation $\dot{I} = \dot{A}_q + \dot{A}_w + \dot{A}_f$. Since $\dot{A}_q = 0$ (adiabatic condition), $\dot{A}_w = -\dot{m}w_t = -(3 \times 10^6)(136.5) = -409.5 \times 10^6$ Btu/hr, and $\dot{A}_f = \dot{m}(a_{f_1} - a_{f_2}) = (3 \times 10^6)(149.4) = 448.2 \times 10^6$ Btu/hr, we have the availability destruction rate as

$$\dot{I} = -409.5 \times 10^6 + 448.2 \times 10^6 = 38.7 \times 10^6 \text{ Btu/hr.}$$

Because the process is assumed to be adiabatic, there is no heat loss from the turbine. It means that the reduction of thermal energy is exactly equal to the generation of mechanical work. From an energy viewpoint the conversion takes place in perfect fashion. However, from an availability point of view, the picture is completely different. As indicated in the above calculations, not all flow availability (a form of work potential) is converted into mechanical work; some of the flow availability is consumed in the process. Thus, the turbine process still has room for improvement. In terms of availability utilization, the process has an efficiency of 91.4%. The first law efficiency for the process is 84.2%. This efficiency has no physical meaning, saying only that the actual turbine work is 84.2% of the work from the ideal process 1-2s. The processes 1-2 and 1-2s are different; therefore, the first law efficiency used to couple them is arbitrary in nature. ■

4-3 NOZZLES AND DIFFUSERS

Nozzles and diffusers are the devices engineers use to change the velocity of fluid flow. For a nozzle, the objective is to convert the thermal energy into kinetic energy and thus to produce a fluid velocity greater than the initial value. The diffuser's function is opposite to that of the nozzle. For a diffuser, the objective is to convert the kinetic energy into thermal energy and, in so doing, to produce a fluid pressure and temperature greater than the initial value.

The fluid flow through a nozzle or diffuser usually is assumed to be adiabatic but not reversible. The irreversibility, due mainly to the friction, increases the entropy of the fluid. These two processes are shown respectively in Figs. 4-3 and 4-4.

The nozzle first law efficiency, based on the concept of energy, is defined as

$$\eta_n = \frac{h_1 - h_2}{h_1 - h_{2s}} \tag{4-8}$$

In terms of the fluid kinetic energy, Eq. (4-8) becomes

$$\eta_n = \frac{V_2^2 - V_1^2}{V_{2s}^2 - V_1^2} \tag{4-9}$$

If the velocity at the nozzle inlet is negligible, Eq. (4-9) is further reduced to

$$\eta_n = \frac{V_2^2}{V_{2s}^2} \tag{4-10}$$

The second law efficiency for nozzles is developed with the concept of availability. The availability equation for the nozzle process 1-2 is $a_q + a_w + a_{f_1} - a_{f_2} =$

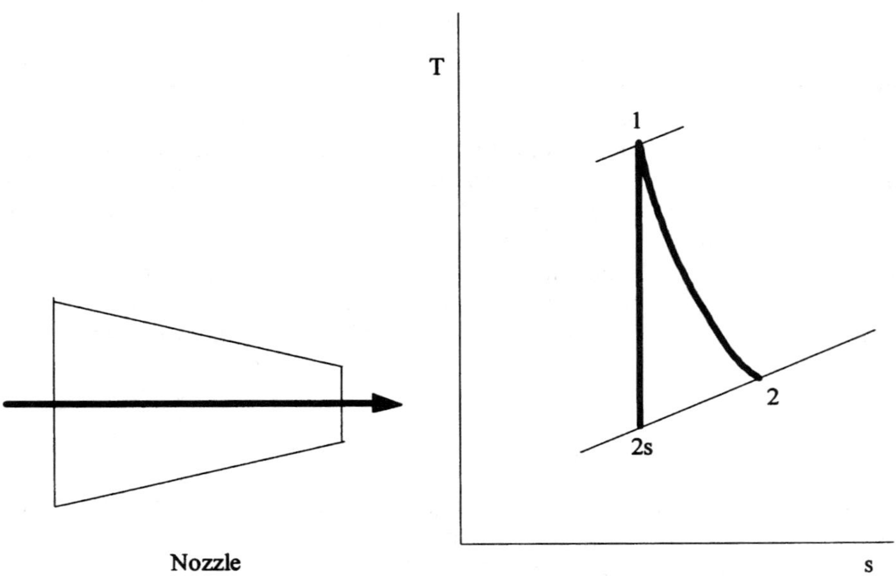

Figure 4-3 A nozzle flow and its T-s diagram.

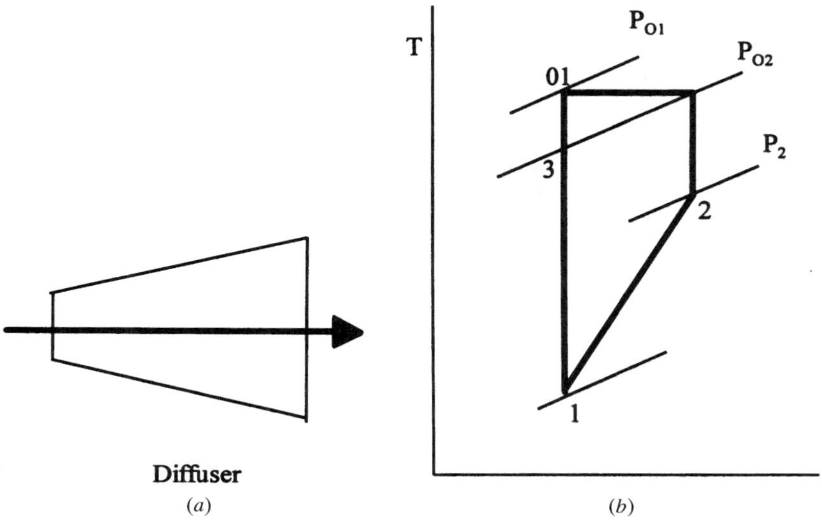

Figure 4-4 A diffuser flow (*a*) and its T-s diagram (*b*).

0. Since $a_q = 0$ and $a_w = 0$, we have

$$i = a_{f_1} - a_{f_2} = [(h_1 - h_2) - T_0(s_1 - s_2)] - \frac{V_2^2 - V_1^2}{2} \tag{4-11}$$

The first term on the right-hand side of the equation represents a decrease of flow availability (the thermal-mechanical type), and the second term represents an increase of kinetic availability (i.e., the availability related to the kinetic energy of the moving fluid). The conversion is not perfect. The difference between these two terms is the availability destruction of the nozzle flow process. The ratio of these terms is defined as the second law efficiency, that is,

$$\varepsilon_n = \frac{(V_2^2 - V_1^2)/2}{(h_1 - h_2) - T_0(s_1 - s_2)} \tag{4-12}$$

When the velocity at the nozzle inlet is negligible, Eq. (4-12) is further reduced to

$$\varepsilon_n = \frac{V_2^2/2}{(h_1 - h_2) - T_0(s_1 - s_2)} \tag{4-13}$$

The first law efficiency for the diffuser shown in Fig. 4-4 is defined as

$$\eta_d = \frac{h_3 - h_1}{h_{o_1} - h_1} \tag{4-14}$$

where h_{o_1} is the stagnation enthalpy at the diffuser inlet, and the term $(h_{o_1} - h_1)$ represents the kinetic energy of the fluid at the same location. In equation form the first law efficiency is written also as

$$\eta_d = \frac{h_3 - h_1}{V_1^2/2} \tag{4-15}$$

The second law diffuser efficiency is developed with the availability equation. The procedure is identical to that used for the nozzle flow. For simplicity we will present the expression for the availability destruction as

$$i = \left(\frac{V_1^2 - V_2^2}{2}\right) - [(h_2 - h_1) - T_0(s_2 - s_1)] \tag{4-16}$$

and the expression for the second law efficiency as

$$\varepsilon_d = \frac{(h_2 - h_1) - T_0(s_2 - s_1)}{(V_1^2 - V_2^2)/2} \tag{4-17}$$

When the velocity at the diffuser exit is negligible, Eq. (4-17) is further reduced to

$$\varepsilon_d = \frac{(h_2 - h_1) - T_0(s_2 - s_1)}{V_1^2/2} \tag{4-18}$$

Example 4-5. An engineer tested an old nozzle and obtained the air data as in Table 4-2. Determine the first law and second law efficiencies and the availability destruction per unit mass. The nozzle process is assumed to be adiabatic.

Table 4-2 Characteristics of air in Example 4-5.

Characteristics	Inlet conditions	Exit conditions
Pressure (psia)	150	14.7
Velocity (ft/s)	100	600
Temperature (°F)	200	—

Solution. The actual and corresponding isentropic processes are shown in Fig. 4-3. We first determine the two temperatures T_2 and T_{2s}. The temperature T_2 is the temperature of the air leaving the nozzle and is calculated by using the first law equation: $h_1 - h_2 = (V_2^2 - V_1^2)/2$ or $c_p(T_1 - T_2) = (V_2^2 - V_1^2)/2$. Substituting the numeric values into the equation, we have

$$0.24(660 - T_2) = \frac{600^2 - 100^2}{2(32.2)(778)}$$

$$T_2 = 630 \text{ R}$$

The temperature T_{2s} is the temperature of air leaving the corresponding ideal nozzle, which usually is assumed frictionless and perfectly insulated. The calculation is done by the isentropic relation

$$\frac{T_{2s}}{T_1} = \left(\frac{P_2}{P_1}\right)^{k-1/k}$$

Substituting the numeric values into this equation, we have

$$T_{2s} = 660\left(\frac{14.7}{150}\right)^{1.4-1/1.4} = 339.9 \text{ R}$$

The first law efficiency is given by

$$\eta_n = \frac{h_1 - h_2}{h_1 - h_{2s}} = \frac{T_1 - T_2}{T_1 - T_{2s}} = \frac{660 - 630}{660 - 339.9} = 0.0937$$

The second law efficiency is given by Eq. (4-12). Since

$$\frac{V_2^2 - V_1^2}{2} = \frac{600^2 - 100^2}{2 \times 32.2 \times 778} = 6.99 \text{ Btu/lbm}$$

and

$$(h_1 - h_2) - T_0(s_1 - s_2) = c_p(T_1 - T_2) - T_0\left(c_p \ln\frac{T_1}{T_2} - R \ln\frac{P_1}{P_2}\right)$$

$$= 0.24(660 - 630) - 537\left(0.24 \ln\frac{660}{630} - \frac{53.3}{778}\ln\frac{150}{14.7}\right)$$

$$= 86.43 \text{ Btu/lbm}$$

the second law efficiency is therefore $\varepsilon_n = 6.99/86.43 = 0.0809$.

The availability destruction per unit mass is $i = a_{f_1} - a_{f_2}$, or

$$i = (h_1 - h_2) - T_0(s_1 - s_2) - \frac{V_2^2 - V_1^2}{2} \tag{4-f}$$

Recognizing $h_1 - h_2 = (V_2^2 - V_1^2)/2$, we have Eq. (4-f) reduced to $i = -T_0(s_1 - s_2)$. With numeric values we have the availability destruction $i = 79.44$ Btu/lbm. This old nozzle has a relatively low efficiency as compared with the modern nozzle. ■

Example 4-6. An air diffuser test has the data shown in Table 4-3. Calculate the second law efficiency.

Solution. The diffuser process 1-2 is shown in Fig. 4-4. The static temperature at the diffuser exit is determined with the first law equation. That is,

$$c_p(T_1 - T_2) = \frac{V_2^2 - V_1^2}{2} \tag{4-g}$$

Substituting the numeric values into Eq. (4-g), we have

$$0.24(480 - T_2) = \frac{200^2 - 600^2}{2(32.2)(778)}$$

$$T_2 = 506.8 \text{ R}$$

The second law efficiency for the diffuser is given by Eq. (4-17). With Eq. (4-g) the second law efficiency is further reduced to

$$\varepsilon_d = 1 - \frac{T_0(s_2 - s_1)}{h_2 - h_1}$$

Since

$$\begin{aligned} T_0(s_2 - s_1) &= T_0\left(c_p \ln\frac{T_2}{T_1} - R\ln\frac{P_2}{P_1}\right) \\ &= 537\left(0.24\ln\frac{506.8}{480} - \frac{53.3}{778}\ln\frac{11.7}{10}\right) \\ &= 1.20 \text{ Btu/lbm} \end{aligned}$$

Table 4-3 Characteristics of air in Example 4-6.

Characteristics	Inlet conditions	Exit conditions
Pressure (psia)	10	11.7
Temperature (°F)	20	—
Velocity (ft/s)	600	200

and

$$h_2 - h_1 = c_p(T_2 - T_1) = 0.24(506.8 - 480) = 643 \text{ Btu/lbm}$$

we have the second law efficiency as $\varepsilon_d = 1 - (1.20/6.43) = 0.813$.

The diffuser has a second law efficiency of 81.3%, as compared with the first law efficiency of 84.4%. The detailed calculation for the first law efficiency is left as an exercise for the reader. ■

4-4 HEAT EXCHANGERS

The heat exchanger is a device in which heat is transferred from a hot substance to a cold substance. It is one of the essential components in many engineering systems. For example, in a coal-fired power plant system, we have heat exchangers known as boilers, feedwater heaters, and condensers. The second law efficiency for heat exchanger is the ratio of the availability received by the cold substance to the availability given up by the hot substance. For the heat exchanger shown in Fig. 4-5, the second law efficiency is

$$\varepsilon_h = \frac{\dot{m}_c(a_{f_e} - a_{f_i})_c}{\dot{m}_h(a_{f_i} - a_{f_e})_h} \tag{4-19}$$

Equation (4-19) indicates how effectively the availability is transferred. Because of the process irreversibility (i.e., the finite temperature difference between the hot and cold substances), the efficiency is expected to be less than unity. This term should not be confused with the heat exchanger effectiveness, which frequently is used in heat transfer textbooks and literature.

Example 4-7. A contact heater in a power plant has the following operating conditions:

Incoming feedwater	10^6 lbm/hr at 150 psia and 200°F
Turbine extraction steam	163,576 lbm/hr at 150 psia and 600°F

Estimate the availability destruction and the second law efficiency for this contact heater operation.

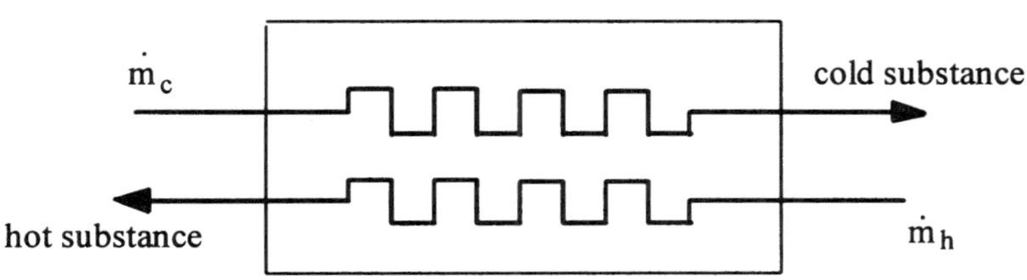

Figure 4-5 Schematic diagram for a heat exchanger.

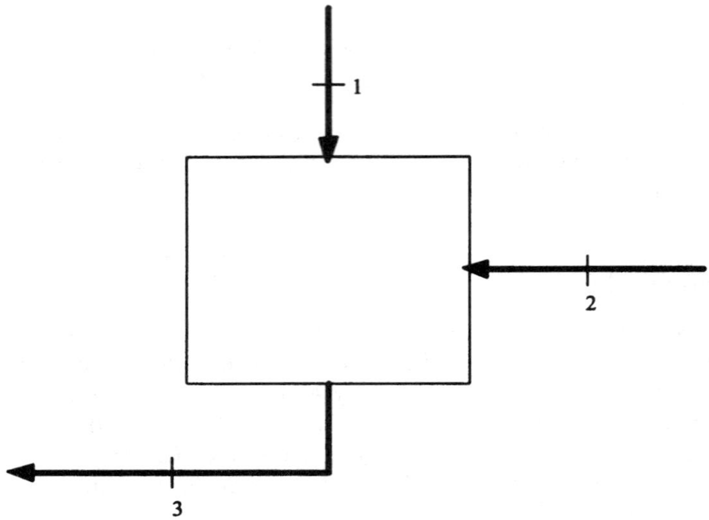

Figure 4-6 Schematic diagram for a contact heater.

Solution. Figure 4-6 indicates a schematic diagram for this contact heater. From the steam table we evaluate the steam properties and flow availabilities at the various locations. The results are summarized in Table 4-4. The flow availabilities are determined with the environmental conditions $T_0 = 77°F$ and $P_0 = 14.7$ psia. The heater exit condition is assumed to be saturated.

The availability destruction in the process is determined with the availability equation. Because there is no work and negligible heat loss from the contact heater, we have $\dot{I} = \dot{m}_1 a_{f_1} + \dot{m}_2 a_{f_2} - \dot{m}_3 a_{f_3}$. We can rearrange this equation as

$$\dot{I} = \dot{m}_1(a_{f_1} - a_{f_3}) - \dot{m}_2(a_{f_3} - a_{f_2}) \tag{4-h}$$

The first term on the right-hand side of the equation represents the availability given up by the turbine extraction steam, and the second term represents the availability received by the feedwater. Substituting the numeric values into Eq.

Table 4-4 Steam properties and flow availabilities for Example 4-7.

Characteristics	1	2	3
Pressure (psia)	150	150	150
Temperature (°F)	600	200	358.5
Enthalpy (Btu/lbm)	1325.7	168	330.8
Entropy (Btu/lbm-R)	1.711	0.294	0.5142
Flow availability (Btu/lbm)	408.9	12.2	56.7
Mass flow rate (lbm/hr)	0.163×10^6	1×10^6	1.163×10^6

(4-h), we have the availability destruction $\dot{I} = 0.163 \times 10^6(408.9 - 56.7) - 10^6(56.7 - 12.2) = 13.12 \times 10^6$ Btu/hr.

The second law efficiency is simply

$$\varepsilon_h = \frac{\dot{m}_2(a_{f_3} - a_{f_2})}{\dot{m}_1(a_{f_1} - a_{f_3})} \tag{4-i}$$

Substituting the numeric values into Eq. (4-i), we have the second law efficiency

$$\varepsilon_h = \frac{10^6(56.7 - 12.2)}{0.163 \times 10^6(408.9 - 56.7)} = 0.772 \quad \blacksquare$$

When the contact heater is treated as a chamber in which two streams of substance are mixed, the second law efficiency for a mixing process may be defined as the ratio of the availability at the chamber exit to the availability at the chamber inlet. In equation form, the second law efficiency is

$$\varepsilon_m = \frac{\sum_e \dot{m}_e a_{f_e}}{\sum_i \dot{m}_i a_{f_i}} \tag{4-20}$$

By Eq. (4-20), the second law efficiency for Example 4-7 is 83.4 percent. Thus it is important to clearly define the second low efficiency in any availability calculation.

Example 4-8. A heat balance of a steam turbine cycle provides the following conditions for the surface feedwater heater:

Steam entering	33,417 lbm/hr, 60 psia, sat. vapor
Heater drain leaving	60 psia, sat. liquid
Feedwater entering	10^6 lbm/hr, 150 psia, 250°F
Feedwater leaving	150 psia, 280°F

Calculate the availability destruction for the condensing steam, the feedwater, and the surface feedwater heater.

Solution. A schematic diagram for the feedwater heater is shown in Fig. 4-7. The availability destruction for the condensing steam is determined with the availability equation:

$$\dot{I} = \int_1^2 \left(1 - \frac{T_0}{T_s}\right) \delta\dot{Q} + \dot{m}_1 a_{f_1} - \dot{m}_2 a_{f_2}$$

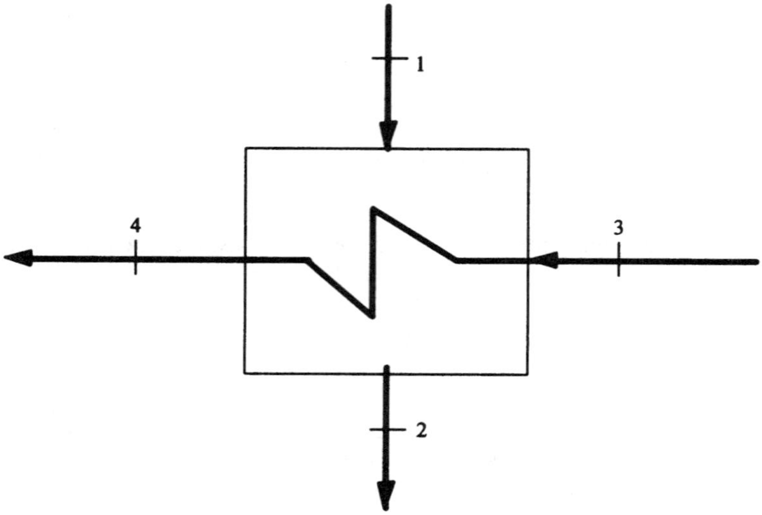

Figure 4-7 Schematic for a feedwater heater.

Since T_s is constant and $\dot{m}_1 = \dot{m}_2$, the equation is reduced to

$$\dot{I} = \left(1 - \frac{T_0}{T_s}\right)\dot{Q} + \dot{m}_1(a_{f_1} - a_{f_2}) \tag{4-j}$$

Also,

$$\dot{Q} = \dot{m}_1(h_2 - h_1) \tag{4-k}$$

and

$$\dot{m}_1(a_{f_1} - a_{f_2}) = \dot{m}_1(h_1 - h_2) - \dot{m}_1 T_0(s_1 - s_2) \tag{4-l}$$

Substituting Eqs. (4-k) and (4-l) into Eq. (4-j) gives

$$\dot{I} = \dot{m}_1(h_2 - h_1) - \frac{T_0}{T_s}\dot{m}_1(h_2 - h_1) + \dot{m}_1(h_1 - h_2) - \dot{m}_1 T_0(s_1 - s_2)$$

Recognizing that $s_2 - s_1 = (h_2 - h_1)/T_s$, we have the availability destruction for the condensing steam as $\dot{I} = 0$. Similarly, the availability destruction for the feedwater is

$$\dot{I} = \int_3^4 \left(1 - \frac{T_0}{T_w}\right)\delta\dot{Q} + \dot{m}_3 a_{f_3} - \dot{m}_4 a_{f_4} \tag{4-m}$$

Table 4-5 Steam properties and flow abilities for Example 4-8.

Characteristics	1	2	3	4
Pressure (psia)	60	60	150	150
Temperature (°F)	292.7	292.7	250	280
Enthalpy (Btu/lbm)	1178	262.3	218.6	249.2
Entropy (But/lbm-R)	1.6444	0.4273	0.3677	0.4099
Flow availability (Btu/lbm)	297	34.9	23.2	31.1
Flow rate (lbm/hr)	33,417	33,417	10^6	10^6

Since $\delta \dot{Q} = \dot{m}_3 c_p \, dT_w$, Eq. (4-m) can be further reduced to

$$\dot{I} = \dot{m}_3 c_p (T_4 - T_3) - T_0 \dot{m}_3 c_p \ln \frac{T_4}{T_3} + \dot{m}_3 c_p (T_3 - T_4) - \dot{m}_3 T_0 (s_3 - s_4)$$

Recognizing that $s_3 - s_4 = c_p \ln(T_3/T_4)$, we have the availability destruction for the feedwater as $\dot{I} = 0$.

To calculate the availability destruction for the feedwater heater, we first determine the steam properties and flow availabilities at the various locations. The results are given in Table 4-5. Because there is no stray heat from the heater, and the availability destruction is simply given by the equation

$$\dot{I} = \dot{m}_1 (a_{f_1} - a_{f_2}) - \dot{m}_3 (a_{f_4} - a_{f_3}) \tag{4-n}$$

Substituting the numeric values into Eq. (4-n), we have the availability destruction for the heater as $\dot{I} = 33{,}417(217 - 34.9) - 10^6(31.1 - 23.2) = 8.76 \times 10^6 - 7.9 \times 10^6 = 0.86 \times 10^6$ Btu/hr.

It is well known from the first law that the energy released by the condensing steam is equal to the energy received by the feedwater. In terms of availability, the principle of conservation does not apply. In this heater the steam gives up 8.76×10^6 Btu/hr of availability, whereas the water receives only 7.9×10^6 Btu/hr of availability. The difference is the amount of availability consumed in the process. The availability consumption is due to the internal irreversibility, which results from the heat-transfer process through the finite temperature difference between the steam and feedwater. There is another observation: In the feedwater heater the steam experiences a drop in flow availability. Numerically, the drop in flow availability is exactly equal to the amount of the heat availability transferred from the steam. Because of this there is no availability destruction in the condensing steam. A similar explanation can be given for the feedwater side. ■

4-5 FLUID FLOW

There are many occasions when an engineer will deal with fluid flow in a pipe. In this section the availability equation will be applied to an open system that is characterized by a pressure drop and sometimes a temperature change. This will be best illustrated by two examples. One example involves water flow through a long pipe, and another involves Freon-12 flow for which the temperature is below the environmental temperature T_0.

Example 4-9. Water at 100°F flows from Point 1 to Point 2 through an 800-ft cast iron pipe of average diameter 12″. Point 2 has an elevation 30 ft above Point 1, and the pressures at Point 1 and Point 2 are 46 psia and 20 psia, respectively. Assuming the heat loss from the pipe is negligible and the flow rate is constant, calculate the availability destruction per unit mass of water flow.

Solution. The availability equation for a unit of mass is $i = a_q + a_w + a_{f_1} - a_{f_2}$. Since there is no work involved and no heat loss from the pipe, the above equation is further reduced to $i = a_{f_1} - a_{f_2}$, or

$$i = (h_1 - h_2) - T_0(s_1 - s_2) + \frac{V_1^2 - V_2^2}{2} + (z_1 - z_2)g \qquad \text{(4-o)}$$

Because the water velocity and temperature are also constant, Eq. (4-o) can be further reduced to

$$i = v_1(P_1 - P_2) + (z_1 - z_2)g \qquad \text{(4-p)}$$

Substituting the numeric values into Eq. (4-p) gives the availability destruction per pound of water flowing in the pipe as

$$i = 0.01613(46 - 20)\frac{144}{778} + \frac{(0 - 30)32.2}{32.2 \times 778} = 0.039 \text{ Btu/lbm}$$

This availability destruction is due entirely to the friction in the pipe. If the friction is decreased, the availability destruction will be reduced. In frictionless flow the availability destruction will vanish and Eq. (4-p) will become the Bernoulli equation.

Equation 4-10. Consider Freon-12 flowing in a horizontal pipe at a constant rate. The fluid enters the pipe at 15 psia and 20°F and leaves at 60°F. Assuming that the pressure drop is negligible, calculate per unit mass of Freon-12 (a) the heat gain and (b) the associated heat availability transfer in the process.

Solution. From the Freon-12 table we evaluate the enthalpies and entropies at the inlet and exit of the pipe. The results are

$$h_1 = 80.71 \text{ Btu/lbm} \qquad h_2 = 86.45 \text{ Btu/lbm}$$
$$s_1 = 0.1835 \text{ Btu/lbm-R} \qquad s_2 = 0.1950 \text{ Btu/lbm-R}$$

(a) The fluid receives heat from the environment as the fluid flows inside the pipe. The heat gain is determined from the first law equation: $q = h_2 - h_1 = 86.45 - 80.71 = 5.74$ Btu/lbm.

(b) The availability transfer associated with the heat gain by Freon-12 is determined from the availability equation $a_q + a_w + a_{f_1} - a_{f_2} - i = 0$.

Since the process is internally reversible and no work is involved in the process, the above equation can be further reduced to $a_q = a_{f_2} - a_{f_1} = (h_2 - h_1) - T_0(s_2 - s_1)$. With the numeric values, we have the heat availability transfer in the process as

$$a_q = (86.45 - 80.71) - 537(0.1950 - 0.1835) = -0.436 \text{ Btu/lbm}.$$

It is important to notice that in this flow process Freon-12 gains heat while it loses heat availability. The direction of heat flow is opposite to that of heat availability transfer. Also, Freon-12 experiences a decrease in flow availability as the fluid flows in the pipe. But this decrease of flow availability is exactly equal to the heat availability transferred from the Freon-12 flow. ■

4-6 ENERGY STORAGE

The need for energy storage arises in many occasions. Take electric-power generation as an example. The demand for electric energy is not constant, and, in fact, it is always either higher or lower than the based-load generating capacity in the utility network system. For reliable, efficient, and economic operation, the utility planners frequently utilize energy storage systems such as pumped-hydro, compressed-air storage gas-turbine plants, and batteries. The objective of energy storage is simple. When the demand is lower than the based-load generating capacity, excess electric energy is stored. When the demand is higher than the base-load generating capacity, the stored energy is released. Use of energy storage usually results in an improved system operation.

In recent years the renewable energy sources have received increasing attention. Because of the nature of energy sources, many systems such as solar and wind frequently require energy storage. The storage scheme usually is of thermal storage types, which may include sensible heat and latent heat. Sometimes chemical energy storage is also utilized.

Another example for use of energy storage is the steam seal regulator frequently required in a coal-fired power plant. The steam seal regulator is used to collect the steam from the turbine gland leakages and to deliver it to various feedwater heaters in the power plant system. The general purpose of the steam seal regulator is to conserve the water as well as its thermal energy.

Presented in this section is an application of the availability equation to the fluid storage unit. For simplicity, the storage unit will be examined under the following operating conditions: (1) charging, (2) discharging, and (3) idling or storaging.

The availability equation for the storage unit is given by Eq. (3-32). For convenience, the equation is reproduced as below:

$$m_2 a_2 - m_1 a_1 = \sum_i \Delta m_i a_{f_i} - \sum_e \Delta m_e a_{f_e} + A_q + A_w - I$$

Since the storage unit generates no work, we can further reduce the above equation to

$$m_2a_2 - m_1a_1 = \sum_i \Delta m_i a_{f_i} - \sum_e \Delta m_e a_{f_e} + A_q - I \qquad (4\text{-}21)$$

In words, the equation shows that the availability change inside the control volume must be equal to the net flow availability in and out of the control volume minus the availability destruction and loss in the process. For a charging operation, the availability equation becomes

$$\sum \Delta m_i a_{f_i} - (m_2a_2 - m_1a_1) = I - A_q$$

The second law efficiency for the charging process simply is the ratio of the availability change inside the storage to the availability supply to the storage. In equation form it is

$$\varepsilon = \frac{m_2a_2 - m_1a_1}{\sum \Delta m_i a_{f_i}} \qquad (4\text{-}22)$$

For a discharging operation, the availability equation is given by

$$(m_1a_1 - m_2a_2) - \sum \Delta m_e a_{f_e} = I - A_q$$

The second law efficiency is the ratio of the availability withdrawn to the availability decrease inside the storage, that is,

$$\varepsilon = \frac{\sum \Delta m_e a_{f_e}}{m_1a_1 - m_2a_2} \qquad (4\text{-}23)$$

Similarly, for an idling operation, the availability equation and its second law efficiency are respectively $m_1a_1 - m_2a_2 = I - A_q$ and

$$\varepsilon = \frac{m_2a_2}{m_1a_1} \qquad (4\text{-}24)$$

Example 4-11. An insulated pressure vessel has a volume of 200 ft^3 and contains saturated steam of 50 psia. Attached to the pressure vessel is a steam pipe in which the conditions of 400 psia and 600°F are maintained. Steam flows into the vessel from the pipe until the vessel pressure reaches 400 psia. Calculate the availability destruction in this charging operation. What is the second law efficiency in this process?

Solution. A schematic diagram for the storage unit is shown in Fig. 4-8. To determine the steam conditions at the end of the process, we apply the first law equation

$$Q_{cv} + \Delta m_i h_i - W_{cv} - \Delta m_e h_e = m_2u_2 - m_1u_1$$

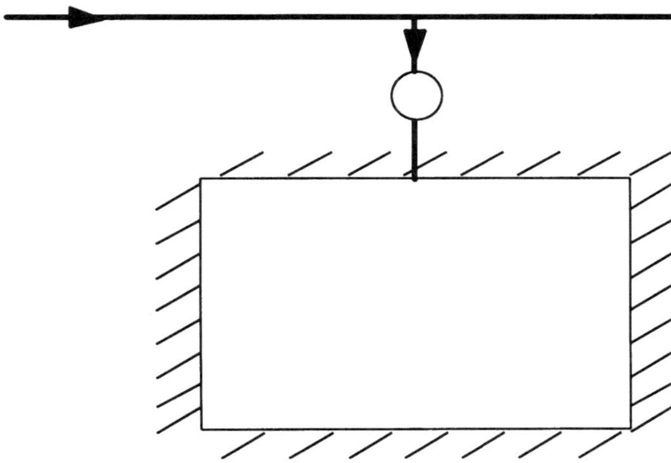

Figure 4-8 Schematic diagram for steam storage.

Because $W_{cv} = 0$, $Q_{cv} = 0$, and $\Delta m_e h_e = 0$, the above equation becomes

$$\Delta m_i h_i = m_2 u_2 - m_1 u_1$$

Also, because $\Delta m_i = m_2 - m_1$, the equation is further reduced to

$$m_2(h_i - u_2) = m_1(h_i - u_1) \tag{4-q}$$

At the end of the process, the pressure is 400 psia, and the following relationship holds:

$$m_2 = \frac{200}{v_2} \tag{4-r}$$

Evidently, Eqs. (4-q) and (4-r) contain only two unknowns: u_2 and v_2. Using the steam properties

$$h_i = 1306.6 \text{ Btu/lbm}$$

$$u_1 = 1095.6 \text{ Btu/lbm}$$

$$v_1 = 8.518 \text{ ft}^3\text{/lbm}$$

$$m_1 = \frac{200}{v_1} = 23.48 \text{ lbm}$$

we calculate the final steam conditions by trial and error. The results are

$$T_2 = 755°\text{F}$$

$$u_2 = 1263.4 \text{ Btu/lbm}$$

$$v_2 = 1.7423 \text{ ft}^3\text{/lbm}$$

$$m_2 = 114.8 \text{ lbm}$$

Now, we start with the availability calculation. The availability destruction is given by the equation

$$I = \Delta m_i a_{f_i} - (m_2 a_2 - m_1 a_1) \tag{4-s}$$

The conditions for the environmental state are 14.7 psia and 77°F. First, we determine the nonflow availability at state 1 as

$$\begin{aligned} a_1 &= (u_1 - u_0) - T_0(s_1 - s_0) + P_0(v_1 - v_0) \\ &= (1095.6 - 45.09) - 537(1.6589 - 0.08775) \\ &\quad + \frac{14.7 \times 144}{778}(8.518 - 0.01606) \\ &= 229.93 \text{ Btu/lbm} \end{aligned}$$

Similarly, the nonflow availability at state 2 is

$$\begin{aligned} a_2 &= (1263.4 - 45.09) - 537(1.6648 - 0.08775) \\ &\quad + \frac{14.7 \times 144}{778}(1.7423 - 0.01606) \\ &= 376.13 \text{ Btu/lbm} \end{aligned}$$

The flow availability at location i is given by the equation

$$a_{f_i} = (h_i - h_0) - T_0(s_i - s_0)$$

Substituting the numeric value into the equation, we have $a_{f_i} = (1306.6 - 45.09) - 537(1.5892 - 0.08775) = 455.23$ Btu/lbm. Finally, by Eq. (4-s), the availability destruction is

$$I = (114.8 - 23.48)(455.23) - (114.8 \times 376.13 - 23.48 \times 229.93) = 3719 \text{ Btu}$$

The second law efficiency for the charging operation is given by

$$\varepsilon = \frac{m_2 a_2 - m_1 a_1}{\Delta m_i a_{f_i}}$$

With the numeric values we have

$$\varepsilon = \frac{114.8 \times 376.13 - 23.48 \times 229.93}{(114.8 - 23.48)(455.23)} = 0.908$$

The charging process is seen highly irreversible. The availability supplied to the storage unit is always greater than the availability change inside. The difference between these two is the availability destruction in the process. For this example, the availability destruction is approximately equal to 10% of the availability flow into the storage unit. ■

Example 4-12. An insulated tank has a volume of 40 ft^3 and initially contains saturated steam at 50 psia. A control valve on the top of the tank is then opened, and steam is allowed to leave until the pressure inside drops to 30 psia. The discharging process is such that only saturated vapor would exit from the tank. Estimate the second law efficiency for this process.

Solution. The second law efficiency for the discharging process is

$$\varepsilon = \frac{\Delta m_e a_{f_e}}{m_1 a_1 - m_2 a_2} \tag{4-t}$$

To estimate the final conditions, we apply the first law equation as

$$Q_{cv} + \Delta m_i h_i - W_{cv} - \Delta m_e h_e = m_2 u_2 - m_1 u_1$$

Because $Q_{cv} = 0$, $W_{cv} = 0$, $\Delta m_i h_i = 0$, and $\Delta m_e = m_1 - m_2$, the above equation becomes

$$-(m_1 - m_2)h_e = m_2 u_2 - m_1 u_1$$

or

$$m_2(h_e - u_2) = m_1(h_e - u_1)$$

With the steam properties

$$m_1 = \frac{V_1}{v_1} = \frac{40}{8.518} = 4.696 \text{ lbm}$$

$$u_1 = 1095.6 \text{ Btu/lbm}$$

$$h_e = \frac{h_{g_1} + h_{g_2}}{2} = \frac{1174.4 + 1164.3}{2} = 1169.4 \text{ Btu/lbm}$$

the equation is further reduced to

$$m_2(1169.4 - u_2) = 4.696(1169.4 - 1095.6) \tag{4-u}$$

Also

$$m_2 = \frac{V_2}{v_2} = \frac{40}{(1 - x_2)(0.017) + x_2(13.748)} \tag{4-v}$$

and

$$u_2 = (1 - x_2)(218.84) + x_2(1088) \tag{4-w}$$

Then we solve Eqs. (4-u), (4-v), and (4-w) simultaneously and obtain $x_2 = 0.9618$, $u_2 = 1054.8$ Btu/lbm, and $m_2 = 3.026$ lbm. For later use, we also determine $v_2 = 13.22$ ft^3/lb and $s_2 = 1.6488$ Btu/lbm-R. Thus the amount of steam discharge from the tank is $m_e = m_1 - m_2 = 4.696 - 3.026 = 1.67$ lbm.

Now we start with the availability calculation. The flow availability at the discharge point is given by the equation

$$a_{f_e} = (h_e - h_0) - T_0(s_e - s_0)$$

Again, it is assumed that

$$s_e = \frac{s_{g_1} + s_{g_2}}{2} = \frac{1.6589 + 1.6996}{2} = 1.6793 \text{ Btu/lbm-R}$$

With the numeric values we have the flow availability at the tank exit as $a_{f_e} = (1169.4 - 45.09) - 537(1.6793 - 0.08775) = 264.1$ Btu/lbm.

The nonflow availability is given by the equation

$$a = (u - u_0) - T_0(s - s_0) + P_0(v - v_0)$$

Numerically, for state 1 we have

$$a_1 = (1095.6 - 45.09) - 537(1.6589 - 0.08775) + \frac{14.7 \times 144}{778}(8.518 - 0.016)$$

$$= 229.9 \text{ Btu/lbm}$$

and for state 2

$$a_2 = (1054.8 - 45.09) - 537(1.6488 - 0.08775) + \frac{14.7 \times 144}{778}(13.22 - 0.016)$$

$$= 207.3 \text{ Btu/lbm}$$

Finally, we have the second law efficiency by Eq. (4-t)

$$\varepsilon = \frac{1.67 \times 264.1}{4.696 \times 229.9 - 3.026 \times 207.3} = 0.9751 \quad \blacksquare$$

It must be emphasized that because of the assumption (using the average value as the tank exit condition), the solution in Example 4-12 is at best approximate. In fact, when the tank is perfectly insulated and the discharge from it takes place very slowly, the process is usually assumed to be isentropic. In that case, there will be no availability destruction and loss. Also, the second law efficiency is expected to be a unity.

The energy storage unit usually is treated as an open system when it undergoes a charging or discharging process. This approach has been demonstrated by both examples in this section. However, the energy storage unit would become a closed system if it is operated in an idling or storaging mode. This is especially true when the leakage from the fluid storage is negligible. Numerical calculations for the idling process can be found in Example 3-3 of the preceding chapter.

4-7 SUMMARY

This chapter presents an application of availability equation to various engineering processes. The availability equation is

$$m_2 a_2 - m_1 a_1 = \sum_j \left(1 - \frac{T_0}{T_j}\right) \delta Q_j - W_{cv} + P_0(V_2 - V_1) + \sum_i \Delta m_i a_{f_i} - \sum_e \Delta m_e a_{f_e} - T_0 \sigma$$

This equation is valid for open systems as well as closed systems. In an availability analysis, one of the objectives is to determine the availability loss and destruction. In many occasions engineers just lumped these two terms together and referred to the combined term as the availability destruction or the process irreversibility.

In an availability analysis, attention is given to the quantity as well as the quality of energy. Different weights are assigned to energies according to their work-producing capabilities. The analysis emphasizes the effects of availability loss and availability destruction on the process performance. In many cases the availability destruction is more dominating than the availability loss.

Another objective of availability analysis is to determine the second law efficiency, or process effectiveness. In this chapter we defined the second law efficiency individually for various engineering processes. Generally, we treat the availability equation for a given process as an availability balance. For each process we define the availability input and availability output. The difference between these two must be the availability loss and destruction in the process. The second law efficiency simply is the ratio of the availability output to the availability input. That is,

$$\varepsilon = \frac{\text{Availability output in the process}}{\text{Availability input in the process}}$$

or

$$\varepsilon = 1 - \frac{\text{Availability loss and destruction}}{\text{Availability input}}$$

As indicated in the above equation, the second law efficiency is based on the concept of availability. It is a true measure of the process efficiency. It should be emphasized that different selections (or decisions) on the process availability input and output will lead to different expressions for the second law efficiency.

4-8 PROBLEMS

4-1 An electric generator driven by a steam turbine has an efficiency of 98 percent. Estimate the availability loss and destruction per unit input for the electric generator.

4-2 In a modern power plant the steam turbine usually consists of high-pressure, intermediate-pressure, and low-pressure cylinders, all coupled with an electric generator. When the coupling efficiency is 98 percent, what is the availability loss and destruction in this coupling process? Express the result per unit input.

4-3 A spray condenser is sometimes used in power plant design. It operates by spraying the cooling water into the steam. At the bottom of the condenser, the mixture is usually at saturated conditions and pumped back to the plant feedwater system and cooling tower. Estimate the availability destruction of this mixing process for the following conditions:

Condenser pressure	1 psia
Cooling water	60°F
Steam conditions	
Flow rate	10^6 lb/hr
Quality	88 percent

4-4 A surface condenser is designed to handle 3×10^6 lbm/hr of 90 percent quality steam at 1 psia, 360,000 lbm/hr of 112°F drain water from the low-pressure feedwater heater, and 1875 lbm/hr of 440°F drain water from the steam-jet air ejector. The cooling water enters the condenser at 70°F and leaves at 90°F. Calculate (a) the availability destruction in the condenser operation using the condenser as a control volume, (b) the availability decrease in the hot side of the condenser, and (c) the availability increase in the cold side of condenser.

4-5 A surface feedwater heater has the following operating data:

Feedwater	3×10^6 lbm/hr, 3000 psia, 370°F in, 398°F out
Bled steam	140,000 lbm/hr, 240 psia, 800°F
Drain from another feedwater heater	300,000 lbm/hr, 520 psia, 410°F
Drain from this heater	440,000 lbm/hr, 380°F

Calculate the availability loss and destruction in this operation, and estimate the second law efficiency.

4-6 A reheater in a modern power plant has the following operating data:

Steam flow rate	4.4×10^6 lbm/hr
Inlet conditions	744 psia, 598°F
Outlet conditions	692 psia, 1000°F

Estimate the availability destruction in the steam reheating process.

4-7 A low-pressure turbine has several steam extractions for feedwater heating. Figure 4-9 shows the typical operation for a low-pressure steam turbine with all mass and heat balance data. Assuming the low-pressure turbine to be a control volume, calculate (a) the availability at the

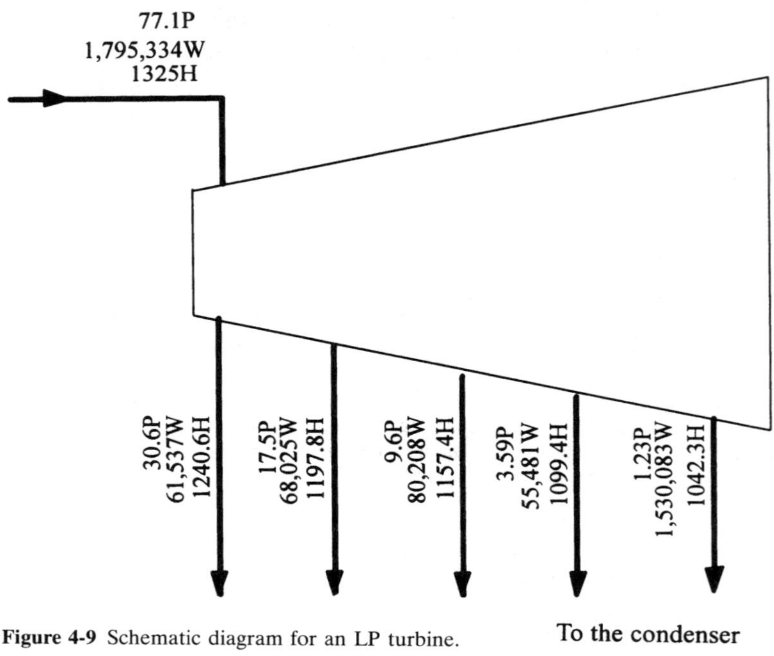

Figure 4-9 Schematic diagram for an LP turbine.

turbine inlet, (b) the availability at the turbine exits, (c) the availability loss and destruction in the process, and (d) the second law efficiency of the low-pressure steam turbine.

4-8 Saturated water enters a boiler feedwater pump at 153 psia and leaves at 4200 psia. The pump has an efficiency of 85 percent and a flow rate of 10^6 lbm/hr. Calculate the availability loss and destruction.

4-9 The boiler feedwater pump frequently is driven by an auxiliary steam turbine. For the pump specified in Problem 4-8, calculate the steam requirement and the turbine second law efficiency for the following conditions:

Turbine inlet conditions	153 psia, 630°F
Turbine exit conditions	2.5″ Hg abs, 88% quality
Mechanical coupling efficiency	99%

4-10 A steam storage has a volume of 60 ft^3 and contains steam at 100 psia and 340°F. Because of insufficient insulation from the storage unit, the steam at one point is cooled to 300°F. Calculate the availability loss associated with the heat loss. What is the second law efficiency for the process?

4-11 An underground compressed-air storage has a volume of one million cubic feet. It contains air initially at 20 psia and 80°F. A control valve is then opened, and the underground air storage is charged with the compressed air at 200 psia and 120°F. The process is assumed to be adiabatic and is terminated at the storage pressure of 200 psia. Determine (a) the initial and final availabilities, (b) the availability destruction, and (c) the second law efficiency for the process.

4-12 An underground air storage has a volume of one million cubic feet and initial conditions 300 psia and 150°F. The discharging takes place until the air pressure inside drops to 200 psia. Assuming negligible heat loss from the storage system, estimate (a) the availability of the air leaving the underground air storage, (b) the change of the availability in the air storage, and (c) the second law efficiency for the discharging process.

4-13 An underground storage contains one million cubic feet of compressed air at 200 psia and 120°F. After 10 hours in idling, the air inside the storage has the pressure of 190 psia and the temperature of 100°F. Estimate the availability loss and destruction. What is the second law efficiency for this process?

4-14 Freon-12 is contained in a 2-ft^3 storage unit at 15 psia and 80°F. It is desired to fill up 80% of the storage unit with liquid (by volume) at this temperature. The storage unit is connected to a line flowing Freon-12 at 100 psia and 100°F. For this process calculate (a) the initial and final nonflow availabilities, (b) the heat availability associated with the heat transfer, and (c) the availability destruction in the process.

CHAPTER
FIVE

DIFFUSION AVAILABILITY

The availability is the maximum work that can be extracted from the substance as it undergoes a reversible process from a given state to the environmental state. The environmental state was defined in an earlier chapter as the state with the temperature and pressure identical to those of the environment. When a substance is in the environmental state, it is in thermal and mechanical equilibrium with the environment and therefore has no work potential. In other words, a substance has a zero availability at the environmental state. The work potential was referred to as the availability. To be more exact, it should have been called the thermomechanical availability.

In the previous chapter we described the environment in terms of its temperature and pressure. No consideration was given to the composition of the environment. This approach is acceptable as long as we are interested only in the thermomechanical availability. However, there are many occasions in which the substance is in thermal and mechanical equilibrium with the environment but not in chemical equilibrium. An interesting question now arises. Is there any work that theoretically can be developed as a substance comes into chemical equilibrium with the environment? If work is developed, what is the maximum work one would expect? In this text, the work potential of this kind will be called either *diffusion availability* or *chemical availability*. The objective of this chapter and the next is to develop the concept of diffusion and chemical availability and the equations for their numeric calculation. Also included are their applications to various engineering processes such as those in heating, ventilating, and air conditioning (HVAC) and in combustion.

5-1 CONCEPT OF DIFFUSION AVAILABILITY

The conditions of the environment play an important role in the calculation of thermomechanical availability. The thermomechanical availability will have different values for different choices of the environmental temperature and pressure. To determine the maximum work that can be developed as the substance comes into thermal, mechanical, and chemical equilibrium with the environment, we need a more detailed description of the environment. In addition to the temperature and pressure, the composition of the environment is needed. In the literature there have been many discussions and suggestions about modeling the environment. For simplicity we will present just one of the popular models for the environment:

Temperature (°F)	77°F
Pressure (psia)	14.7 psia
Components	Mole Fraction (x^0)
Nitrogen	0.7567
Oxygen	0.2035
Water vapor	0.0303
Carbon dioxide	0.0003
Other gases	0.0092

To determine the combined availability (i.e., the combination of thermomechanical and diffusion availabilities), we start with the equation for the available work of a substance in the process between the states 1 and 2:

$$w_{\text{ava}} = (h_1 - h_2) - T_0(s_1 - s_2)$$

The substance is one of the components in the environment. When the final state (state 2) is the dead state, the available work will become the combined availability of the substance in state 1. In a general form, we simply write the combined availability of a substance as

$$a_f = (h - h_d) - T_d(s - s_d) \tag{5-1}$$

There should be some explanation about the dead state. The *dead state* is defined as the state in which the substance is in thermal, mechanical, and chemical equilibrium with the environment. For the case of a pure substance, equilibrium means that the substance has a temperature equal to the environmental temperature T_0 and a pressure equal to the partial pressure of the same component in the environment. That is,

$$T_d = T_0 \tag{5-2}$$

$$P_d = x^0 P_0 \tag{5-3}$$

To satisfy the condition that the substance be in chemical equilibrium with the environment, in a physical sense there must be no difference in the concentration level between the substance and the corresponding component in the environment; thus there should be no net diffusion between them. Mathematically, Eq. (5-3) is the condition for the pure substance in chemical equilibrium with the environment.

In this text the dead state is defined differently than the environmental state. In the environmental state, there is no requirement for chemical equilibrium with the environment. It can also be interpreted that there is no interaction between the substance and the contents of the environment. We can visualize the substance as being contained in a big envelope where the inside temperature and pressure are equal to those in the environment. In the dead state, however, we can imagine that the substance is in direct contact with the contents of the environment and that, because of chemical equilibrium, there will be no further diffusion or chemical reaction between them. For the case of the pure substance, the substance at the dead state must have a pressure equal to the partial pressure of the same component in the environment. When the substance under consideration is an ideal gas, we can further reduce Eq. (5-1) to

$$a_f = c_p(T - T_0) - T_0\left(c_p \ln\frac{T}{T_0} - R\ln\frac{P}{x^0 P_0}\right)$$

or

$$a_f = a_f^{\text{th}} + a_f^{\text{di}} \tag{5-4}$$

where

$$a_f^{\text{th}} = c_p(T - T_0) - T_0\left(c_p \ln\frac{T}{T_0} - R\ln\frac{P}{P_0}\right) \tag{5-5}$$

$$a_f^{\text{di}} = T_0 R \ln\frac{1}{x^0} \tag{5-6}$$

Eq. (5-4) indicates that the total availability of a substance at a given state is the thermomechanical availability a_f^{th} plus the diffusion availability a_f^{di} at that state. We can imagine taking the reversible process between the given state and the dead state and breaking it into two processes. The first process will take the substance into thermal and mechanical equilibrium with the environment. The first process terminates at the environmental state, where the temperature and pressure are respectively equal to T_0 and P_0. The second process starts at the environment state (T_0, P_0) and eventually brings the substance into chemical equilibrium with the environment. In the dead state the substance has the temperature T_0 and the pressure equal to the partial pressure of the corresponding component in the environment. The available work in the second process is defined as the diffusion availability for a substance. In an equation form, the available work is

$$a_f^{\text{di}} = w_{\text{ava}} = c_p(T_0 - T_0) - T_0\left(c_p \ln\frac{T_0}{T_0} - R\ln\frac{P_0}{x^0 P_0}\right)$$

Rearranging this equation provides an expression for the diffusion availability as

$$a_f^{\mathrm{di}} = T_0 R \ln \frac{1}{x^0} \tag{5-7}$$

which is identical to the diffusion flow availability as defined in Eq. (5-6). There are two observations to be made here. First, the diffusion availability for a pure substance does not depend on the thermodynamic properties at a given state. In fact, the diffusion availability is constant in value. Second, the diffusion availability is very much related to the model of the environment. Different choices of the environment model will result in different calculated values.

Similarly, we can write the total nonflow availability as the sum of the thermomechanical and diffusion availability. That is,

$$a = a^{\mathrm{th}} + a^{\mathrm{di}} \tag{5-8}$$

where a^{th} is the thermomechanical nonflow availability and a^{di} is the diffusion nonflow availability. The thermomechanical nonflow availability has been discussed in Chapter 3. The diffusion nonflow availability is the available work between the environmental state (T_0, P_0) and the dead state. As before, the dead state is the state at which the substance is in chemical equilibrium with the environment. For a pure substance the pressure of the substance at the dead state is equal to the partial pressure of the same material in the environment. In equation form, the diffusion nonflow availability is

$$a^{\mathrm{di}} = T_0 R \ln \frac{1}{x^0} \tag{5-9}$$

which is identical to the diffusion flow availability.

Steam is a pure substance of considerable interest. The total availability of steam at any state (T, P) can be developed and expressed as

$$a_f = \left[h(T,P) - h_g(T_0)\right] - T_0\left[s(T,P) - s_g(T_0)\right] + T_0 R_v \ln \frac{P_g(T_0)}{x_v^0 P_0} \tag{5-10}$$

or

$$a_f = a_f^{\mathrm{th}} + a_f^{\mathrm{di}}$$

where

$$a_f^{\mathrm{th}} = \left[h(T,P) - h_g(T_0)\right] - T_0\left[s(T,P) - s_g(T_0)\right] \tag{5-11}$$

$$a_f^{\mathrm{di}} = T_0 R_v \ln \frac{P_g(T_0)}{x_v^0 P_0} \tag{5-12}$$

Equation (5-10) is valid for both steam and water. For the water only, Eq. (5-10) can be rewritten in a more convenient form. For this purpose, we make use of the conditions

$$h(T, P) = h_f(T) + v_f\left[P - P_g(T)\right] \tag{5-13}$$

and

$$s(T, P) = s_f(T) \tag{5-14}$$

Then we substitute Eqs. (5-13) and (5-14) into Eq. (5-10) to get total availability of water at any given state (T, P) as

$$a_f = \left[h_f(T) - h_g(T_0)\right] + v_f\left[P - P_g(T)\right] - T_0\left[s_f(T) - s_g(T_0)\right] + T_0 R_v \ln\frac{P_g(T_0)}{x_v^0 P_0} \tag{5-15}$$

where the subscript f denotes saturated water at the temperature T. While Eqs. (5-10) and (5-15) are developed for steam and water, they are very general in nature and valid for other liquids.

There should be some discussion about Eq. (5-11). In this equation we set the enthalpy and entropy at the environmental state to be the same as those of saturated vapor at the temperature T_0. Because of the relationship

$$h_g(T_0) - T_0 s_g(T_0) = h_f(T_0) - T_0 s_f(T_0) \tag{5-16}$$

Eq. (5-11) can be further reduced to

$$a_f^{\text{th}} = \left[h(T, P) - h_f(T_0)\right] - T_0\left[s(T, P) - s_f(T_0)\right] \tag{5-17}$$

Equation (5-17) is identical to the equation for the thermomechanical availability developed in Chapter 3.

In summary, we have developed the concept of diffusion availability for a pure substance that exists as one of the components in the environment. Physically, the diffusion availability is the available work that can be extracted from the substance as it undergoes a reversible process from the environment state (T_0, P_0) to the dead state $(T_0, x^0 P_0)$. In the dead state the substance is expected to have the temperature T_0 and the pressure equal to the partial pressure of that substance in the environment. For calculations, Eqs. (5-6) and (5-12) are used for ideal gases and vapors, respectively.

Example 5-1. Consider steam flowing inside a long pipe. The steam experiences a pressure drop from 2000 psia to 1800 psia and a temperature drop from 1040°F to 1000°F. Calculate the combined flow availability at both states.

Solution. Using the steam table, we evaluate the steam properties for both states. The results summarized below:

$$\begin{aligned} P_1 &= 2000 \text{ psia} & P_2 &= 1800 \text{ psia} \\ T_1 &= 1040°\text{F} & T_2 &= 1000°\text{F} \\ h_1 &= 1499.6 \text{ Btu/lbm} & h_2 &= 1480.7 \text{ Btu/lbm} \\ s_1 &= 1.5771 \text{ Btu/lbm-R} & s_2 &= 1.5749 \text{ Btu/lbm-R} \end{aligned}$$

The combined flow availability is given by the equation

$$a_f = a_f^{\text{th}} + a_f^{\text{di}} \tag{5-a}$$

where

$$a_f^{\text{th}} = (h - h_0) - T_0(s - s_0) \tag{5-b}$$

and

$$a_f^{\text{di}} = T_0 R_v \ln \frac{P_g(T_0)}{x_v^0 P_0} \tag{5-c}$$

As before, we have $h_0 = 45.09$ Btu/lbm and $s_0 = 0.0877$ Btu/lbm-R. Substituting the numeric values into the above equations, we get the thermomechanical and diffusion availabilities as state 1:

$$a_{f_1}^{\text{th}} = (1499.6 - 45.09) - 537(1.5771 - 0.0877) = 654.7 \text{ Btu/lbm}$$

$$a_{f_1}^{\text{di}} = \frac{(537)(85.76)}{778} \ln \frac{0.46}{(0.0303)(14.7)} = 1.91 \text{ Btu/lbm}$$

By Eq. (5-a), the combined flow availability at state 1 is $a_{f_1} = 654.7 + 1.91 = 656.6$ Btu/lbm. Similarly, the combined flow availability at state 2 is $a_{f_2} = a_{f_2}^{\text{th}} + a_{f_2}^{\text{di}} = 637.0 + 1.91 = 638.9$ Btu/lbm.

The diffusion availability for this steam flow is small relative to the thermomechanical availability. Also, the diffusion availability remains unchanged in the process, as expected. Because of this, the change of combined availability is exactly equal to that of thermomechanical availability. ■

Example 5-2. Oxygen is at the conditions of 14.7 psia and 100°F. Calculate the thermomechanical and diffusion availability per unit mass of oxygen.

Solution. The thermomechanical availability is given by

$$a_f^{\text{th}} = (h - h_0) - T_0(s - s_0) \tag{5-d}$$

where h_0 and s_0 are the enthalpy and entropy of the substance at the environmental state (T_0, P_0). When oxygen is treated as an ideal gas, the equation becomes

$$a_f^{\text{th}} = c_p(T - T_0) - T_0\left(c_p \ln\frac{T}{T_0} - R\ln\frac{p}{p_0}\right)$$

With numeric values we have

$$a_f^{\text{th}} = 0.219(560 - 537) - (537)(0.219)\ln\frac{560}{537} = 0.11 \text{ Btu/lbm}$$

The diffusion availability is

$$a_f^{\text{di}} = T_0 R \ln\frac{1}{x_{O_2}^0} \tag{5-e}$$

With numeric values, we have

$$a_f^{\text{di}} = \frac{(48.28)(537)}{778}\ln\frac{1}{0.2035} = 53.1 \text{ Btu/lbm}$$

The diffusion availability is much more dominating than the thermomechanical availability for the oxygen under these conditions. Omitting the diffusion availability in availability calculations would result in a significant mistake under these conditions. ■

5-2 AVAILABILITY OF MIXTURES

There are many engineering applications in which the working substance is a mixture. The products of combustion are a good example. In this section we will deal only with the mixture of ideal gases. The mixture itself also is assumed to be an ideal gas.

Consider a mixture of n components. The mixture has a volumetric analysis as $(x_1, x_2, \ldots, x_n)$. From elementary thermodynamics the internal energy, enthalpy, and entropy of the mixture can be expressed respectively as below:

$$\bar{u} = \sum_{k=1}^{n} x_k \bar{u}_k \tag{5-18}$$

$$\bar{h} = \sum_{k=1}^{n} x_k \bar{h}_k \tag{5-19}$$

$$\bar{s} = \sum_{k=1}^{n} x_i \bar{s}_k \tag{5-20}$$

For ideal gas components the terms $\bar{u}_k$ and $\bar{h}_k$ are evaluated at the component temperature, which is equal to the temperature of the mixture. The term $\bar{s}_k$ is a function of component temperature and pressure. The component pressure is the partial pressure of the component in the mixture.

The total flow availability for a mixture is important to many availability analyses. It consists of thermomechanical and diffusion availabilities. To determine these two availabilities, we follow the procedure used for the mixture internal energy, enthalpy, and entropy. In this section attention will be focused on the mixture for which all components exist in the environment. For the thermomechanical availability of a mixture at any given state (T, P), we simply present the equation

$$\bar{a}_f^{\text{th}} = \sum_{k=1}^{n} x_k \bar{a}_{f_k}^{\text{th}} \tag{5-21}$$

or

$$\bar{a}_f^{\text{th}} = \sum_{k=1}^{n} x_k \left[\bar{h}_k(T) - \bar{h}_k(T_0) - T_0[\bar{s}_k(T, x_k P) - \bar{s}_k(T_0, x_k P_0)] \right] \tag{5-22}$$

In case the thermodynamic properties are to be evaluated with the thermodynamic property table, Eq. (5-22) should be reduced to

$$\bar{a}_f^{\text{th}} = \sum_{k=1}^{n} x_k \left[\bar{h}_k(T) - \bar{h}_k(T_0) - T_0[\bar{s}'_k(T) - \bar{s}'_k(T_0)] \right] + T_0 \bar{R} \ln \frac{P}{P_0} \tag{5-23}$$

In the same manner we can develop the equation for the diffusion availability of a mixture:

$$\bar{a}_f^{\text{di}} = \sum_{k=1}^{n} x_k \bar{a}_{f_k}^{\text{di}} \tag{5-24}$$

or

$$\bar{a}_f^{\text{di}} = T_0 \bar{R} \sum_{k=1}^{n} x_k \ln \frac{x_k}{x_k^0} \tag{5-25}$$

Example 5-3. The products of combustion have the volumetric analysis as below:

CO_2	6.47%
H_2O	7.29%
O_2	10.12%
N_2	76.12%

The products have a pressure of 20 psia and a temperature of 2040°F. Calculate the diffusion availability and thermomechanical availability per mole of the mixture.

Solution. The diffusion availability is given by the equation

$$\bar{a}_f^{\text{di}} = T_0\bar{R}\left[x_{CO_2}\ln\frac{x_{CO_2}}{x^0_{CO_2}} + x_{H_2O}\ln\frac{x_{H_2O}}{x^0_{H_2O}} + x_{O_2}\frac{x_{O_2}}{x^0_{O_2}} + x_{N_2}\ln\frac{x_{N_2}}{x^0_{N_2}}\right] \tag{5-f}$$

Using the environmental model specified in Section 5-1, we have

$$x^0_{CO_2} = 0.0003, \qquad x^0_{H_2O} = 0.0303$$

$$x^0_{O_2} = 0.2035, \qquad x^0_{N_2} = 0.7567$$

Substituting the numeric values into Eq. (5-f), we have the diffusion availability

$$\begin{aligned}\bar{a}_f^{\text{di}} &= \frac{(537)(1545)}{778}\left[0.0647\ln\frac{0.0647}{0.0003} + 0.0729\ln\frac{0.0729}{0.0303}\right.\\ &\quad\left. +0.1212\ln\frac{0.1012}{0.2035} + 0.7612\ln\frac{0.7612}{0.7567}\right]\\ &= 368.5\text{ Btu/lbmole}\end{aligned}$$

The thermomechanical availability is given by Eq. (5-23). Substituting the numeric values into the equation, we have

$$\begin{aligned}\bar{a}_f^{\text{th}} &= 0.0647[(27{,}801 - 4{,}030) - 537(68.7 - 51.0)]\\ &\quad + 0.0729[(22{,}735 - 4{,}258) - 537(59.0 - 45.1)]\\ &\quad + 0.1012[(19{,}443 - 3{,}725) - 537(60.9 - 49.0)]\\ &\quad + 0.7612[(18{,}590 - 3{,}730) - 537(57.1 - 45.7)]\\ &\quad + \frac{(537)(1545)}{778}\ln\frac{20}{14.7} = 9670\text{ Btu/lbmole}\end{aligned}$$

At a temperature of 2040°F, the diffusion availability of the combustion products is small relative to the thermomechanical availability. As the temperature decreases, the diffusion availability will become more dominant. It can be shown that the diffusion availability is a function of the mixture composition when the conditions at the dead state are specified. ■

5-3 AVAILABILITY OF MOIST AIR

Moist air is one of the substances frequently encountered in many engineering applications. These may include heating, ventilating, and air conditioning. Moist air is regarded as an ideal gas mixture of water vapor and dry air. Water vapor is highly superheated in the moist air and therefore can be approximated as an ideal gas. Dry air itself is a mixture but for simplicity is usually treated as a pure substance.

In this section moist air is treated as an ideal gas mixture. The composition of the mixture is described by the mole fractions of water vapor and dry air x_v, x_a). The mole fraction for the dry air component is

$$x_a = \frac{n_a}{n_a + n_v}$$

or

$$x_a = \frac{1}{1 + \left(\dfrac{n_v}{n_a}\right)} \tag{5-26}$$

Since

$$\frac{n_v}{n_a} = \left(\frac{M_a}{M_v}\right)\left(\frac{m_v}{m_a}\right) = c\omega$$

Eq. (5-26) is reduced to

$$x_a = \frac{1}{1 + c\omega} \tag{5-27}$$

where ω is the humidity ratio, and c is the ratio of molecular weights, M_a/M_v, which is numerically equal to 1.6078. Using

$$x_a + x_v = 1$$

we have the mole fraction for the water vapor component as

$$x_v = \frac{c\omega}{1 + c\omega} \tag{5-28}$$

Since moist air is treated as an ideal gas mixture of two components, both its thermomechanical and diffusion availabilities can be determined by Eqs. (5-22) and (5-25). For the purpose of the availability calculation, a new set of conditions at the dead state is used. They are

Temperature	90°F
Pressure	14.7 psia
Components	mole fraction (x^0)
dry air	0.976
water vapor	0.024

These conditions are equivalent to the weather conditions at 14.7 psia, 90°F, and relative humidity (RH) 50%. The corresponding humidity ratio ω^0 is approximately 0.0153 pound of water vapor per pound of dry air.

Example 5-4. Consider the moist air at 14.7 psia, 110°F, and 50% RH. What is the total flow availability?

Solution. We first determine the mole fractions for dry air and water vapor. Using the equations

$$P_v = \phi P_g$$

$$\omega = 0.622 \frac{P_v}{P - P_v}$$

we have

$$P_v = 0.5(1.2763) = 0.638 \text{ psia}$$

$$\omega = 0.622 \frac{0.638}{14.7 - 0.638} = 0.0282 \text{ lbm(v)/lbm(a)*}$$

Using Eqs. (5-27) and (5-28), we obtain the mole fractions for dry air and water vapor as $x_a = 0.9566$ and $x_v = 0.0434$, respectively.

The total flow availability for moist air consists of thermomechanical and diffusion availability. The thermomechanical availability is given by Eq. (5-22). Using the ideal gas model, we have

$$\bar{a}_f^{\text{th}} = x_a \left[\bar{c}_{p_a}(T - T_0) - T_0 \left(\bar{c}_{p_a} \ln \frac{T}{T_0} - \bar{R} \ln \frac{x_a P}{x_a P_0} \right) \right]$$

$$+ x_v \left[\bar{c}_{p_v}(T - T_0) - T_0 \left(\bar{c}_{p_v} \ln \frac{T}{T_0} - \bar{R} \ln \frac{x_v P}{x_v P_0} \right) \right]$$

Using the numeric values $P = P_0 = 14.7$ psia, $\bar{c}_{p_a} = 6.94$ Btu/lbmole-R, and $\bar{c}_{p_v} = 8.10$ Btu/lbmole-R, we have the thermomechanical availability as

$$\bar{a}_f^{\text{th}} = 0.9566 \left[6.94(570 - 550) - 550(6.94) \ln \frac{570}{550} \right]$$

$$+ 0.0434 \left[8.10(570 - 550) - 550(8.10) \ln \frac{570}{550} \right]$$

$$\bar{a}_f^{\text{th}} = 2.48 \text{ Btu/lbmole}$$

The diffusion availability for moist air is given by Eq. (5-25). That is,

$$\bar{a}_f^{\text{di}} = T_0 \bar{R} \left[x_a \ln \frac{x_a}{x_a^0} + x_v \ln \frac{x_v}{x_v^0} \right]$$

$$= (550)(1.986) \left[0.9566 \ln \frac{0.9566}{0.9760} + 0.0434 \ln \frac{0.0434}{0.0240} \right]$$

$$= 7.10 \text{ Btu/lbmole}$$

Finally, the total flow availability is $\bar{a}_f = \bar{a}_f^{\text{th}} + \bar{a}_f^{\text{di}} = 2.48 + 7.10 = 9.58$ Btu/lbmole. ■

*lbm(v)/lbm(a) = pound of water vapor per pound of dry air

It is of interest to know that the diffusion availability is substantially greater than the thermomechanical availability for the moist air under these conditions. This is usually true for moist air with temperature and pressure very close to the environmental temperature and pressure.

In practice, engineers often present the availability of moist air per pound of dry air. Using the similar pattern for enthalpy, that is, $H = h_a + \omega h_v$, we have the thermomechanical availability of moist air as

$$a_f^{\text{th}} = a_{f_a}^{\text{th}} + \omega a_{f_v}^{\text{th}} \tag{5-29}$$

Since the dry air and water vapor are assumed ideal gases, we have

$$a_{f_a}^{\text{th}} = c_{p_a}(T - T_0) - T_0\left[c_{p_a} \ln\frac{T}{T_0} - R_a \ln\frac{x_a P}{x_a P_0}\right]$$

and

$$a_{f_v}^{\text{th}} = c_{p_v}(T - T_0) - T_0\left[c_{p_v} \ln\frac{T}{T_0} - R_v \ln\frac{x_v P}{x_v P_0}\right]$$

Substituting these two expressions into Eq. (5-29), we have the thermomechanical availability of moist air in Btu per pound of dry air as

$$a_f^{\text{th}} = \hat{c}_p(T - T_0) - T_0\left(\hat{c}_p \ln\frac{T}{T_0} - \hat{R} \ln\frac{P}{P_0}\right) \tag{5-30}$$

where

$$\hat{c}_p = c_{p_a} + \omega c_{p_v} \tag{5-31}$$

and

$$\hat{R} = R_a + \omega R_v \tag{5-32}$$

Similarly, the diffusion availability of moist air is

$$a_f^{\text{di}} = a_{f_a}^{\text{di}} + \omega a_{f_v}^{\text{di}} \tag{5-33}$$

where $a_{f_a}^{\text{di}}$ and $a_{f_v}^{\text{di}}$ are the diffusion availabilities of dry air and water vapor, respectively. These two expressions can be shown as

$$a_{f_i}^{\text{di}} = T_0 R_a \ln\frac{x_a}{x_a^0}$$

and

$$a_{f_v}^{\text{di}} = T_0 R_v \ln\frac{x_v}{x_v^0}$$

Substituting them into Eq. (5-33), we have

$$a_f^{\text{di}} = T_0 R_a \ln\frac{x_a}{x_a^0} + T_0 R_v \omega \ln\frac{x_v}{x_v^0}$$

To further modify the above equation, we make use of Eqs. (5-27) and (5-28). Since

$$\frac{x_a}{x_a^0} = \frac{1 + c\omega^0}{1 + c\omega}$$

and

$$\frac{x_v}{x_v^0} = \frac{\omega}{\omega^0} \times \frac{1 + c\omega^0}{1 + c\omega}$$

we have the diffusion availability in Btu per pound of dry air as

$$a_f^{\mathrm{di}} = T_0 R_a \ln\frac{1 + c\omega^0}{1 + c\omega} + T_0 R_v \omega \ln\left(\frac{\omega}{\omega^0} \times \frac{1 + c\omega^0}{1 + c\omega}\right)$$

or

$$a_f^{\mathrm{di}} = T_0\left(\hat{R} \ln\frac{1 + c\omega^0}{1 + c\omega} + c\omega R_a \ln\frac{\omega}{\omega^0}\right) \tag{5-34}$$

In summary, the thermomechanical and diffusion availabilities have been derived for moist air per pound of dry air. They are given by Eqs. (5-30) and (5-34), respectively. It is seen that the thermomechanical availability of moist air is a function of its temperature, pressure, and the humidity ratio (or relative humidity). Diffusion availability also is strongly influenced by the humidity ratio of moist air. However, in both cases the conditions of the dead state must be defined in order to make any availability calculations.

5-4 AVAILABILITY ANALYSIS FOR AIR CONDITIONING PROCESSES

There are many processes engineers can utilize to provide comfortable air within a building. These processes may include heating, cooling, humidifying, and dehumidifying. In addition, the conditions of air can be changed by mixing the building air directly with another air stream, for example, the air outside the building. The objectives of this section are to apply the concepts of availability and to present the availability analyses for several air conditioning processes.

Three processes are considered in this section. The first is the adiabatic mixing of two streams. As shown in Fig. 5-1, the air streams at states 1 and 2 have different dry-bulb temperatures and relative humidities; after mixing, the air stream will be at state 3. Because heat transfer to the environment usually is negligible, the process generally is assumed to be adiabatic. Figure 5-2 shows another process called *evaporative cooling*. This process frequently is used to make hot and dry air more comfortable. This process is accomplished by passing the air stream through a water-spray section. The end effects are a reduction of air temperature and an increase of the water content in the air. In the

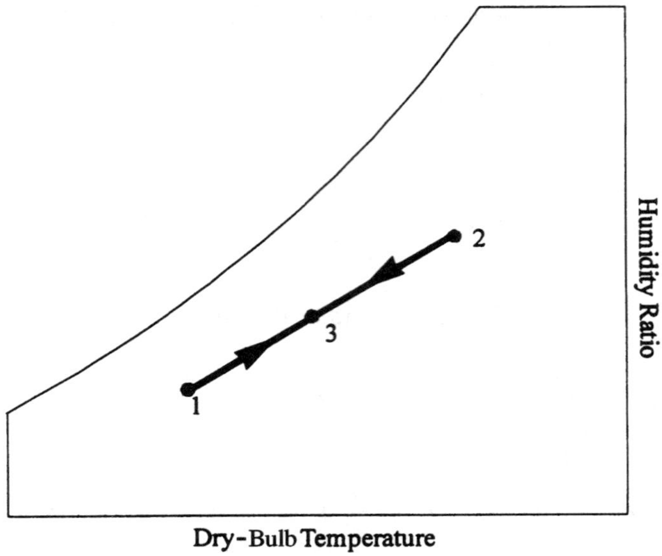

Figure 5-1 A mixing process on a psychrometric chart.

psychrometric diagram, this process is characterized by a constant wet-bulb temperature.

Figure 5-3 presents the air conditioning process for hot and humid air. First the air temperature is reduced to the dew-point temperature. During the process 1-a, the humidity ratio remains constant. As the air continues to be cooled, water vapor will condense out from the air and the air will remain saturated. As shown in Fig. 5-3, both the air temperature and the humidity ratio are lowered in the process a-b. At state b, the air generally has a comfortable humidity ratio, but the temperature is too low. This situation is overcome by passing the air stream through a heating section. This process b-2 is of constant humidity ratio.

The following are examples of availability analyses for these air conditioning processes. The dead state conditions used for these analyses are those presented in Section 5-3: the pressure is 14.7 psia, the temperature is 90°F, and the relative humidity is 50%.

Example 5-5. An air stream (50°F, 80% RH) enters an adiabatic mixing chamber at a rate of 1500 cfm and mixes with another air stream (90°F, 60% RH) at the rate of 1000 cfm. By the first law analysis, the air stream leaving the chamber would have a temperature of 65.2°F and a relative humidity of 80.8%. For convenience, all properties and other information are summarized in Table 5-1. Calculate the availability destruction and the second law efficiency for the mixing process. The pressure of the mixing chamber is assumed to be 14.7 psia.

(*a*)

(*b*)

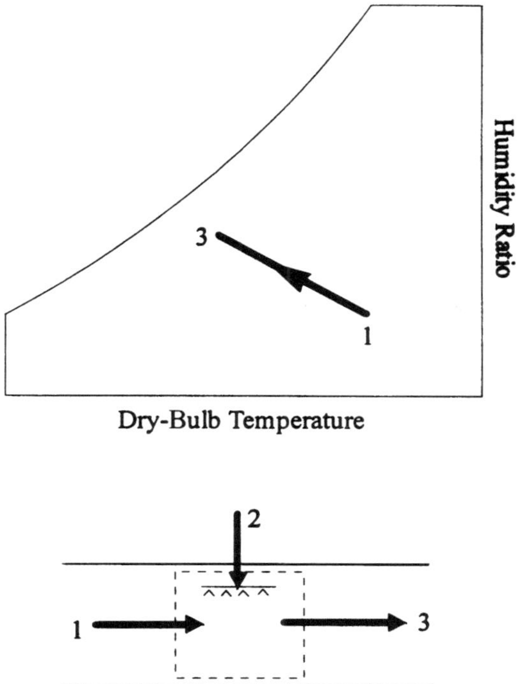

Figure 5-2 Evaporative cooling: (*a*) process in a psychrometric chart; (*b*) schematic diagram for its equipment.

Solution. First we evaluate the thermomechanical and diffusion availabilities at these three locations using Eqs. (5-30) through (5-32) and (5-34). For this example, we use the following values:

$$c_{p_a} = 0.240 \text{ Btu/lbm-R} \qquad R_a = 53.34 \text{ lbf-ft/lbm-R}$$
$$c_{p_v} = 0.445 \text{ Btu/lbm-R} \qquad R_v = 85.76 \text{ lbf-ft/lbm-R}$$
$$\omega^0 = 0.0153 \text{ lbm(v)/lbm(a)}$$

The calculations are straightforward, and the results are summarized in Table 5-2.

The availability destruction for this mixing process is given by the availability equation:

$$\begin{aligned} \dot{I} &= \dot{m}_{a_1} a_{f_1} + \dot{m}_{a_2} a_{f_2} - \dot{m}_{a_3} a_{f_3} \\ &= 115.7(0.5838) + 70.2(0.0158) - 185.9(0.1881) \\ &= 33.69 \text{ Btu/min} \end{aligned}$$

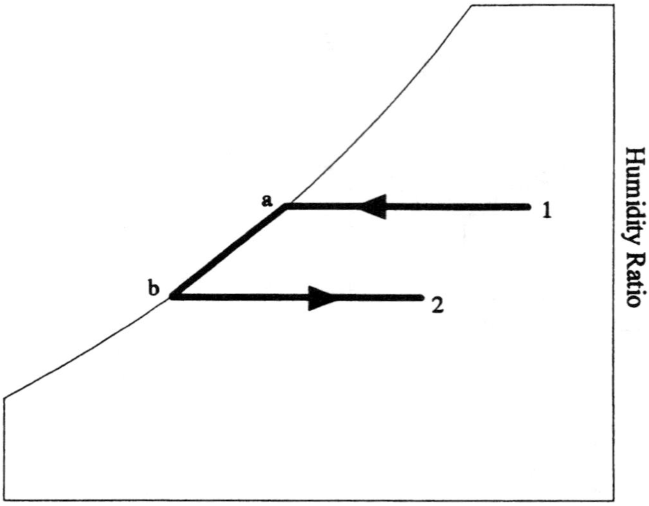

Figure 5-3 Dehumidification process with heating on a psychrometric chart.

Table 5-1 Properties of the air stream in Example 5-5.

Location	1	2	3
Temperature (°F)	50	90	65.2
Relative humidity (%)	80	60	80.8
Humidity ratio (lbm(v)/lbm(a))	0.00608	0.01826	0.0107
Volume flow rate (cfm)	1500	1000	—
Dry air flow rate (lbm(a)/min)	115.7	70.2	185.9

Table 5-2 Availability calculations for Example 5-5.

Location	1	2	3
$\hat{c}_p$ (Btu/lbm(a)-R)	0.2427	0.2481	0.2448
$\hat{R}$ (lb_f-ft/lbm(a)-R)	53.86	54.90	54.26
a_f^{th} (Btu/lbm(a))	0.3691	0	0.1422
a_f^{di} (Btu/lbm(a))	0.2147	0.0158	0.0459
a_f (Btu/lbm(a))	0.5838	0.0158	0.1881

Table 5-3 Data for the environment in Example 5-6.

Location	1	2	3
Temperature (°F)	98	70	70
Relative humidity (%)	10	—	65.3
Humidity ratio (lbm(v)/lbm(a))	0.00381	—	0.0102

The second law efficiency is

$$\varepsilon = \frac{\dot{m}_{a_3} a_{f_3}}{\dot{m}_{a_1} a_{f_1} + \dot{m}_{a_2} a_{f_2}} = \frac{185.9(0.1881)}{115.7(0.5838) + 70.2(0.0158)} = 0.5093$$

The inclusion of the diffusion availability in the calculation is important. Omitting the diffusion term would result in a significant mistake. For this particular example, the availability destruction would have been 16.27 Btu/min and the second law efficiency 61.8 percent. ■

Example 5-6. The evaporative cooler in Fig. 5-2 operates in a desert environment. The operating data are presented in Table 5-3. The spray water conditions are those at location 2. Calculate the total availability per unit mass at these three locations as well as the availability destruction for the process.

Solution. First we evaluate the availability for water at 70°F. For the thermomechanical availability, we have by Eq. (5-17)

$$\begin{aligned} a_{f_2}^{\text{th}} &= \left[h_2 - h_f(T_0)\right] - T_0\left[s_2 - s_f(T_0)\right] \\ &= (38.09 - 58.07) - 550(0.07463 - 0.11165) \\ &= 0.38 \text{ Btu/lbm} \end{aligned}$$

For the diffusion availability we have by Eq. (5-12)

$$a_{f_2}^{\text{di}} = T_0 R_v \ln \frac{P_g(T_0)}{x_v^0 P_0} = \frac{(550)(85.76)}{778} \ln \frac{0.6988}{(0.024)(14.7)} = 41.44 \text{ Btu/lbm}$$

Thus, the total availability for water at 70°F is $a_{f_2} = a_{f_2}^{\text{th}} + a_{f_2}^{\text{di}} = 0.38 + 41.44 = 41.82$ Btu/lbm.

The thermomechanical and diffusion availabilities for moist air are determined by Eqs. (5-30) and (5-34). Using the same procedure as in Example 5-5, we have the unit availability for moist air at locations 1 and 3. The results, including those for water, are given in Table 5-4.

Assuming that the evaporative cooling process is adiabatic, we have the availability destruction per unit mass of dry air as

$$\begin{aligned} i &= a_{f_1} + (\omega_3 - \omega_1) a_{f_2} - a_{f_3} \\ &= 0.3796 + (0.0102 - 0.00381)(41.82) - 0.1484 \\ &= 0.496 \text{ Btu/lbm(a)} \end{aligned}$$

Table 5-4 Availability calculations for Example 5-6.

Location	1	2	3
$\hat{c}_p$ (Btu/lbm(a)-R)	0.2417	—	0.2445
$\hat{R}$ (lbf-ft/lbm(a)-R)	53.67	—	54.21
a_f^{th} (Btu/lbm(a))	0.0103	0.38	0.0911
a_f^{di} (Btu/lbm(a))	0.3693	41.44	0.0573
a_f (Btu/lbm(a))	0.3796	41.82	0.1484

Since evaporative cooling is highly irreversible, the availability destruction is unavoidable, as indicated in the example. In the calculation, the availability that will be used must be equal to the total availability. Otherwise the result on the process irreversibility would be misleading. ■

Example 5-7. Air enters an air conditioning unit at 80°F and 50% RH and leaves at 58°F and 80% RH. The process is shown in Fig. 5-3. Determine the unit availability at four locations (1, a, b, and 2) and the net heat availability transfer for the entire process. The condensate is assumed to be at 58°F.

Solution. The availability consists of two parts: the thermomechanical availability and the diffusion availability. The thermomechanical availability is given by the Eq. (5-30), and the diffusion availability is given by Eq. (5-34), where $\hat{c}_p = c_{p_a} + \omega c_{p_v}$ and $\hat{R} = R_a + \omega R_v$. In this example we use the numeric values

$$c_{p_a} = 0.240 \text{ Btu/lbm-R} \qquad R_a = 53.34 \text{ lbf-ft/lbm-R}$$

$$c_{p_v} = 0.445 \text{ Btu/lbm-R} \qquad R_v = 85.76 \text{ lbf-ft/lbm-R}$$

Since the calculations are quite straightforward, we present the unit availabilities in Table 5-5.

To determine the unit availability for water at 58°F, we use Eq. (5-15).

$$a_f = \left[h_f(T) - h_g(T_0)\right] + v_f\left[P - P_g(T)\right] - T_0\left[s_f(T) - s_g(T_0)\right] + T_0 R_v \ln\frac{P_g(T_0)}{x_v^0 P_0}$$

Table 5-5 Unit availabilities for the process in Example 5-7.

Location	1	a	b	2
T (°F)	80	59.8	52	58
ω (lbm(v)/lbm(a))	77/7000	77/7000	58/7000	58/7000
$\hat{c}_p$ (Btu/lbm(a)-R)	0.2449	0.2449	0.2437	0.2437
$\hat{R}$ (lb_f-ft/lbm(a)-R)	54.28	54.28	54.05	54.05
a_f^{th} (Btu/lbm(a))	0.0225	0.2108	0.3355	0.2361
a_f^{di} (Btu/lbm(a))	0.0398	0.0398	0.1148	0.1148
a_f (Btu/lbm(a))	0.0623	0.2506	0.4503	0.3509

Substituting the numeric values into the above equation, we have

$$a_f = (26.08 - 1100.7) + \frac{0.0160 \times 144}{778}(14.7 - 0.6988)$$
$$- 550(0.05169 - 2.0083) + \frac{550 \times 85.76}{778} \ln \frac{0.6988}{0.024 \times 14.7}$$
$$= 43.02 \text{ Btu/lbm}$$

Based on one pound of dry air, the flow availability of the condensate is given by

$$a_{f_w} = (\omega_a - \omega_b)a_f = \frac{77 - 58}{7000} \times 43.02 = 0.1168 \text{ Btu/lbm(a)}$$

The net heat availability transfer for the entire process is given by the equation

$$a_q = a_{q_{1\text{-a}}} + a_{q_{\text{a-b}}} + a_{q_{\text{b-2}}} \tag{5-g}$$

To determine the heat availability transfer $a_{q_{1\text{-a}}}$ in process 1-a, we apply the availability equation as

$$a_{q_{1\text{-a}}} + a_w + a_{f_1} - a_{f_a} - i = 0 \tag{5-h}$$

Since the process is internally reversible and no work is produced, Eq. (5-h) is further reduced to

$$a_{q_{1\text{-a}}} = a_{f_a} - a_{f_1}$$

With the numeric values, we have $a_{q_{1\text{-a}}} = 0.2506 - 0.0623 = 0.1883$ Btu/lbm(a).

Similarly, the heat availability transfer in process a-b is

$$a_{q_{\text{a-b}}} = a_{f_b} + a_{f_w} - a_{f_a}$$
$$= 0.4503 + 0.1168 - 0.2506$$
$$= 0.3165 \text{ Btu/lbm(a)}$$

and the heat availability in process b-2 is

$$a_{q_{\text{b-2}}} = a_{f_2} - a_{f_b} = 0.3509 - 0.4503 = -0.0994 \text{ Btu/lbm(a)}$$

Finally, by applying Eq. (5-g) we determine the net heat availability transfer for the overall process as $a_q = 0.1883 + 0.3165 - 0.0994 = 0.4054$ Btu/lbm(a).

A few comments are in order regarding the heat availability transfer in this example. As the heat availability transfer takes place in the process, there is a change in the flow availability. Because the process is internally reversible, the amount of heat availability transfer is exactly equal to the increase of flow availability in the moist air. It is also of interest to notice that the heat availability flows in the direction opposite that of the heat energy in this example. In process 1-a, for instance, the moist air loses the heat energy but receives the heat availability. Similar observations can be made for processes a-b and b-2.

5-5 PROBLEMS

5-1 Calculate the thermomechanical and diffusion availabilities for water at the following conditions:

a. 2400 psia, 1000°F
b. 10 psia, 80% quality
c. 100 psia, 120°F
d. 100 psia saturated vapor
e. 100 psia saturated liquid

5-2 Carbon dioxide is compressed in a piston-cylinder assembly. The initial conditions are 15 psia, 100°F, and 1 ft^3. During the process the pressure and specific volume are related by the equation

$$Pv^{1.2} = \text{constant}$$

For the final pressure of 150 psia, calculate (a) the combined nonflow availability at the initial and final states, and (b) the availability loss and destruction in the process.

5-3 A well-insulated tank contains carbon dioxide at the conditions of 20 psia and 120°F. The tank has a volume of 2 ft^3 and is equipped with a paddle wheel. During the process, 30 Btu of energy is transferred to the carbon dioxide through the paddle wheel. Determine (a) the combined nonflow availability at the initial and final states, (b) the availability destruction in the process, and (c) the availability loss from the carbon dioxide.

5-4 Saturated water from a boiler drum is throttled into a blowdown tank. The boiler drum and blowdown tank are at the pressures 1800 psia and 200 psia, respectively. Calculate, per pound of water, (a) the combined flow availability for water before and after throttling, (b) the second law efficiency based on the combined flow availability, and (c) the second law efficiency based on the thermomechanical flow availability.

5-5 A power plant stack is composed of the following gases:

Nitrogen = 80%	Carbon dioxide = 8%
Oxygen = 6%	Water vapor = 6%

The gases are at the pressure of 15 psia and the temperature of 400°F. Calculate the diffusion availability and thermomechanical flow availability per pound of gases. What is the percentage of diffusion availability?

5-6 Determine the effect of relative humidity on the diffusion availability for moist air at a constant temperature of 90°F.

5-7 Determine the effect of temperature on the combined flow availability for moist air at a constant relative humidity of 50%.

5-8 Moist air is heated at a constant pressure of 14.7 psia. The initial and final temperatures are 60°F and 90°F, respectively, while the humidity ratio remains constant at 0.0057 lbm(v)/lbm(a). Calculate in Btu per pound of dry air (a) the flow availabilities at the initial and final states and (b) the availability destruction and loss.

5-9 Moist air at 90°F and 60% relative humidity is cooled to 58°F and 80% relative humidity. The volume flow rate is 1000 cfm, and the condensate leaves at 58°F. Determine the change of flow availability in moist air during the process.

5-10 Moist air at 60°F and 20% relative humidity flows through a heater and humidifier at the rate of 1000 cfm. Heat is transferred to the air at the rate of 100,000 Btu/hr, and saturated water at 80°F is injected at the rate of 60 lbm/hr. Present the availability balance on the heater-humidifier combined system.

CHAPTER

SIX

CHEMICAL AND FUEL AVAILABILITIES

In this chapter we will develop the concepts of chemical availability and fuel availability and the equations for their numerical calculations. The approach is similar to that in Chapter 5 and basically involves the use of available work in a reversible process. The chemical availability is the available work from the substance at the state (T_0, P_0) both as the substance reacts chemically with those drawn from the environment and as the reaction products eventually come to the dead state. As in Chapter 5, the dead state is defined as the state which is in thermal, mechanical, and chemical equilibrium with the environment.

6-1 GIBBS FUNCTION

This section presents a brief review of the Gibbs function, which is defined by

$$G = H - TS \tag{6-1}$$

when Eq. (6-1) is divided by the mass of the substance, we have a specific Gibbs function as

$$g = \frac{G}{m} = h - Ts \tag{6-2}$$

Similarly, the Gibbs function per mole of the substance is

$$\bar{g} = \bar{h} - T\bar{s} \tag{6-3}$$

As shown by the definition, the Gibbs function is a compound property. Like other thermodynamic properties, the Gibbs function is a function of two independent properties such as temperature and pressure. For an ideal gas mixture, the Gibbs function is written as

$$\begin{aligned}\bar{g}(T,P) &= \sum_i x_i \bar{g}_i(T, x_i P) \\ &= \sum_i x_i \bar{h}_i(T) - T \sum_i x_i \bar{s}_i(T, x_i P)\end{aligned} \tag{6-4}$$

The Gibbs function frequently is used when a process involves a chemical reaction. The Gibbs function represents the chemical potential of the substance in a given state and becomes very useful in the calculation of available work in a chemical reaction process. These can be seen later in this chapter.

In the following we derive an expression for the Gibbs function of a substance at a pressure different from the atmospheric pressure. Starting with a substance at the state (T_0, P_0), we have the Gibbs function

$$\bar{g}(T_0, P_0) = \bar{h}(T_0, P_0) - T_0 \bar{s}(T_0, P_0) \tag{6-5}$$

Similarly, for the substance at (T_0, P), we have

$$\bar{g}(T_0, P) = \bar{h}(T_0, P) - T_0 \bar{s}(T_0, P) \tag{6-6}$$

Combining these two equations yields

$$\bar{g}(T_0, P) - \bar{g}(T_0, P_0) = -T_0[\bar{s}(T_0, P) - \bar{s}(T_0, P_0)]$$

or

$$\bar{g}(T_0, P) = \bar{g}(T_0, P_0) + \bar{R}T_0 \ln \frac{P}{P_0} \tag{6-7}$$

Equation (6-7) indicates that the Gibbs function at any pressure can be determined with the Gibbs function at the state (T_0, P_0). $\bar{g}(T_0, P_0)$ frequently is referred to as the Gibbs function at the standard state and is readily available in thermodynamic tables.

When the substance is one of the components in the environment, the substance would have the temperature T_0 and the pressure $x^0 P_0$, where x^0 is the mole fraction of the substance in the environment. Under these conditions the Gibbs function for the substance would become

$$\bar{g}(T_0, x^0 P_0) = \bar{g}(T_0, P_0) + \bar{R}T_0 \ln x^0 \tag{6-8}$$

Equation (6-8) is a very useful equation and frequently is used in the calculation of chemical and fuel availabilities.

6-2 AVAILABLE WORK IN A CHEMICAL REACTION

The available work for a steady-state, steady-flow process in the absence of changes in kinetic and potential energy is given by

$$w_{\text{ava}} = \sum_i m_i(h_i - T_0 s_i) - \sum_e m_e(h_e - T_0 s_e) \tag{6-9}$$

Applying this equation to a process involving a chemical reaction, we have the available work

$$w_{\text{ava}} = \sum_r n_r\left(\bar{h}_r - T_0\bar{s}_r\right) - \sum_p n_p\left(\bar{h}_p - T_0\bar{s}_p\right) \tag{6-10}$$

where the subscripts r and p denote the reactants and products, respectively. For the case where the reactants and products are in temperature and pressure equilibrium with the surroundings, Eq. (6-10) can be modified and expressed in terms of a Gibbs function as follows:

$$w_{\text{ava}} = \sum_r n_r\bar{g}_r(T_0, x_r P_0) - \sum_p n_p\bar{g}_p(T_0, x_p P_0) \tag{6-11}$$

where x_r and x_p are the mole fractions of reactants and products, respectively. Equation (6-11) indicates that the available work in a chemical reaction process is the Gibbs function of all reactants minus the Gibbs function of all products. When all individual reactants and products are assumed to be in thermal and mechanical equilibrium with the environment, Eq. (6-11) can be approximated by

$$w_{\text{ava}} = \sum_r n_r\bar{g}_r(T_0, P_0) - \sum_p n_p\bar{g}_p(T_0, P_0)$$

or

$$w_{\text{ava}} = -\Delta G(T_0, P_0) \tag{6-12}$$

Note that the available work obtained by Eq. (6-12) is an approximation. While the atmospheric air used in a chemical reaction is in the standard conditions (T_0, P_0), the components of the atmospheric air experience only the partial pressure $(x^0 P_0)$. Similar observations can be made for the reaction products.

To demonstrate the approximate nature of Eq. (6-12), let us take the chemical reaction of methane as an example. Figure 6-1 shows methane(g) entering the reactor at 14.7 psia and 77°F and reacting with atmospheric air also at 14.7 psia and 77°F. After the reaction, the products leave the reactor at 14.7 psia and 77°F. For 100 percent theoretical air, the equation for this chemical

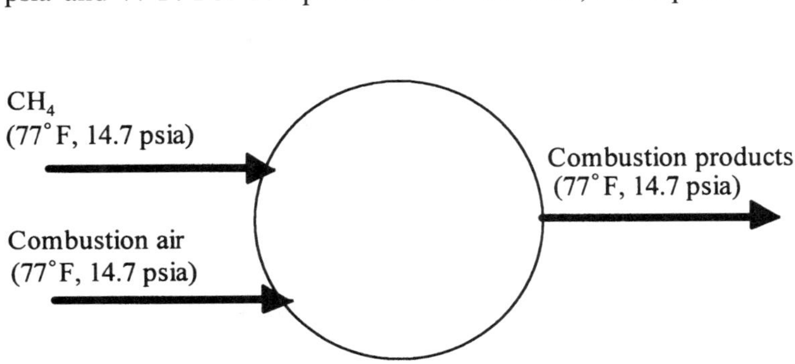

Figure 6-1 A schematic diagram for a reactor.

reaction is

$$CH_4(g) + \alpha_1 O_2 + \alpha_2 N_2 \rightarrow \beta_1 CO_2 + \beta_2 H_2O + \beta_3 N_2 \tag{6-13}$$

where

$$\begin{aligned} \alpha_1 &= 2 & \beta_1 &= 1 \\ \alpha_2 &= 2(3.76) & \beta_2 &= 2 \\ & & \beta_3 &= 2(3.76) \end{aligned}$$

Applying Eq. (6-11), we have the available work

$$\begin{aligned} w_{\text{ava}} = {} & \bar{g}_{CH_4}(T_0, P_0) + \alpha_1 \bar{g}_{O_2}\left(T_0, x^0_{O_2} P_0\right) \\ & + \alpha_2 \bar{g}_{N_2}\left(T_0, x^0_{N_2} P_0\right) - \beta_1 \bar{g}_{CO_2}(T_0, x_{CO_2} P_0) \\ & - \beta_2 \bar{g}_{H_2O}(T_0, x_{H_2O} P_0) - \beta_3 \bar{g}_{N_2}(T_0, x_{N_2} P_0) \end{aligned} \tag{6-14}$$

where x_i^0 is the mole fraction of component i in the environment, and x_i is the mole fraction of the component i in the reaction products. Making use of Eq. (6-7), we can rewrite Eq. (6-14) as

$$\begin{aligned} w_{\text{ava}} = {} & \bar{g}_{CH_4}(T_0, P_0) + \alpha_1 \bar{g}_{O_2}(T_0, P_0) + \alpha_1 \overline{R} T_0 \ln x^0_{O_2} \\ & + \alpha_2 \bar{g}_{N_2}(T_0, P_0) + \alpha_2 \overline{R} T_0 \ln x^0_{N_2} \\ & - \beta_1 \bar{g}_{CO_2}(T_0, P_0) - \beta_1 \overline{R} T_0 \ln x_{CO_2} \\ & - \beta_2 \bar{g}_{H_2O}(T_0, P_0) - \beta_2 \overline{R} T_0 \ln x_{H_2O} \\ & - \beta_3 \bar{g}_{N_2}(T_0, P_0) - \beta_3 \overline{R} T_0 \ln x_{N_2} \end{aligned}$$

Rearranging the above equation gives

$$w_{\text{ava}} = -\Delta G(T_0, P_0) + \overline{R} T_0 \ln \frac{\left(x^0_{O_2}\right)^{\alpha_1} \left(x^0_{N_2}\right)^{\alpha_2}}{(x_{CO_2})^{\beta_1} (x_{H_2O})^{\beta_2} (x_{N_2})^{\beta_3}} \tag{6-15}$$

Compared with Eq. (6-12), Eq. (6-15) indicates the existence of a new term (i.e., the second term on the right-hand side of the equation) which accounts for the effects of the partial pressure of the components in both atmospheric air and reaction products. With numeric values, we can calculate the first and second terms as

$$\begin{aligned} -\Delta G(T_0, P_0) &= \bar{g}_{CH_4}(T_0, P_0) + \alpha_1 \bar{g}_{O_2}(T_0, P_0) + \alpha_2 \bar{g}_{N_2}(T_0, P_0) \\ & \quad - \beta_1 \bar{g}_{CO_2}(T_0, P_0) - \beta_2 \bar{g}_{H_2O}(T_0, P_0) - \beta_3 \bar{g}_{N_2}(T_0, P_0) \\ &= -21{,}861 + 2(0) + 2(3.76)(0) - (-169{,}677) - 2(-98{,}345) \\ & \quad - 2(3.76)(0) \\ &= 344{,}506 \text{ Btu/mole of } CH_4 \end{aligned}$$

and

$$\bar{R}T_0 \ln \frac{\left(X^0_{O_2}\right)\left(x^0_{N_2}\right)^{\alpha_2}}{(x_{CO_2})^{\beta_1}(x_{H_2O})^{\beta_2}(x_{N_2})^{\beta_3}}$$

$$= \frac{(1545)(537)}{778} \ln \frac{(0.2035)^2(0.7567)^{7.52}}{(0.0951)^{1.0}(0.1901)^{2.0}(0.7148)^{7.52}} = 3111 \text{ Btu/mole of } CH_4$$

Thus, the available work for this chemical reaction is

$$w_{ava} = 344{,}506 + 3111 = 347{,}617 \text{ Btu/mole of } CH_4$$

There are several observations to be made about this calculation. First, the available work by Eq. (6-12) is quite close to that of Eq. (6-11). In this numerical calculation, the difference is less than one percent. Second, the substance methane is at thermal and mechanical equilibrium with the environment (i.e., $T = T_0$ and $P = P_0$) and therefore has a zero thermomechanical availability. However, as shown in this section, the substance still has a great deal of work potential, and thus it is expected to have another kind of availability at the environmental state (T_0, P_0). This new availability will be formally introduced in the next section. Third, the partial pressures of the individual components in the products are different from those of the same components in the environment. Because of this, the reaction products are not in chemical equilibrium with the environment. In the above calculation, the diffusion availability is not taken into account.

6-3 CHEMICAL AVAILABILITY

In this section we introduce the chemical availability. First, we define the environment as we did for the calculations of the thermomechanical and diffusion availabilities. For the present purpose, the environment conditions are selected as identical to those in Chapter 5:

Pressure	14.7 psia
Temperaure	77°F
Components	Mole fraction (x^0)
Nitrogen	0.7567
Oxygen	0.2035
Water vapor	0.0303
Carbon dioxide	0.0003
Other gases	0.0092

Second, we confine ourselves to those substances for which the products of reaction are also the components of the environment. In other words, the reaction products must be nitrogen, oxygen, water vapor, and carbon dioxide.

The equation for the chemical availability of a substance at (T_0, P_0) can be developed with the concept of available work in a reversible process. The procedure is similar to that for thermomechanical availability. For a process involving a chemical reaction, the available work is the maximum work that theoretically can be extracted as the substance at (T_0, P_0) reacts chemically with the components of the environment and as the reaction products eventually come to the dead state in a reversible process. The dead state refers to the state in which all products of reaction would be at temperature T_0 and in which the partial pressures would be equal to those of similar substances in the environment.

Here we develop the equation for chemical availability of a typical hydrocarbon fuel C_aH_b. The fuel reacts with oxygen from the environment, and the chemical reaction produces carbon dioxide and water vapor. The chemical equation is

$$C_aH_b + \left(a + \frac{b}{4}\right)O_2 \rightarrow (a)CO_2 + \left(\frac{b}{2}\right)H_2O(v)$$

Applying Eq. (6-11), we have the available work for this chemical process as

$$w_{\text{ava}} = \bar{g}_{C_aH_b}(T_0, P_0) + \left(a + \frac{b}{4}\right)\bar{g}_{O_2}\left(T_0, x^0_{O_2}P_0\right)$$
$$-(a)\bar{g}_{CO_2}\left(T_0, x^0_{CO_2}P_0\right) - \left(\frac{b}{2}\right)\bar{g}_{H_2O}\left(T_0, x^0_{H_2O}P_0\right) \qquad (6\text{-}16)$$

where x_i^0 is the mole fraction of component i within the environment. When the conditions of the environment are specified, the numeric values for these mole fractions will be determined. Applying Eq. (6-8), we have the following relations:

$$\bar{g}_{O_2}\left(T_0, x^0_{O_2}P_0\right) = \bar{g}_{O_2}(T_0, P_0) + \bar{R}T_0 \ln x^0_{O_2}$$
$$\bar{g}_{CO_2}\left(T_0, x^0_{CO_2}P_0\right) = \bar{g}_{CO_2}(T_0, P_0) + \bar{R}T_0 \ln x^0_{CO_2}$$
$$\bar{g}_{H_2O}\left(T_0, x^0_{H_2O}P_0\right) = \bar{g}_{H_2O}(T_0, P_0) + \bar{R}T_0 \ln x^0_{H_2O}$$

Substituting these Gibbs functions into Eq. (6-16), we have the available work for the chemical process, which is also defined as the chemical availability of hydrocarbon fuel C_aH_b:

$$\bar{a}^{\text{ch}}_{C_aH_b} = w_{\text{ava}} = -\Delta G(T_0, P_0) + \bar{R}T_0 \ln\left(\frac{\left(x^0_{O_2}\right)^{a+b/4}}{\left(x^0_{CO_2}\right)^a \left(x^0_{H_2O}\right)^{b/2}}\right) \qquad (6\text{-}17)$$

where

$$-\Delta G(T_0, P_0) = \bar{g}_{C_aH_b}(T_0, P_0) + \left(a + \frac{b}{4}\right)\bar{g}_{O_2}(T_0, P_0)$$
$$-(a)\bar{g}_{CO_2}(T_0, P_0) - \left(\frac{b}{2}\right)\bar{g}_{H_2O}(T_0, P_0) \qquad (6\text{-}18)$$

Equation (6-17) indicates the maximum work one could extract from one mole of hydrocarbon fuel C_aH_b at the state (T_0, P_0). The equation has been developed for the condition in which the hydrocarbon fuel reacts with oxygen. This equation is valid also when atmospheric air is used for the chemical reaction.

It is important to recognize that the available work calculated by Eq. (6-17) is independent of the excess oxygen or excess air. The reason for it is quite simple. The atmospheric air and reaction products have no work potential when they are at equilibrium with the environment. For the same reasons, the available work by Eq. (6-17) must be numerically equal to the availability of hydrocarbon fuel C_aH_b at the standard state (T_0, P_0).

Example 6-1. Calculate the chemical availability of methane gas at 14.7 psia and 77°F.

Solution. The reaction of methane gas with oxygen from the environment produces carbon dioxide and water vapor. The chemical equation is given by

$$CH_4 + 2O_2 \rightarrow CO_2 + 2H_2O$$

The change of Gibbs function for this reaction is

$$\begin{aligned} -\Delta G(T_0, P_0) &= \bar{g}_{CH_4}(T_0, P_0) + 2\bar{g}_{O_2}(T_0, P_0) \\ &\quad - \bar{g}_{CO_2}(T_0, P_0) - 2\bar{g}_{H_2O}(T_0, P_0) \\ &= -21{,}861 + 2(0) - (169{,}677) - 2(-98{,}345) \\ &= 344{,}506 \text{ Btu/mole of } CH_4 \end{aligned}$$

Inserting numerical values into Eq. (6-17), we have chemical availability per mole of methane gas

$$\begin{aligned} \bar{a}_f^{ch} &= 344{,}506 + \frac{(1545)(537)}{778} \ln \frac{(0.2035)^2}{(0.0003)(0.0303)^2} \\ &= 344{,}506 + 12{,}712 \\ &= 357{,}218 \text{ Btu/mole of } CH_4 \end{aligned}$$

While the methane gas has no thermomechanical availability at 14.7 psia and 77°F, it has a substantial amount of chemical availability. Numerically, the chemical availability can also be approximated by $-\Delta G$. In this example, the error due to this approximation is around 3.5 percent. ■

For hydrocarbons (C_aH_b), the chemical availabilities can be approximated in terms of the lower heating value (LHV). These equations are

$$\frac{\bar{a}_f^{ch}}{\text{LHV}} = 1.033 + 0.0169\frac{b}{a} - \frac{0.0698}{a}$$

Table 6-1 Chemical availability of some selected hydrocarbons[a] [A-10]

Fuel	Low heating value (LHV)	High heating value (HHV)	$-\Delta G$	$\bar{a}^{ch}$
Methane(g) CH_4	191.76	212.80	191.38	198.44
Ethane(g) C_2H_6	341.27	372.82	350.73	357.04
Propane(g) C_3H_8	488.53	530.60	503.93	513.62
Butane(g) C_4H_{10}	635.39	687.98	656.74	669.82
Octane(g) C_8H_{18}	1222.79	1317.45	1268.43	1295.07
Octane(l) C_8H_{18}	1212.87	1307.53	1266.06	1292.70
Ethylene(g) C_2H_4	316.20	337.23	318.18	324.96
Propylene(g) C_3H_6	460.44	491.99	467.81	477.98
Benzene(g) C_6H_6	757.54	789.09	766.62	788.37

[a]All values are in kcal/gmol relative to the dead state defined in this text. To obtain the value in Btu/lbmole, the table value must be multiplied by 1800.

Source: Data extracted from National Bureau of Standards Circular 500, "Selected Values of Chemical Thermodynamic Properties," February 1952.

for gaseous hydrocarbons and

$$\frac{\bar{a}_f^{ch}}{\text{LHV}} = 1.04224 + 0.011925\frac{b}{a} - \frac{0.042}{a}$$

for liquid hydrocarbons.

There are many situations in which the hydrocarbon fuel C_aH_b is a member of an ideal gas mixture. When the mixture is at the state (T_0, P_0), the hydrocarbon fuel C_aH_b is in the state (T_0, x_iP_0). The chemical availability of the hydrocarbon fuel at this condition has a different value. The following relation can be developed for numerical calculation:

$$\bar{a}_f^{ch}(T_0, x_iP_0) = \bar{a}_f^{ch}(T_0, P_0) + \bar{R}T_0 \ln x_i \tag{6-19}$$

where x_i is the mole fraction of a particular hydrocarbon fuel in the mixture. The derivation of Eq. (6-19) is left as an exercise.

For quick reference we present the chemical availability of selected hydrocarbons at the standard conditions in Table 6-1. The chemical availability of some combustible elements in coals are presented in Table 6-2.

Table 6-2 Chemical availability of some elements in coals[a]

Elements	$-\Delta G$	$\bar{a}^{ch}$
Carbon (C)	94.26	98.12
Hydrogen (H_2)	56.69	56.22
Sulfur (S)	71.68	—

[a]All values are in kcal/gmol relative to the dead state defined in the text. To obtain the value in Btu/lbmole, the table value must be multiplied by 1800.

Example 6-2. Incomplete combustion frequently results in the formation of carbon monoxide gas in the products of combustion. Calculate the chemical availability of carbon monoxide in the combustion products at 14.7 psia and 77°F. The mole fraction for carbon monoxide is assumed to be 2 percent.

Solution. First we determine the chemical availability of carbon monoxide gas at the standard conditions. For this purpose, we use the chemical equation $CO + (1/2)O_2 \rightarrow CO_2$. The change of Gibbs function for this reaction is

$$-\Delta G(T_0, P_0) = \bar{g}_{CO}(T_0, P_0) + \frac{1}{2}\bar{g}_{O_2}(T_0, P_0) - \bar{g}_{CO_2}(T_0, P_0)$$

Substituting the standard Gibbs functions for various components into the above equation, we have

$$\begin{aligned}-\Delta G(T_0, P_0) &= (-59{,}010) + \frac{1}{2}(0) - (-169{,}680)\\ &= 110{,}670 \text{ Btu/mole of CO}\end{aligned}$$

The chemical availability of carbon monoxide at the standard conditions is given by the equation

$$\bar{a}^{ch}_{f_{CO}} = -\Delta G(T_0, P_0) + \bar{R}T_0 \ln \frac{\left(x^0_{O_2}\right)^{0.5}}{\left(x^0_{CO_2}\right)}$$

Inserting the numeric values into this equation we have

$$\bar{a}^{ch}_{f_{CO}} = 110{,}670 + \frac{(1545)(537)}{778} \ln \frac{0.2035^{0.5}}{0.0003} = 118{,}471 \text{ Btu/mole of CO}$$

Finally the chemical availability of carbon monoxide in the combustion products is

$$\begin{aligned}\bar{a}^{ch}_{f_{CO}}(T_0, x_i P_0) &= \bar{a}^{ch}_{f_{CO}}(T_0, P_0) + \bar{R}T_0 \ln x_i\\ &= 118{,}471 + \frac{(1545)(537)}{778} \ln 0.02\\ &= 114{,}299 \text{ Btu/mole of CO}\end{aligned}$$

Example 6-3. Calculate the chemical availability for the producer gas. The producer gas at the standard conditions (14.7 psia, 77°F) has the volumetric analysis as follows:

Methane	3.0%
Hydrogen	14.0
Nitrogen	50.9
Oxygen	0.6
Carbon monoxide	27.0
Carbon dioxide	4.5

Solution. The producer gas has zero thermomechanical availability at the standard conditions. The total availability for the producer gas is therefore just the chemical availability, which is given by the equation

$$\bar{a}_f^{\text{ch}} = \sum_i x_i \bar{a}_{f_i}^{\text{ch}} \tag{6-a}$$

For the components methane (CH_4), carbon monoxide (CO), and hydrogen (H_2), we apply Eq. (6-19) in the form of

$$\bar{a}_{f_i}^{\text{ch}}(T_0, x_i P_0) = \bar{a}_{f_i}^{\text{ch}}(T_0, P_0) + \bar{R}T_0 \ln x_i \tag{6-b}$$

For methane (CH_4),

$$\bar{a}_{f_{\text{CH}_4}}^{\text{ch}} = (198.42)(1800) + (1.986)(537)\ln(0.03) = 353{,}416 \text{ Btu/mole}$$

For hydrogen (H_2),

$$\bar{a}_{f_{\text{H}_2}}^{\text{ch}} = (56.22)(1800) + (1.986)(537)\ln(0.14) = 99{,}099 \text{ Btu/mole}$$

For carbon monoxide (CO),

$$\bar{a}_{f_{\text{CO}}}^{\text{ch}} = (65.82)(1800) + (1.986)(537)\ln(0.27) = 117{,}080 \text{ Btu/mole}$$

For the components nitrogen (N_2), oxygen (O_2), and carbon dioxide (CO_2), we apply the equation

$$\bar{a}_{f_i}^{\text{ch}} = \bar{R}T_0 \ln \frac{x_i}{x_i^0} \tag{6-c}$$

for nitrogen (N_2),

$$\bar{a}_{f_{\text{N}_2}}^{\text{ch}} = (1.986)(537)\ln\frac{0.5090}{0.7567} = -423 \text{ Btu/mole}$$

For oxygen (O_2),

$$\bar{a}_{f_{\text{O}_2}}^{\text{ch}} = (1.986)(537)\ln\frac{0.0060}{0.2035} = -3758 \text{ Btu/mole}$$

For carbon dioxide (CO_2),

$$\bar{a}^{ch}_{f_{CO_2}} = (1.986)(537)\ln\frac{0.0450}{0.0003} = 5344 \text{ Btu/mole}$$

Inserting the numeric values into Eq. (6-a), we have the chemical availability for the producer gas as

$$\bar{a}^{ch}_f = (0.03)(353{,}416) + (0.14)(99{,}099) + (0.509) - (-423)$$

$$+(0.006)(-3758) + (0.27)(117{,}080) + (0.045)(5344)$$

$$\bar{a}^{ch}_f = 56{,}089 \text{ Btu/mole of producer gas} \quad \blacksquare$$

6-4 FUEL AVAILABILITY

There are many types of fuels engineers may encounter in practice. These fuels are in solid, liquid, and gaseous forms. Table 6-3 presents some bituminous coals with their ultimate analyses and heating values. Similar information on petroleums and gaseous fuels are presented in Table 6-4 and Table 6-5, respectively.

The chemical availability of fuel, sometimes referred to as the fuel availability, is of interest to engineers. The fuel availability is needed in the availability analysis of the combustion process. As shown later in the text, it also is needed in energy system analysis. The procedures presented in the last section generally can be used to estimate the fuel availability. These procedures are summarized below:

1. Prepare a chemical equation for the fuel combustion process.

Table 6-3 Ultimate analyses of some selected bituminous coals [A-12][a]

		Percentage by weight						
Bituminous coals	HHV (Btu/lbm)	C	H	O	N	S	A	W
Alabama	14,157	79.71	4.99	4.72	1.42	1.26	5.19	2.71
Arkansas	13,561	75.45	4.75	2.32	1.31	3.71	10.62	1.84
California	12,130	65.21	5.01	8.92	1.24	1.09	7.51	8.02
Illinois	10,856	59.83	4.13	7.47	1.14	3.66	10.73	13.04
Indiana	11,882	65.36	4.56	7.21	1.44	2.63	7.96	10.84
Kansas	13,082	71.81	4.66	5.63	1.20	3.34	8.27	5.09
Kentucky	14,200	79.81	5.06	6.82	1.53	0.65	3.13	3.00
Missouri	11,327	62.92	3.99	6.70	0.95	3.71	10.57	11.16
New Mexico	12,307	68.36	4.90	7.66	1.32	0.82	13.75	3.19
Ohio	13,390	73.68	4.95	5.33	1.85	3.21	7.38	3.60
Oklahoma	14,148	80.85	4.42	2.92	1.80	0.99	6.00	3.02
Pennsylvania	13,217	73.77	4.93	5.67	1.45	1.12	10.20	2.86
West Virginia	14,789	83.86	4.77	3.47	1.68	0.50	3.15	2.57

[a] Reprinted with permission from Krieger Publishing.

Table 6-4 Ultimate analyses of some selected petroleums [A-12][a]

Petroleums	HHV (Btu/lbm)	Specific gravity at 60°F	Percentage by weight C	H	O	N	S
California Crude	18,920	0.917	84.00	12.70	1.20	1.35	0.75
Kansas Crude	19,100	0.921	84.15	13.00	—	0.45	1.90
Ohio Crude	21,600	0.887	80.20	17.10	2.70	—	—
Oklahoma Crude	19,500	0.869	85.70	13.11	—	0.30	0.40
Pennsylvania Crude	19,500	0.813	86.06	13.88	—	—	0.60
Texas Crude	19,450	0.875	85.05	12.30	—	0.70	1.75
Virginia Crude	19,200	0.860	87.10	11.70	1.20	—	—
West Virginia Crude	21,240	—	86.60	12.90	0.50	—	—
Wyoming	19,700	0.996	82.00	14.20	3.60	—	—

[a]Reprinted with permission from Krieger Publishing.

2. The fuel and combustion air are assumed to be at the standard conditions and the products are completely in thermal and chemical equilibrium with the environment.
3. Calculate the available work for the above combustion process. Note that the available work is numerically equal to the availability of the fuel under consideration.

Table 6-5 Composition of some selected natural and artificial gases [A-12][b]

Gases	HHV[a] (Btu/ft^3)	Percentage of volume CH_4	C_2H_6	CO_2	H_2	CO	O_2	N_2
NATURAL								
Alabama	983	97.6	—	0.3	—	—	—	2.1
Arkansas	1000	99.2	—	0.2	—	—	—	0.6
California	1062	77.5	16.0	6.5	—	—	—	—
Illinois	963	95.6	—	0.5	—	—	—	3.9
Indiana	1171	75.4	23.4	—	—	—	—	1.2
Kansas	988	98.0	—	1.2	—	—	—	0.8
Kentucky	1178	75.0	24.0	—	—	—	—	1.0
Missouri	965	84.1	6.7	0.8	—	—	—	8.4
New York	1111	84.0	15.0	—	—	—	—	1.0
Ohio	1066	83.5	12.8	—	—	—	—	3.7
Oklahoma	1064	73.5	18.4	—	—	—	—	8.1
Pennsylvania	1065	90.0	9.0	0.2	—	—	—	0.8
West Virginia	1169	76.8	22.5	—	—	—	—	0.7
ARTIFICIAL								
Producer gas								
(from anthracite)	138	—	—	5.5	15.5	22.7	0.3	56.0
(from bituminous)	151	3.7	0.1	4.8	11.6	24.4	0.6	54.8
Illuminating gas	504	23.6	10.5	7.2	11.7	13.7	0.7	32.6
Blast furnace gas	98.3	0.2	—	12.7	3.6	26.5	—	57.0

[a]Measured at 14.7 psia and 60°F.
[b]Reprinted with permission from Krieger Publishing.

In Appendix B there is an example illustrating the use of these procedures for a coal availability calculation. The procedures are quite lengthy and cumbersome. However, it must be emphasized that the above approach is approximate in nature. The approach does not take into consideration the effects of chemical bonds of all elements in the industrial fuels. It is especially true for solid and liquid fuels. Because of this, the approach would at best produce approximate values of chemical availability for solid and liquid fuels. Other approaches have been proposed in recent years. The following are correlation equations suggested by Rodriguez [B-22].

For solid fuel Szargut and Styrylska proposed the correlation equations:

$$\frac{a_f^{\mathrm{ch}}(T_0, P_0)}{(\mathrm{LVH})_d} = 1.0438 + 0.0013\frac{\mathrm{H}}{\mathrm{C}} + 0.1083\frac{\mathrm{O}}{\mathrm{C}} + 0.0549\frac{\mathrm{N}}{\mathrm{C}}$$

if

$$\frac{\mathrm{O}}{\mathrm{C}} \leq 0.666 \tag{6-20}$$

and

$$\frac{a_f^{\mathrm{ch}}(T_0, P_0)}{(\mathrm{LHV})_d} = \frac{1.0438 + 0.0013\frac{\mathrm{H}}{\mathrm{C}} - 0.4453\frac{\mathrm{O}}{\mathrm{C}}\left(1 + 0.0051\frac{\mathrm{H}}{\mathrm{C}}\right) + 0.0521\frac{\mathrm{N}}{\mathrm{C}}}{1 - 0.5385\frac{\mathrm{O}}{\mathrm{C}}}$$

if

$$2.67 > \frac{\mathrm{O}}{\mathrm{C}} > 0.666 \tag{6-21}$$

where H/C, O/C, and N/C respectively denote the mass ratios of hydrogen to carbon, of oxygen to carbon, and of nitrogen to carbon. The quantity $(\mathrm{LHV})_d$ is the lower heating value for the solid fuel with the free moisture removed and expressed per unit of moist fuel. Equations (6-20) and (6-21) are estimated to be accurate to within 1%. For the solid fuel that contains sulfur, an additional corrective term must be added to the value calculated from the above equations. Rodriguez [B-22] suggested the correction term as

$$\Delta a_f^{\mathrm{ch}} = 6740(\mathrm{S})\ \mathrm{kJ/kg} \tag{6-22}$$

or

$$\Delta a_f^{\mathrm{ch}} = 2898(\mathrm{S})\ \mathrm{Btu/lbm}$$

where S represents the mass fraction of sulfur in the fuel.

For liquid fuels, the correlation equation is

$$\frac{a_f^{\mathrm{ch}}(T_0, P_0)}{(\mathrm{LHV})_d} = 1.0374 + 0.0013\frac{\mathrm{H}}{\mathrm{C}} + 0.0756\frac{\mathrm{O}}{\mathrm{C}} + 1.5960\frac{\mathrm{S}}{\mathrm{C}}\left(1 - 0.0145\frac{\mathrm{H}}{\mathrm{C}}\right) \tag{6-23}$$

and for gaseous fuels, the correlation equation is

$$\frac{a_f^{\mathrm{ch}}(T_0, P_0)}{(\mathrm{LHV})_d} = 1.0334 + 0.0015\frac{\mathrm{H}}{\mathrm{C}} - 0.0058\frac{1}{\mathrm{C}} \tag{6-24}$$

Equations (6-23) and (6-24) yield average derivations of $\pm 0.38\%$ and $\pm 0.15\%$, respectively.

Example 6-4. Calculate the fuel availability for an anthracite coal. The coal has the following ultimate analysis:

$C = 69.10$	$N_2 = 1.50$	$A = 13.36$
$H_2 = 3.71$	$S = 0.57$	$W = 3.00$
$O_2 = 8.76$		

and a lower heating value of 12,000 Btu/lbm.

Solution. We first calculate the mass ratio of oxygen to carbon as $O/C = 8.76/69.10 = 0.13$. Second, we apply Eq. (6-20) and have

$$\frac{a_f^{ch}(T_0, P_0)}{(\mathrm{LHV})_d} = 1.0438 + 0.0013\left(\frac{3.71}{69.10}\right) + 0.1083\left(\frac{8.76}{69.10}\right) + 0.0549\left(\frac{1.5}{69.10}\right)$$
$$= 1.0588$$

Finally, the fuel availability for the anthracite coal is $a_f^{ch}(T_0, P_0) = 1.0588(12{,}000) = 12{,}705$ Btu/lbm.

Since the coal contains 0.57% of sulfur, the correction must be determined with Eq. (6-22) and added to the above calculated value. However, the correction is relatively small. For practical purposes, the term frequently is neglected in calculations. ■

Example 6-5. Calculate the fuel availability for an oil. The oil has the following composition:

$C = 0.861$	$O_2 = 0.0$
$H_2 = 0.132$	$S = 0.007$

and a lower heating value of 18,400 Btu/lbm.

Solution. By Eq. (6-23) we have

$$\frac{a_f^{ch}(T_0, P_0)}{(\mathrm{LHV})_d} = 1.0374 + 0.0013\left(\frac{0.132}{0.861}\right) + 0.0756\left(\frac{0}{0.861}\right)$$
$$+ 1.5960\left(\frac{0.007}{0.861}\right)\left(1 - 0.0145\frac{0.132}{0.861}\right)$$
$$= 1.0505$$

This fuel contains no moisture, and the effect of sulfur is included in the correlation. Thus, the fuel availability for the oil is $a_f^{ch}(T_0, P_0) = 1.0505(\mathrm{LHV})_d = 1.0505(18{,}400) = 19{,}329$ Btu/lbm. ■

For a quick estimate of the fuel availability, one can use the empirical equation

$$a_f^{\text{ch}}(T_0, P_0) = c_1(\text{LHV})$$

where the typical values of c_1 are 1.06–1.10 for coals, 1.04–1.08 for oils, and 1.04–1.06 for natural gases.

6-5 AVAILABILITY ANALYSES FOR COMBUSTION PROCESSES

The combustion process is one of many processes frequently used in energy conversion systems. For a steady-state and steady-flow combustion process, the energy equation is written in the form

$$Q_{\text{cv}} + \sum_r n_r \bar{h}_r = W_{\text{cv}} + \sum_p n_p \bar{h}_p \tag{6-25}$$

The equation often is prepared for one mole of fuel. For simplicity, kinetic and potential energies have been omitted. For the combustion process the availability equation is given by

$$\left(1 - \frac{T_0}{T_{\text{av}}}\right) Q_{\text{cv}} - W_{\text{cv}} + \sum_r n_r \bar{a}_{f_r} - \sum_p n_p \bar{a}_{f_p} - I = 0 \tag{6-26}$$

Equations (6-25) and (6-26) are quite general and can be further simplified under certain conditions. For instance, when the combustion process is adiabatic and produces no mechanical work, the energy and availability equations will become respectively

$$\sum_r n_r \bar{h}_r - \sum_p n_p \bar{h}_p = 0 \tag{6-27}$$

and

$$\sum_r n_r \bar{a}_{f_r} - \sum_p n_p \bar{a}_{f_p} = I \tag{6-28}$$

It is of interest to compare these two equations. While Eq. (6-27) indicates the nature of energy conservation in combustion, Eq. (6-28) reveals an energy degradation. In Eq. (6-28), the irreversibility I represents availability destruction in the combustion process. In numerical calculations, first the exit temperature of the combustion products must be determined with Eq. (6-27) and then, at this temperature, the availabilities of the combustion products are evaluated. The following is an illustrative example.

Example 6-6. Liquid octane at 14.7 psia and 77°F enters a combustion chamber and burns with 400 percent theoretical air at 14.7 psia and 77°F. Assuming the combustion process is adiabatic, determine the availability destruction in this combustion process.

Solution. The equation for the combustion process is

$$C_8H_{18}(l) + 4(12.5)O_2 + 4(12.5)(3.76)N_2 \rightarrow 8CO_2 + 9H_2O(g) + 37.5O_2 + 188.0N_2$$

To determine the temperature at which the combustion products leave the chamber, we apply the equation

$$\sum_r n_r \bar{h}_r - \sum_p n_p \bar{h}_p = 0$$

For this combustion process we have the enthalpy of the reactants as

$$\sum_r n_r \bar{h}_r = \bar{h}^0_{f_{C_8H_{18}}} + 4(12.5)\bar{h}^0_{f_{O_2}} + 4(12.5)(3.76)\bar{h}^0_{f_{N_2}} \tag{6-d}$$

Because all reactants are in standard conditions (14.7 psia and 77°F), the enthalpy of formation for O_2 and N_2 is zero. Therefore, we have $\Sigma_r n_r \bar{h}_r = \bar{h}^0_{f_{C_8H_{18}}} = -107{,}532$ Btu/mole.

Similarly, we write the enthalpy of combustion products at the exit gas temperature T_p in the following form:

$$\sum_p n_p \bar{h}_p = 8\left(\bar{h}^0_f + \Delta\bar{h}\right)_{CO_2} + 9\left(\bar{h}^0_f + \Delta\bar{h}\right)_{H_2O} + 37.5\left(\bar{h}^0_f + \Delta\bar{h}\right)_{O_2} + 188.0\left(\bar{h}^0_f + \Delta\bar{h}\right)_{N_2} \tag{6-e}$$

Substituting the standard enthalpy of formation into the above equation we have

$$\sum_p n_p \bar{h}_p = 8(-169{,}297 + \Delta\bar{h})_{CO_2} + 9(-104{,}036 + \Delta\bar{h})_{H_2O} + 37.5(0 + \Delta\bar{h})_{O_2} + 188.0(0 + \Delta\bar{h})_{N_2}$$

or

$$\sum_p n_p \bar{h}_p = -2.290 \times 10^6 + \left(8\Delta\bar{h}_{CO_2} + 9\Delta\bar{h}_{H_2O} + 37.5\Delta\bar{h}_{O_2} + 188\Delta\bar{h}_{N_2}\right) \tag{6-f}$$

In the above equation, the terms $\Delta\bar{h}$ depend on the exit temperature of the combustion products. Table 6-6 gives the values of those terms at two different exit temperatures, 1620°R and 1800°R. Substituting these values into Eq. (6-f) at 1620°R, we have $\Sigma n_p \bar{h}_p = -325{,}039$ Btu/mole of fuel, and at 1800°R we have

Table 6-6 Temperature of combustion products

Product	1620 (°R)	1800 (°R)
$\Delta\bar{h}_{CO_2}$	12,063	14,371
Δh_{H_2O}	9,432	11,176
$\Delta\bar{h}_{O_2}$	8,281	9,769
$\Delta\bar{h}_{N_2}$	7,839	9,232

$\Sigma n_p \bar{h}_p = 26{,}807$ Btu/mole of fuel. By interpolation, we have the exit temperature approximately 1732°R.

$$\sum_p n_p \bar{h}_p = \sum_r n_r \bar{h}_r = -107{,}532 \text{ Btu/mole of fuel}$$

Thus, the exit gas temperature for this combustion process is 1732°R. This temperature frequently is referred to as the *adiabatic flame temperature*. To determine the availability destruction in this combustion process, we apply the availability equation as

$$I = \sum_r n_r \bar{a}_{f_r} - \sum_p n_p \bar{a}_{f_p} \tag{6-g}$$

For this combustion process we have the availability of all reactants

$$\sum_r n_r \bar{a}_{f_r} = \bar{a}_{f_{C_8H_{18}}} + 4(12.5)\bar{a}_{f_{O_2}} + 4(12.5)(3.76)\bar{a}_{f_{N_2}} \tag{6-h}$$

because $\bar{a}_{f_{O_2}}$ and $\bar{a}_{f_{N_2}}$ are zero at the given conditions. The availability of all reactants is simply the availability of liquid octane, which is

$$\sum_r n_r \bar{a}_{f_r} = \bar{a}_{f_{C_8H_{18}}} = (1800)(1292.7)$$

$$= 2.326 \times 10^6 \text{ Btu/mole of fuel}$$

Similarly, we write the availability of combustion products as

$$\sum_p n_p \bar{a}_{f_p} = 8\bar{a}_{f_{CO_2}} + 9\bar{a}_{f_{H_2O}} + 37.5\bar{a}_{f_{O_2}} + 188\bar{a}_{f_{N_2}} \tag{6-i}$$

The flow availability $\bar{a}_{f_i}$ is the total availability, which is $\bar{a}_{f_i} = \bar{a}_{f_i}^{\text{th}} + \bar{a}_{f_i}^{\text{di}}$, where

$$\bar{a}_{f_i}^{\text{th}} = \left(\bar{h} - \bar{h}_0\right)_i - T_0(\bar{s} - \bar{s}_0)_i$$

and

$$\bar{a}_{f_i}^{\text{di}} = T_0 \bar{R} \ln \frac{x_i}{x_i^0}$$

Table 6-7 Results for the availability products in Example 6-6.

Products	x_i	$(\bar{h}-\bar{h}_0)_i$ $\left(\frac{\text{Btu}}{\text{lbmole}}\right)$	$(\bar{s}-\bar{s}_0)_i$ $\left(\frac{\text{Btu}}{\text{lbmole-R}}\right)$	$\bar{a}_{fi}^{\text{th}}$, $\left(\frac{\text{Btu}}{\text{lbmole}}\right)$	x_i^0	$\bar{a}_{f_i}^{\text{di}}$ $\left(\frac{\text{Btu}}{\text{lbmole}}\right)$	$\bar{a}_{f_i}$ $\left(\frac{\text{Btu}}{\text{lbmole}}\right)$
CO_2	0.0330	13,499	12.76	6,647	0.0003	5013	11,660
H_2O	0.0371	10,515	10.10	5,091	0.0303	216	5,307
O_2	0.1546	9,207	8.86	4,449	0.2035	−293	4,156
N_2	0.7753	8,706	8.43	4,179	0.7567	26	4,205

The numerical calculations are quite straightforward. The results for the availability of the combustion products are summarized in Table 6-7. Substituting the above numeric values into Eq. (6-i), we have

$$\Sigma n_p \bar{a}_{f_p} = 8(11{,}660) + 9(5{,}307) + 37.5(4{,}156) + 188(4{,}205)$$

$$= 1.087 \times 10^6 \text{ Btu/mole of fuel}$$

Finally, by Eq. (6-g), the availability destruction for the combustion process is $I = 2.326 \times 10^6 - 1.087 \times 10^6 = 1.239 \times 10^6$ Btu/mole of fuel. ■

The availability destruction in the combustion process is quite substantial and equals approximately 53 percent of the flow availability before combustion. As indicated later in the text, the combustion process is the most inefficient element in many energy conversion systems. An examination also reveals the importance of diffusion availability. At the combustion chamber exit, the combustion products at 14.7 psia and 1732°R have a total availability of 1.087×10^6 Btu per mole of fuel, of which approximately 3.3 percent is the diffusion availability. An omission of the diffusion term will result in a mistake. ■

Excess air plays an important role in combustion. An increase in excess air will lower the adiabatic flame temperature. This temperature affects not only the process irreversibility but also the distribution of the thermomechanical and diffusion availability in the combustion products.

In an adiabatic combustion process, the availability destruction also can be calculated in term of entropy production:

$$I = T_0\left(\sum_p n_p \bar{s}_p - \sum_r n_r \bar{s}_r \right) \tag{6-29}$$

The derivation is left as an exercise.

The adiabatic combustion process is an ideal process. The actual combustion process is much more complicated because of both heat loss from the combustion chamber and incomplete combustion in the process. Incomplete combustion means an appearance of carbon monoxide in the products of combustion when the fuel is a hydrocarbon. The example below will provide an illustration of an incomplete combustion.

Example 6-7. Liquid octane enters the combustion chamber of a gas turbine system at 77°F, and air enters at 440°F. The pressure in the combustion chamber is 150 psia, and the excess air for the combustion is 300 percent. For some reasons the combustion is incomplete, and 2 percent of the carbon in the fuel will result in carbon monoxide. What is the availability destruction of this combustion process if the temperature of the combustion products is 1520°F?

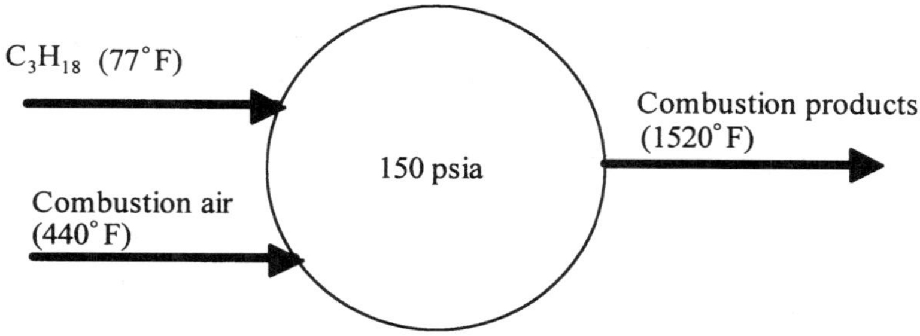

Figure 6-2 A sketch for example 6-7.

Solution. Figure 6-2 shows a sketch of the combustion chamber. The pressure drop across the chamber is assumed to be negligible, and both reactants and products are therefore at 150 psia. The combustion equation with excess air of 300% is

$$C_8H_{18}(l) + 4(12.5)O_2 + 4(12.5)(3.76)N_2$$
$$\rightarrow 7.84CO_2 + 0.16CO + 9H_2O + 37.58O_2 + 188N_2$$

The availability destruction for the combustion process is calculated by using the equation

$$I = \left(1 - \frac{T_0}{T_{av}}\right)Q_{cv} + \Sigma n_r \bar{a}_{f_r} - \Sigma n_p \bar{a}_{f_p} \tag{6-j}$$

where T_{av} is the average temperature of this combustion process. For simplicity, we simply take the average value of 440°F and 1520°F. To determine the heat loss (Q_{cv}) from the combustion chamber, we apply the first law equation as follows:

$$Q_{cv} = \Sigma n_p \bar{h}_p - \Sigma n_r \bar{h}_r \tag{6-k}$$

The enthalpy for all reactants is

$$\Sigma n_r \bar{h}_r = \bar{h}_{C_8H_{18}} + 4(12.5)\bar{h}_{O_2} + 4(12.5)(3.76)\bar{h}_{N_2}$$

Inserting the numeric values into the equation, we have

$$\begin{aligned}\Sigma n_r \bar{h}_r &= -107{,}537 + 4(12.5)(2{,}619) + 4(12.5)(3.76)(2{,}543)\\ &= 501{,}502 \text{ Btu/mole of fuel}\end{aligned}$$

Similarly, the enthalpy for combustion products at 1520°F is

$$\begin{aligned}\Sigma n_p \bar{h}_p &= 7.84\bar{h}_{CO_2} + 0.16\bar{h}_{CO} + 9\bar{h}_{H_2O} + 37.58\bar{h}_{O_2} + 188\bar{h}_{N_2}\\ &= 7.84(-169{,}297 + 16{,}733) + 0.16(-47{,}551 + 10{,}769)\\ &\quad + 9(-104{,}036 + 12{,}978) + 37.58(11{,}279) + 188(10{,}651)\\ &= 404{,}744 \text{ Btu/mole of fuel}\end{aligned}$$

Substituting the numeric values into Eq. (6-k), we have the heat loss from the combustion chamber as $Q_{cv} = 404{,}744 - 501{,}502 = -96{,}758$ Btu/mole of fuel. The availability loss associated with the heat loss is, therefore,

$$\left(1 - \frac{T_0}{T_{av}}\right)Q_{cv} = \left(1 - \frac{537}{1440}\right)(-96{,}758)$$
$$= -60{,}675 \text{ Btu/mole of fuel}$$

The flow availability of the reactants and combustion products for the combustion process are respectively

$$\Sigma n_r \bar{a}_{f_r} = \bar{a}_{f_{C_8H_{18}}} + 4(12.5)\bar{a}_{f_{O_2}} + 4(12.5)(3.76)\bar{a}_{f_{N_2}} \tag{6-l}$$

and

$$\Sigma n_p \bar{a}_{f_p} = 7.84\bar{a}_{f_{CO_2}} + 0.16\bar{a}_{f_{CO}} + 9\bar{a}_{f_{H_2O}} + 37.58\bar{a}_{f_{O_2}} + 188\bar{a}_{f_{N_2}} \tag{6-m}$$

The flow availability in the above equations consists of a thermomechanical component and a diffusion (or chemical) component. That is, $\bar{a}_{f_i} = \bar{a}^{th}_{f_i} + \bar{a}^{di}_{f_i}$, where

$$\bar{a}^{th}_{f_i} = (\bar{h} - \bar{h}_0)_i - T_0(\bar{s} - \bar{s}_0)_i$$

and

$$\bar{a}^{di}_{f_i} = \bar{R}T_0 \ln\frac{x_i}{x_i^0}$$

The numerical calculations for these flow availabilities are summarized in Table 6-8. Substituting the numeric values into Eq. (6-l), we have the availability of the fuel and combustion air as

$$\Sigma n_r \bar{a}_{f_r} = 2.3268 \times 10^6 + 50(3132) + 188(3123)$$
$$= 3{,}070{,}524 \text{ Btu/mole of fuel}$$

Table 6-8 Flow availability calculations for Example 6-7.

	x_i	$(\bar{h}-\bar{h}_0)_i$ (Btu/lbmole)	$(\bar{s}-\bar{s}_0)_i$ (Btu/lbmole-R)	$\bar{a}^{th}_{f_i}$ (Btu/lbmole)	x_i^0	$\bar{a}^{di}_{f_i}$ (Btu/lbmole)	$\bar{a}_{f_i}$ (Btu/lbmole)
Reactant							
C_3H_{18}(l)	—	—	—	—	—	—	2.3268×10^6
O_2	0.2101	2,619	−0.892	3,098	0.2035	34	3132
N_2	0.7899	2,543	−0.994	3,077	0.7567	46	3123
Products							
CO_2	0.0323	16,733	9.912	11,410	0.0003	4990	16,400
CO	0.0006	10,769	4.966	8,102	—	110,560[a]	118,662
H_2O	0.0371	12,978	6.829	9,311	0.0303	216	9,527
O_2	0.1549	11,279	5.377	8,392	0.2035	−291	8,101
N_2	0.7751	10,651	4.878	8,032	0.7567	26	8,058

[a]See similar calculation in Example 6-2.

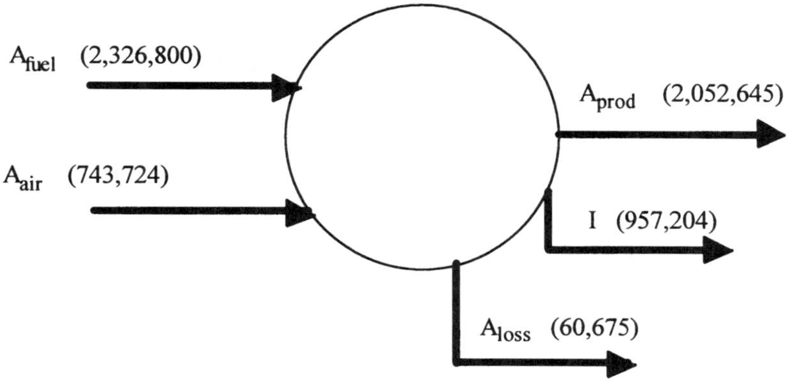

Figure 6-3 An availability balance for Example 6-7. (All units are in Btu/mole of fuel.)

Similarly, the availability of the combustion products by Eq. (6-m) is

$$\Sigma n_p a_{f_p} = 7.84(16{,}400) + 0.16(118{,}662) + 9(9{,}527) + 37.58(8{,}101) + 188(8{,}058) = 2{,}052{,}645 \text{ Btu/mole of fuel}$$

Finally, using Eq. (6-j), the availability destruction for the combustion process is

$$I = -60{,}675 + 3{,}070{,}524 - 2{,}052{,}645 = 957{,}204 \text{ Btu/mole of fuel}$$

Figure 6-3 presents the availability balance for the combustion chamber. It is seen that the availability of the fuel and combustion air is much greater than the availability of combustion products. The differences are the availability loss associated with the heat transfer and the availability destruction due to the irreversible nature of combustion. ■

6-6 AVAILABILITY ANALYSIS OF A STEAM GENERATOR

The steam generator, also referred to as a boiler, is a complicated system from which high-pressure, high-temperature steam is produced. In a fossil-fuel power station, a boiler is used to generate superheated steam and to reheat steam after its partial expansion in a turbine. The first law efficiency of a boiler is simply the ratio of the heat received by the water and steam to the energy supplied to the boiler. Neglecting the energy of combustion air, the boiler first law efficiency is given by the equation

$$\eta_b = \frac{\dot{Q}_{st}}{\dot{m}_f(\text{LHV})} \tag{6-30}$$

where

$$\dot{Q}_{st} = \dot{m}_b(h_e - h_i)_b + \dot{m}_r(h_e - h_i)_r \tag{6-31}$$

The subscripts b and r denote boiler (i.e., economizer, evaporator, and superheater) and reheater, respectively. The boiler energy loss is mainly in the form of thermal energy in the combustion products which eventually leave the system. The boiler first law efficiency is relatively high, usually around 90 percent for boilers in central power stations. However, the first law efficiency does not present a true picture for boiler operation. The objectives of this section are to present an availability balance for a boiler and to determine the availability loss and destruction by using a simplified model.

In terms of availability, the boiler output is simply the availability increase in the water and steam. It is given by the equation

$$\dot{A}_{\text{st}} = \dot{m}_b(a_{f_e} - a_{f_i})_b + \dot{m}_r(a_{f_e} - a_{f_i})_r \tag{6-32}$$

The term a_f is the flow availability of steam or water and can be easily determined at a location where the temperature and pressure are known. The input to the boiler is the availability of the fuel and combustion air. It is given by the equation $\dot{A}_{\text{input}} = \dot{A}_{\text{fuel}} + \dot{A}_{\text{air}}$, or

$$\dot{A}_{\text{input}} = \dot{m}_f a_{\text{fuel}} + \dot{m}_a a_{f_a} \tag{6-33}$$

In an availability analysis the difference between the boiler input and output is the availability loss and destruction. The ratio of the boiler output to input is defined as the second law efficiency:

$$\varepsilon_b = \frac{\dot{A}_{\text{st}}}{\dot{m}_f a_{\text{fuel}} + \dot{m}_a a_{f_a}} \tag{6-34}$$

When the availability of combustion air is small relative to the fuel availability, Eq. (6-34) is further simplified as

$$\varepsilon_b = \frac{\dot{A}_{\text{st}}}{\dot{m}_f a_{\text{fuel}}} \tag{6-35}$$

In an availability analysis, attention is given not only to the boiler second law efficiency but also to the distribution of availability loss and destruction in the overall process. In a boiler operation there are many processes taking place at the same time. These include at least combustion and heat transfer processes. The availability destruction in the boiler is due mainly to the irreversible nature of combustion and heat transfer processes. In addition, there are availability losses such as those associated with incomplete combustion, exhaust loss of combustion products, radiation, and additional heat loss from the boiler structure.

Figure 6-4 presents a simplified model for a boiler availability analysis. In the model, the boiler operation is approximated by three processes: (1) the adiabatic combustion process, (2) the heat transfer process, and (3) the exhaust process.

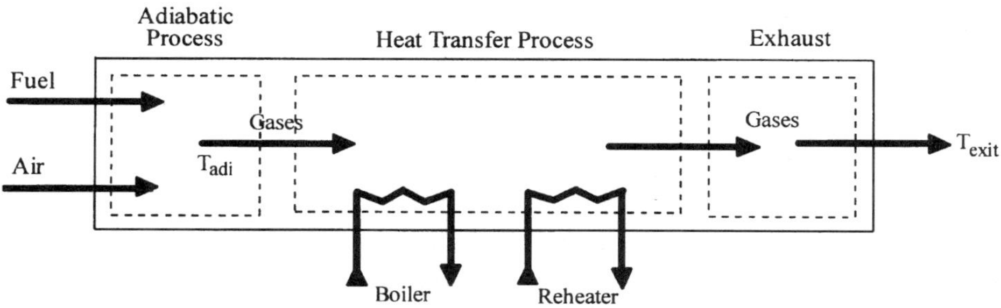

Figure 6-4 A simplified model for a stream generator.

The first is a combustion process which results in a substantial amount of availability destruction. Numerically it approximately equals the availability of the fuel and combustion air minus the availability of the combustion products at the adiabatic flame temperature. That is,

$$\dot{A}_{\text{comb}} = \left(\dot{A}_{\text{fuel}} + \dot{A}_{\text{air}}\right) - \dot{A}_{\text{prod at } T_{\text{adi}}} \tag{6-36}$$

The second is essentially a heat transfer process. The availability increase is on the steam and water side. This increase can be determined by Eq. (6-32). On the gas side the availability is expected to decrease. Numerically, the decrease must be equal to the gas availability at T_{adi} minus the gas availability at T_{exit}. Thus, the availability loss and destruction for this process is given by

$$\dot{A}_{\text{ht}} = \left(\dot{A}_{\text{prod at } T_{\text{adi}}} - \dot{A}_{\text{prod at } T_{\text{exit}}}\right) - \dot{A}_{\text{st}} \tag{6-37}$$

The third in the model is an exhaust process in which the combustion products will leave the steam-generating system. At the exit location, the products of combustion are at the temperature T_{exit}, and their availability will not be utilized. Thus, the availability of combustion products is the availability loss of the exhaust process, or

$$\dot{A}_{\text{exh}} = \dot{A}_{\text{prod at } T_{\text{exit}}} \tag{6-38}$$

On the availability balance for a steam-generating system, we must take into account the availability loss and destruction as given by Eqs. (6-36), (6-37), and (6-38). The sum of these three terms is expected to equal the boiler inputs minus the boiler output. Thus, we have the boiler availability balance equation

$$\left(\dot{A}_{\text{fuel}} + \dot{A}_{\text{air}}\right) - \dot{A}_{\text{st}} = \dot{A}_{\text{comb}} + \dot{A}_{\text{ht}} + \dot{A}_{\text{exh}} \tag{6-39}$$

Equation (6-39) indicates that there are three major losses in a boiler operation. In an availability analysis, the combustion loss $\dot{A}_{\text{comb}}$ is expected to be much greater than the exhaust loss $\dot{A}_{\text{exh}}$. In fact, the combustion loss always is a dominating factor in the boiler performance and results in a relatively low second law efficiency. For boilers in central power stations, the second law efficiency is usually around 50 percent.

6-7 PROBLEMS

6-1 Determine the chemical availability of propane gas (C_3H_8) at the standard conditions.

6-2 Estimate the chemical availability of sulfur at the conditions of 14.7 psia and 77°F.

6-3 Determine the chemical availability of California natural gas at the conditions of 14.7 psia and 77°F. The gas has the following volumetric compositions:

$$CH_4 = 77.5\% \qquad CO_2 = 6.5\% \qquad C_2H_6 = 16.0\%$$

6-4 A bituminous coal has the following ultimate analysis:

C = 79.81	N_2 = 1.53	A = 3.13
H_2 = 5.06	S = 0.65	W = 3.00
O_2 = 6.82		

Estimate the coal chemical availability and compare it with the coal lower heating value.

6-5 A West Virginia crude has the following composition in percentage by weight:

$$C = 86.6 \qquad H_2 = 12.9 \qquad O_2 = 0.5$$

Determine the crude chemical availability and lower heating value.

6-6 Propane gas enters a combustion chamber and burns with 300 percent theoretical air. Both propane gas and air are at the standard conditions, and the combustion products are at 1500°R. Determine, per pound of propane gas, (a) the heat loss from the combustion chamber, and (b) the net change of entropy production in the combustion process.

6-7 Prepare an availability analysis and determine the second law efficiency for the combustion process in Problem 6-6. The heat loss from the combustion chamber takes place at the location where the temperature is 1000°R.

6-8 A central power station boiler has the following operation information:

Steam side

Boiler and Superheater Conditions
Inlet 4000 psia, 510°F, 5×10^6 lbm/hr
Outlet 3500 psia, 1000°F

Reheater conditions
Inlet 750 psia, 600°F, 4.4×10^6 lbm/hr
Outlet 700 psia, 1000°F

Steam generator overall efficiency 98%

Coal and Combustion

Coal ultimate analysis on the dry basis

C = 71.81	N_2 = 1.20	A = 8.27
H_2 = 5.23	S = 3.34	
O_2 = 10.15		

Lower heating value	12,000 Btu/lbm
Excess Air	15%
Combustion air temperature	100 F

Determine the following:
a. the coal consumption per hour
b. the temperature of combustion products leaving the boiler stack
c. the availability output in the boiler
d. the availability input in the boiler
e. the availability loss including the exhaust loss in the boiler
f. the second law efficiency.

6-9 The gas dissociation usually takes place at high temperature. The following is the dissociation reaction for carbon dioxide:

$$CO_2 \rightarrow CO + \frac{1}{2}O_2$$

Assuming the dissociation occurs at 14.7 psia and 5400°F, calculate the availability change and the heat availability transfer in the dissociation process.

6-10 The chemical reaction

$$CH_4 + H_2O \rightarrow CO + 3H_2$$

is an endothermic process. Assuming that the process takes place at 20 psia and 500°F, determine the availability change and the heat availability transfer.

CHAPTER

SEVEN

AVAILABILITY ANALYSES OF ENGINEERING CYCLES

The purpose of this and the next two chapters is to introduce the use of the availability concepts and equations in various energy system analyses. These systems may include steam turbines, gas turbines, heat pumps, refrigerators, and others. The analysis generally produces the second law efficiency, which is a true measure of fuel utilization. In addition, the analysis will show the distribution of availability loss and destruction among the components of the system. With an availability analysis, one can easily identify the component of the system that suffers the most in terms of availability loss and destruction. To improve system performance, attention must be focused to this critical component.

There are many variables affecting the performance of an energy-conversion system. Some of these variables are dominating, while some are not. This chapter is intended to identify these variables and show how they affect the energy-conversion systems. To make the task more manageable, the energy-conversion systems are approximated with idealized cycles in which all processes are internally reversible. The cycles to be considered are Carnot, Rankine, Brayton, Otto, and Diesel.

7-1 BASIC EQUATIONS

All the cycles to be considered in this chapter are internally reversible. Before the specific cycle is discussed, certain features about a reversible cycle are presented. First we start with the availability equation for a steady-state and steady-flow process:

$$a_q + a_w + a_{f_i} - a_{f_e} - i = o \tag{7-1}$$

When the equation is applied to a reversible cyclic process, we have

$$a_{f_i} = a_{f_e} \tag{7-2}$$

and

$$i = o \tag{7-3}$$

Substituting Eqs. (7-2) and (7-3) into Eq. (7-1), we have

$$a_q = -a_w \tag{7-4}$$

This equation indicates that the availability input associated with the heat transfer to the cycle is equal to the availability output associated with the work produced in the cycle. This statement is correct provided that the cycle is internally reversible. We can also verify Eq. (7-4) as follows.

The term a_w is given by $a_w = -w_{cy}$, where w_{cy} is the net work produced by the cycle. According to the first law of thermodynamics, the cycle network is

$$w_{cy} = \oint \delta w = \oint \delta q$$

Combining these two equations, the term a_w is given by

$$a_w = -\oint \delta q \tag{7-5}$$

Also, the term a_q is given by

$$a_q = \oint \left(1 - \frac{T_0}{T}\right) \delta q$$

Rearranging this equation would give

$$a_q = \oint \delta q - \oint T_0 \frac{\delta q}{T}$$

Because $ds = \delta q/T$ for reversible process, the second term in the above equation vanishes, and we are left with

$$a_q = \oint \delta q \tag{7-6}$$

Finally, the combination of Eqs. (7-5) and (7-6) will lead to an equation identical to Eq. (7-4).

The cycle efficiency is important in engineering cycle analysis. The equation for the cycle efficiency can be developed by applying the availability equation $a_q + a_w = o$, or

$$\oint \left(1 - \frac{T_0}{T}\right) \delta q - \oint \delta w = o \tag{7-7}$$

Assuming that there is one process in the cycle for heat supply and another process for heat rejection, we can rewrite the above equation as

$$\int\left(1-\frac{T_0}{T_h}\right)\delta q_h+\int\left(1-\frac{T_0}{T_l}\right)\delta q_l-w_{cv}=o$$

or

$$\int\left(1-\frac{T_0}{T_h}\right)\delta q_h=-\int\left(1-\frac{T_0}{T_l}\right)\delta q_l+w_{cy} \tag{7-8}$$

$$\begin{bmatrix}\text{availability}\\ \text{input to}\\ \text{the cycle}\end{bmatrix}=\begin{bmatrix}\text{availability}\\ \text{loss from}\\ \text{the cycle}\end{bmatrix}+\begin{bmatrix}\text{availability}\\ \text{output from}\\ \text{the cycle}\end{bmatrix}$$

Basically, Eq. (7-8) is the availability equation applied to the reversible cycle and can be expressed in terms of availability input, output, and loss as indicated above. Because the cycle is reversible, there is no availability destruction in the cyclic process. Traditionally, the second law cycle efficiency is defined as a ratio of the output to the input. That is,

$$\varepsilon_{cy}=\frac{w_{cy}}{\int\left(1-\frac{T_0}{T_h}\right)\delta q_h} \tag{7-9}$$

Since the cycle first law efficiency is

$$\eta_{cy}=\frac{w_{cy}}{q_h} \tag{7-10}$$

the cycle second law efficiency can be presented also as

$$\varepsilon_{cy}=\eta_{cy}\frac{q_h}{\int\left(1-\frac{T_0}{T_h}\right)\delta q_h} \tag{7-11}$$

When the energy conversion system is used as a refrigerator or heat pump, the second law efficiency is still the ratio of output to input for the cycle. For a refrigeration system, the input is the compressor work and the output is the availability transfer associated with the heat removed from the refrigerated space. To develop the second law efficiency for a refrigeration cycle, we arrange the availability equation in terms of input, output, and loss as follows:

$$-w_{cy}=-\oint\left(1-\frac{T_0}{T}\right)\delta q$$

or

$$-w_{cy} = -\int\left(1 - \frac{T_0}{T_h}\right)\delta q_h - \int\left(1 - \frac{T_0}{T_l}\right)\delta q_l \tag{7-12}$$

$$\begin{bmatrix}\text{availability}\\ \text{input to}\\ \text{the cycle}\end{bmatrix} = \begin{bmatrix}\text{Availability}\\ \text{loss from}\\ \text{the cycle}\end{bmatrix} + \begin{bmatrix}\text{Availability}\\ \text{output from}\\ \text{the cycle}\end{bmatrix}$$

Thus, the second law efficiency for a refrigeration cycle is

$$\beta_{ref} = \frac{\int\left(1 - \frac{T_0}{T_l}\right)\delta q_l}{w_{cy}} \tag{7-13}$$

By a similar procedure, the second law efficiency for a heat pump cycle can be developed and expressed as

$$\beta_{hp} = \frac{\int\left(1 - \frac{T_0}{T_h}\right)\delta q_h}{w_{cy}} \tag{7-14}$$

The coefficient of performance (COP) is used frequently in the energy analysis of refrigeration and heat pump cycles. They are defined respectively as

$$\mathrm{COP}_{ref} = \frac{q_l}{-w_{cy}} \tag{7-15}$$

and

$$\mathrm{COP}_{hp} = \frac{q_h}{w_{cy}} \tag{7-16}$$

Combining Eqs. (7-13) and (7-15) yields a relationship between the second law efficiency and the coefficient of performance:

$$\beta_{ref} = \mathrm{COP}_{ref}\frac{-\int\left(1 - \frac{T_0}{T_l}\right)\delta q_l}{q_l} \tag{7-17}$$

Similarly, we combine Eqs. (7-14) and (7-16) to obtain the second law efficiency for a heat pump in terms of its coefficient of performance:

$$\beta_{hp} = \mathrm{COP}_{hp}\frac{\int\left(1 - \frac{T_0}{T_h}\right)\delta q_h}{q_h} \tag{7-18}$$

7-2 THE CARNOT CYCLE

The Carnot cycle consists of the following four processes as shown in Fig. 7-1.

1. Process 1-2 is reversible and isothermal. Heat is transferred from the high-temperature reservoir.
2. Process 2-3 is reversible and adiabatic. The working substance expands and has its temperature decreased to that of the low-temperature reservoir.
3. Process 3-4 is reversible and isothermal. Heat is transferred to the low-temperature reservoir.
4. Process 4-1 is reversible and adiabatic. The working substance is compressed and has its temperature increased back to that of the high-temperature reservoir.

It must be pointed out that the Carnot cycle is not only internally reversible but also externally reversible. When heat transfer takes place between the reservoir and the substance, the temperature between them must be infinitesimally small. It is well known that the first law efficiency for the Carnot cycle is

$$n_{\text{cy}} = 1 - \frac{q_l}{q_h} = 1 - \frac{T_l}{T_h} \tag{7-19}$$

When the concept of availability is applied, the second law efficiency can be developed with Eq. (7-9), giving

$$\varepsilon_{\text{cy}} = \frac{w_{\text{cy}}}{\int \left(1 - \frac{T_0}{T_h}\right) \delta q_h}$$

Since T_h is constant for a given Carnot cycle, the above equation is further modified and shown as

$$\varepsilon_{\text{cy}} = \frac{w_{\text{cy}}}{\left(1 - \frac{T_0}{T_h}\right) q_h} \tag{7-20}$$

or

$$\varepsilon_{\text{cy}} = \frac{\eta_{\text{cy}}}{1 - \frac{T_0}{T_h}} \tag{7-21}$$

Equation (7-21) indicates a relationship between the first law and second law cycle efficiencies. As the temperature T_h increases, the second law cycle efficiency will approach the value of the first law cycle efficiency. Equation (7-21) also indicates that the temperatures T_h and T_l will affect the second law efficiency in the same manner that they affect the first law efficiency.

It should be pointed out that the Carnot cycle may have different working substances. A change of substance will not affect the cycle performance. In the case where the gas is used in the Carnot cycle as shown in Fig. 7-1, the first law cycle efficiency can be developed and expressed in a different form. That is,

$$\eta_{cy} = 1 - \frac{T_4}{T_1} \tag{7-22}$$

Since $\frac{T_1}{T_4} = \left(\frac{p_1}{p_4}\right)^{k-1/k} = \rho_p$ Eq. (7-22) is given by

$$\eta_{cy} = 1 - \frac{1}{\rho_p} \tag{7-23}$$

where ρ_p is the isentropic temperature ratio. As the isentropic temperature ratio increases, the first law efficiency will increase. Because of the linear relationship shown in Eq. (7-21), the isentropic temperature ratio has similar influences on the second law cycle efficiency.

7-3 THE RANKINE CYCLE

The Rankine cycle consists of four processes as shown in Fig. 7-2. These processes are:

1. Reversible and adiabatic compression process in the pump (1-2).
2. Constant pressure heat addition process in the boiler (2-3).
3. Reversible and adiabatic expansion process in the turbine (3-4).
4. Constant pressure heat rejection process in the condenser (4-1).

Figure 7-2 indicates that the Rankine cycle is similar to the Carnot cycle with one exception in the condensation process. In the Rankine cycle, the condensation process terminates at the saturated liquid state. Because of this, a simple liquid pump can replace the two-phase compressor. It is evident in the *T-s* diagram that the Rankine cycle is less efficient than a Carnot cycle for the same maximum and minimum temperatures. The net output for the Rankine cycle is the turbine work minus the pump work, $w_{cy} = w_t - w_p$, or $w_{cy} = (h_3 - h_4) - (h_2 - h_1)$. The cycle net input is the amount of heat received by the water and steam in the boiler. In terms of enthalpy, it is given by the equation $q_h = h_3 - h_2$.

The first law cycle efficiency usually is defined as the ratio of output to input in the cycle. That is, $\eta_{cy} = w_{cy}/q_h$, or

$$\eta_{cy} = \frac{(h_3 - h_4) - (h_2 - h_1)}{h_3 - h_2} \tag{7-24}$$

It is of interest to note that Eq. (7-24) can be further modified and approximated by the equation

$$\eta_{cy} = \frac{h_3 - h_4}{h_3 - h_1} \tag{7-25}$$

when the pump work $(h_2 - h_1)$ is negligible.

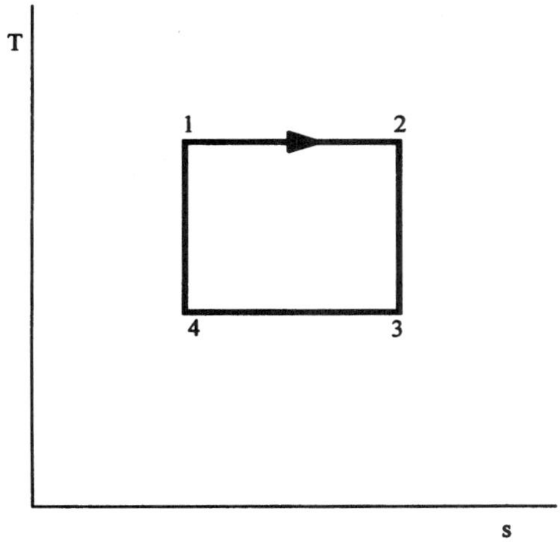

Figure 7-1 *T-s* diagram for a Carnot cycle.

The second law cycle efficiency is important in cycle analysis. For the Rankine cycle, the second law efficiency is given by

$$\varepsilon_{cy} = \frac{w_{cy}}{\int_2^3 \left(1 - \frac{T_0}{T}\right) \delta q} \tag{7-26}$$

The denominator in the above equation represents the availability transfer associated with the heat transfer in the boiler. It is also the availability input to

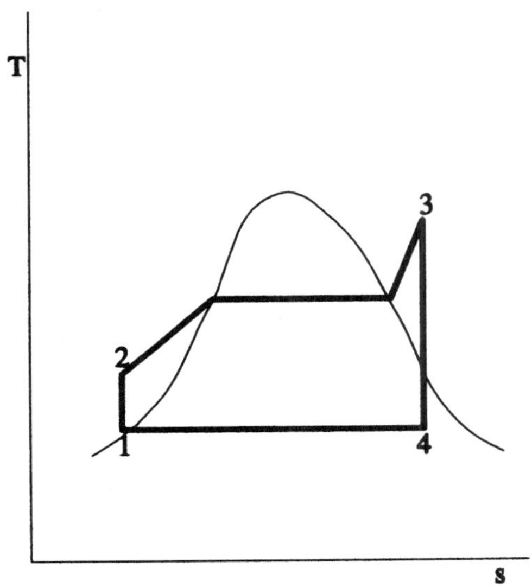

Figure 7-2 *T-s* diagram for a Rankine cycle.

the Rankine cycle. Applying the availability equation to Process 2-3, we have

$$\int_2^3 \left(1 - \frac{T_0}{T}\right) \delta q + a_w + a_{f_2} - a_{f_3} - i = o$$

Since $a_w = o$ and $i = o$, the above equation becomes

$$\int_2^3 \left(1 - \frac{T_0}{T}\right) \delta q = a_{f_3} - a_{f_2} = (h_3 - h_2) - T_0(s_3 - s_2) \qquad (7\text{-}27)$$

Substituting Eq. (7-27) into Eq. (7-26), we have the second law cycle efficiency as

$$\varepsilon_{cy} = \frac{(h_3 - h_4) - (h_2 - h_1)}{(h_3 - h_2) - T_0(s_3 - s_2)}$$

or

$$\varepsilon_{cy} = \frac{\eta_{cy}}{1 - T_0 \dfrac{s_3 - s_2}{h_3 - h_2}} \qquad (7\text{-}28)$$

For convenience, Eq. (7-28) frequently is approximated by

$$\varepsilon_{cy} = \frac{\eta_{cy}}{1 - T_0 \dfrac{s_3 - s_1}{h_3 - h_1}} \qquad (7\text{-}29)$$

when the pump work $(h_2 - h_1)$ is negligible. Some observations should be made at this point. The variables affecting the first law efficiency are the same as those affecting the second law efficiency. These include the steam pressure and temperature at the turbine inlet, and the condenser pressure. It is well known from elementary thermodynamics that increasing the steam pressure and temperature will result in an improvement of cycle efficiency. Also, it is true that lowering the condenser pressure will achieve similar effects. In power plant design, the selection of these variables must be tempered with consideration of some other technical and economic factors.

Reheating frequently is utilized for improvement of the Rankine cycle performance. Figure 7-3 shows a Rankine cycle with one reheating process between the two turbine processes. The heating process 4-5 is reversible and at constant pressure in the idealized cycle. For the reheat Rankine cycle, the net output is still the turbine work minus the pump work and is given by $w_{cy} = (h_3 - h_4) + (h_5 - h_6) - (h_2 - h_1)$, and the net input is the amount of heat received by the water in the boiler and reheater (physically, the boiler and reheater are an integrated unit) and is given by $q_h = (h_3 - h_2) + (h_5 - h_4)$.

The first law cycle efficiency is simply

$$\eta_{cy} = \frac{(h_3 - h_4) + (h_5 - h_6) - (h_2 - h_1)}{(h_3 - h_2) + (h_5 - h_4)} \qquad (7\text{-}30)$$

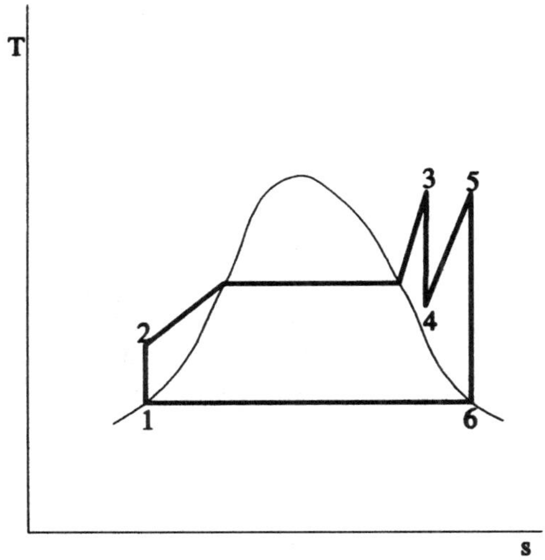

Figure 7-3 *T-s* diagram for a reheat Rankine cycle.

When the pump work is negligible and omitted from consideration, the above equation can be changed to

$$\eta_{cy} = \frac{(h_3 - h_4) + (h_5 - h_6)}{(h_3 - h_1) + (h_5 - h_4)} \tag{7-31}$$

To develop the equation for the second law cycle efficiency for a reheat Rankine cycle, we follow a similar procedure. That is,

$$\varepsilon_{cy} = \frac{w_{cy}}{a_{q_{2\text{-}3}} + a_{q_{4\text{-}5}}} \tag{7-32}$$

where $a_{q_{2\text{-}3}}$ and $a_{q_{4\text{-}5}}$ represent the heat availability transfer in the boiler and reheater, respectively. As indicated before, they are numerically equal to the increase of flow availability in the process. With these relationships, we can rewrite Eq. (7-32) as

$$\varepsilon_{cy} = \frac{(h_3 - h_4) + (h_5 - h_6) - (h_2 - h_1)}{(h_3 - h_2) - T_0(s_3 - s_2) + (h_5 - h_4) - T_0(s_5 - s_4)}$$

or

$$\varepsilon_{cy} = \frac{(h_3 - h_4) + (h_5 - h_6) - (h_2 - h_1)}{(h_3 - h_2) + (h_5 - h_4) - T_0(s_5 - s_2)} \tag{7-33}$$

In terms of the first law efficiency, Eq. (7-33) can be further changed and expressed as

$$\varepsilon_{cy} = \frac{\eta_{cy}}{1 - \dfrac{T_0(s_5 - s_2)}{(h_3 - h_2) + (h_5 - h_4)}} \tag{7-34}$$

Similarly, the second law efficiency can be derived for the Rankine cycle with more than one reheat.

Another approach for improvement of the Rankine cycle is the use of a regenerative process. In the regenerative cycle, effort is made to raise the water temperature to the saturation level at the boiler inlet, and therefore the average temperature during the heat addition process in the cycle will be increased. The second law cycle efficiency for this arrangement is left as an exercise.

7-4 THE BRAYTON CYCLE

A simple gas turbine system, shown in Fig. 7-4, consists of a compressor, combustor and turbine. Air is first compressed and then mixed with fuel and burned in the combustor. After leaving the combustor, the products of combustion enter the turbine and produce mechanical work in an expansion process. Some of the turbine work is used to drive the compressor, and the remainder is delivered outside to meet the user's demand. Because of the combustion process, there is a change in the composition of the working substance. It should be noted that in an actual system the working substance does not go through a thermodynamic cycle. In order to analyze the gas turbine system in a convenient form, the assumptions listed below are frequently used.

1. The working substance is treated as air of fixed composition. The air is an ideal gas and has a constant specific heat.

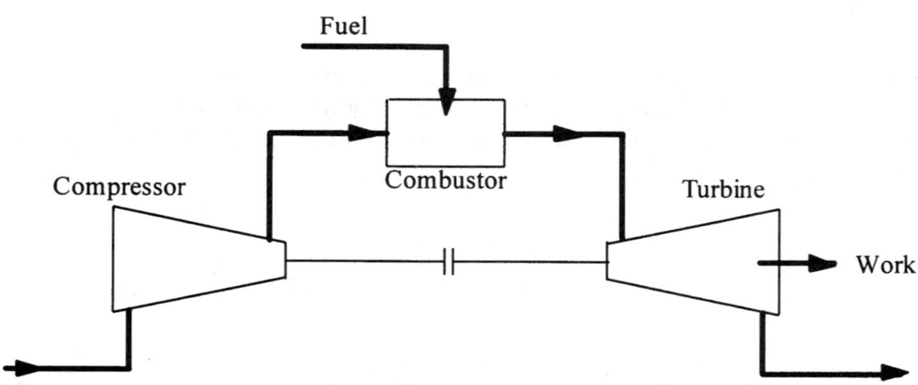

Figure 7-4 A schematic diagram for a simple gas turbine system.

2. The combustion process is replaced by a constant-pressure process in which heat is transferred from an external source. in other words, the mass flow rate remains unchanged.
3. The inlet and exhaust processes are replaced by one constant-pressure process in which the heat is transferred from the working substance; at the end, the cycle is completed.
4. All processes are internally reversible.

The set of these assumptions is called the *air-standard cycle approach*. The idealized gas turbine cycle is referred to as the Brayton cycle. Figure 7-5 shows the Brayton cycle and its *T-s* diagram. To evaluate the first law cycle efficiency, we determine the cycle net work, which is also the net heat supply to the cycle. That is,

$$w_{cy} = c_p(T_3 - T_2) - c_p(T_4 - T_1) = c_pT_1\left(\frac{T_3}{T_1} - \frac{T_2}{T_1} - \frac{T_4}{T_1} + 1\right)$$

Let $k_1 = T_3/T_1$ and let $\rho_p = T_2/T_1$; thus

$$\frac{T_4}{T_1} = \frac{T_4}{T_3} \times \frac{T_3}{T_1} = \frac{k_1}{\rho_p}$$

and we have the cycle net work given by

$$w_{cy} = c_pT_1\left(k_1 - \rho_p - \frac{k_1}{\rho_p} + 1\right)$$

or

$$w_{cy} = c_pT_1\left(1 - \frac{1}{\rho_p}\right)(k_1 - \rho_p) \tag{7-35}$$

The heat supply to the cycle takes place in Process 2-3. It is given by the equation

$$q_h = c_p(T_3 - T_2)$$

Making use of definitions for k_1 and ρ_p, we have

$$q_h = c_pT_1(k_1 - \rho_p) \tag{7-36}$$

With Eqs. (7-35) and (7-36), the first law cycle efficiency is

$$\eta_{cy} = 1 - \frac{1}{\rho_p} \tag{7-37}$$

The equation indicates that the cycle efficiency for the air standard Brayton cycle is simply a function of the isentropic temperature ratio ρ_p. To evaluate the second law efficiency, we determine the availability transfer associated with the

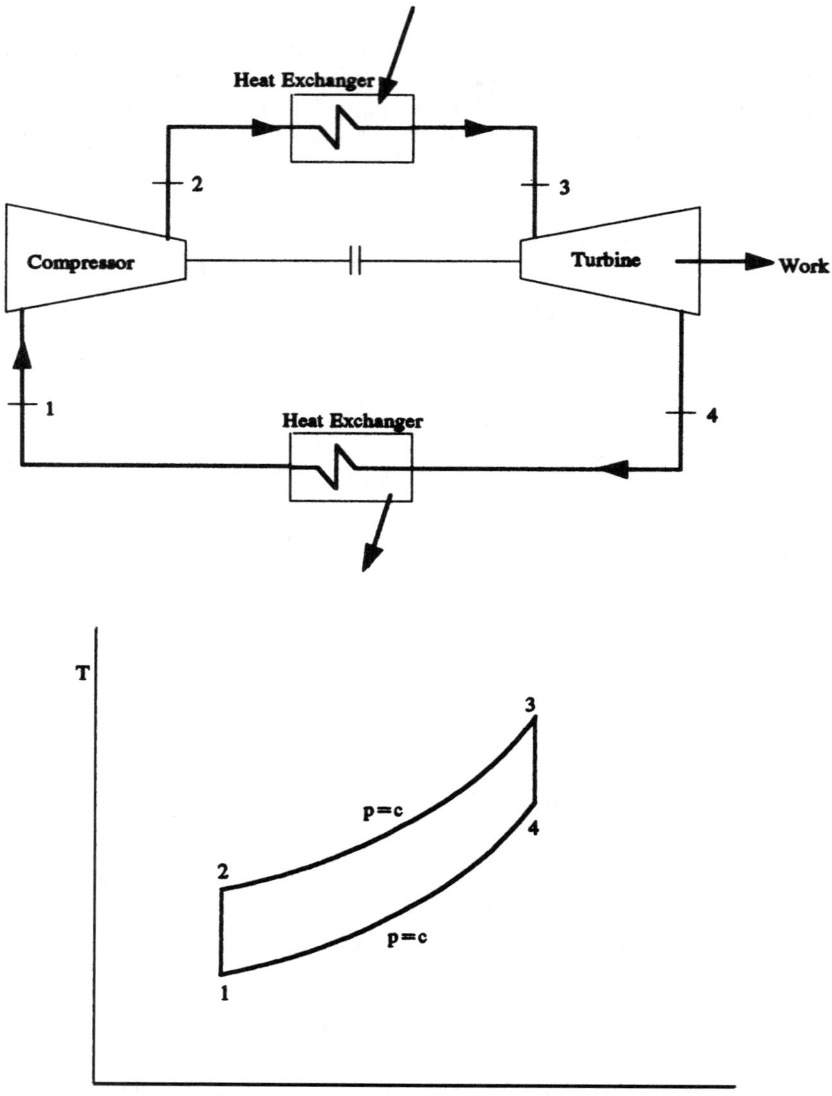

Figure 7-5 A Brayton cycle and its *T-s* diagram.

heat supply in Process 2-3. That is,

$$a_{q_h} = \int_2^3 \left(1 - \frac{T_0}{T}\right) c_p \, dT$$

Integrating the above equation, we have

$$a_{q_h} = c_p(T_3 - T_2) - T_0 c_p \ln \frac{T_3}{T_2}$$

Using the definition of k_1 and ρ_p, this equation becomes

$$a_{q_h} = c_p T_1 \left(k_1 - \rho_p - \frac{T_0}{T_1} \ln \frac{k_1}{\rho_p} \right) \tag{7-38}$$

With Eqs. (7-35) and (7-38), the second law cycle efficiency can be expressed as

$$\varepsilon_{cy} = \frac{w_{cy}}{a_{q_h}} = \frac{\left(1 - \dfrac{1}{\rho_p}\right)(k_1 - \rho_p)}{(k_1 - \rho_p) - \dfrac{T_0}{T_1} \ln \dfrac{k_1}{\rho_p}} \tag{7-39}$$

In terms of the first law cycle efficiency, Eq. (7-39) is further changed to

$$\varepsilon_{cy} = \frac{\eta_{cy}}{1 - \dfrac{T_0}{T_1} \dfrac{\ln(k_1/\rho_p)}{k_1 - \rho_p}} \tag{7-40}$$

It is of interest to note that the second law efficiency is function of three variables: the isentropic temperature ratio ρ_p, the normalized turbine inlet temperature (T_3/T_1) and the normalized reference temperature (T_0/T_1). Figure 7-6 shows the effects of the isentropic temperature ratio on both efficiencies. The normalized turbine inlet temperature also is important. As the inlet temperature increases, the cycle efficiency will increase. The turbine inlet temperature usually is fixed by the metallurgical temperature limit of the turbine blade. This temperature ranges between 2000°F and 2600°F in gas turbine designs.

The cycle efficiency of the Brayton cycle can be improved by introducing a regenerator. Figure 7-7 shows a regenerative Brayton cycle and the corresponding *T-s* diagram. In a regenerator, heat is transferred from the turbine exhaust gas to the compressed air. When the regenerator is an ideal heat exchanger, the temperature of the air leaving the regenerator T_x is equal to the temperature T_4, the temperature of the substance leaving the turbine. In the regenerative Brayton cycle, the heat-transfer process from an external source will take place between state x and state 3. Obviously, the amount of heat supplied to the cycle

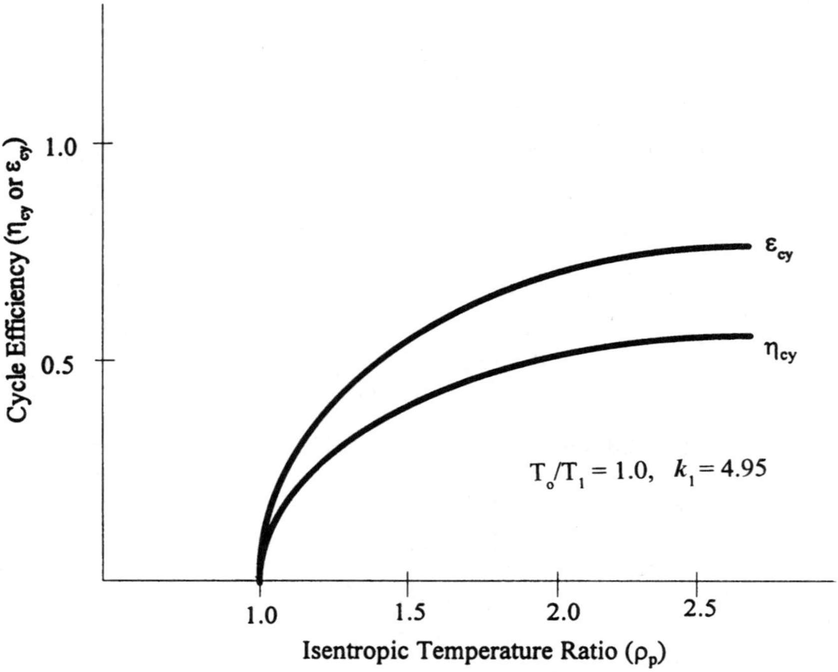

Figure 7-6 Cycle efficiency for a Brayton cycle.

is reduced. In equation form, it is

$$q_h = c_p(T_3 - T_x) = c_p(T_3 - T_4)$$

Making use of the definitions of k_1 and ρ_p, the cycle heat supply is expressed as

$$q_h = c_p T_1 \left(k_1 - \frac{k_1}{\rho_p} \right) \tag{7-41}$$

Thus, the first law cycle efficiency becomes

$$\eta_{cy} = \frac{w_{cy}}{q_h} = \frac{c_p T_1 \left(1 - \dfrac{1}{\rho_p} \right) (k_1 - \rho_p)}{c_p T_1 \left(k_1 - \dfrac{k_1}{\rho_p} \right)}$$

or

$$\eta_{cy} = 1 - \frac{\rho_p}{k_1} \tag{7-42}$$

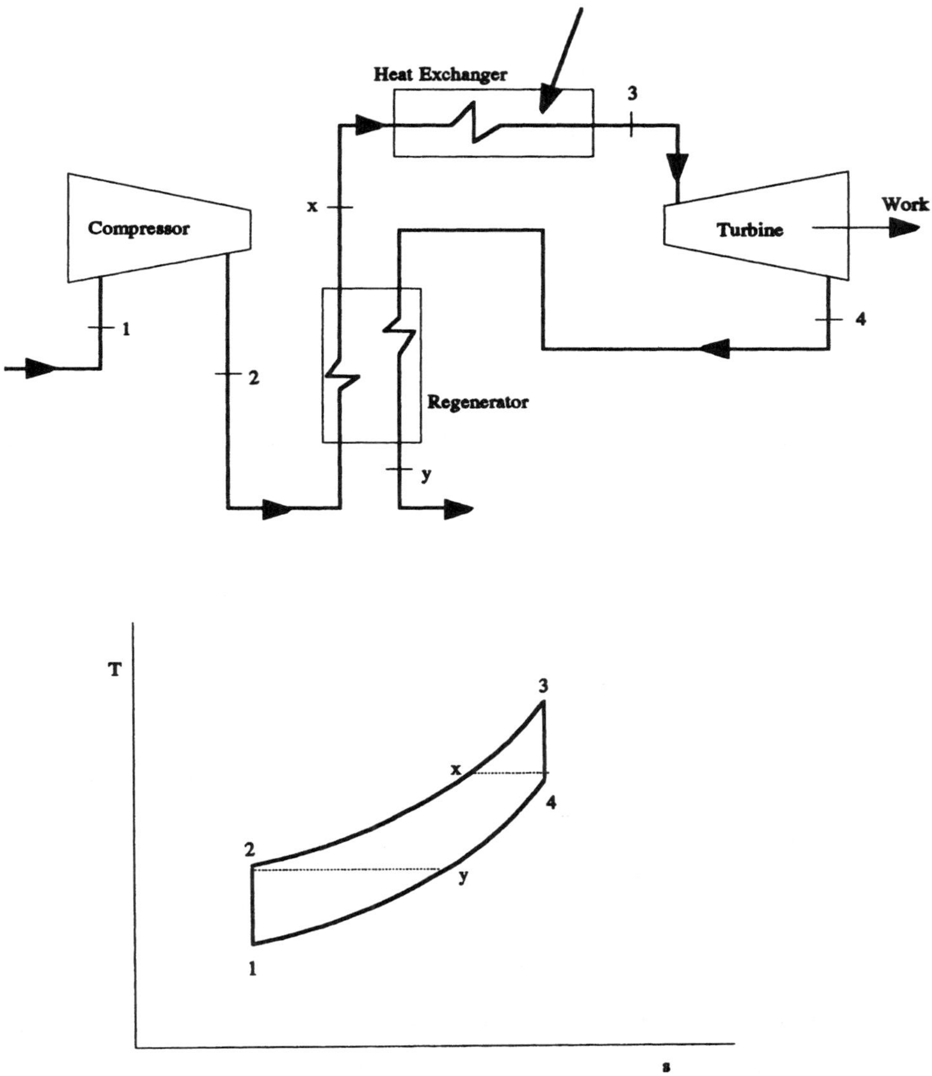

Figure 7-7 A regenerative Brayton cycle and its *T-s* diagram.

The heat availability transfer in Process x-3 is

$$a_{q_h} = \int_{x}^{3} \left(1 - \frac{T_0}{T}\right) c_p \, dT$$

Integrating this equation gives

$$a_{q_h} = c_p(T_3 - T_x) - c_p T_0 \ln \frac{T_3}{T_x}$$

With the definitions of k_1 and ρ_p, the above equation can be changed and expressed as

$$a_{q_h} = c_p T_1 \left[k_1 - \frac{k_1}{\rho_p} - \frac{T_0}{T_1} \ln \rho_p \right] \tag{7-43}$$

Combining Eqs. (7-35) and (7-43), we have the second law efficiency for the idealized regenerative Brayton cycle as

$$\varepsilon_{cy} = \frac{\left(1 - \dfrac{1}{\rho_p}\right)(k_1 - \rho_p)}{\left(k_1 - \dfrac{k_1}{\rho_p}\right) - \dfrac{T_0}{T_1} \ln \rho_p} \tag{7-44}$$

The influences of the isentropic temperature ratio are shown in Fig. 7-8. It is seen also that the optimal compressor pressure ratio is always smaller for a regenerative Brayton cycle than that for a simple Brayton cycle.

The introduction of a regenerator increases the cycle efficiency. This happens mainly because of both an increase in the average temperature of heat reception and a simultaneous decrease in the average temperature of heat rejection. Another practical way to improve the cycle efficiency is to use intercooling between compressor stages and reheating between turbine cylinders. Figure 7-9 shows a cycle with one intercooler and one reheater. In most cases the temperature of the compressed air leaving the intercooler is equal to the temperature of the air entering the first stage of the compressor. The pressure ratio across these two compressors usually is the same for optimal operation. On the expansion side, the temperature after reheating usually is equal to the temperature at the turbine inlet. The pressure at which the reheating takes place is approximately $\sqrt{p_i p_e}$, where p_i and p_e are the turbine inlet and outlet pressures, respectively.

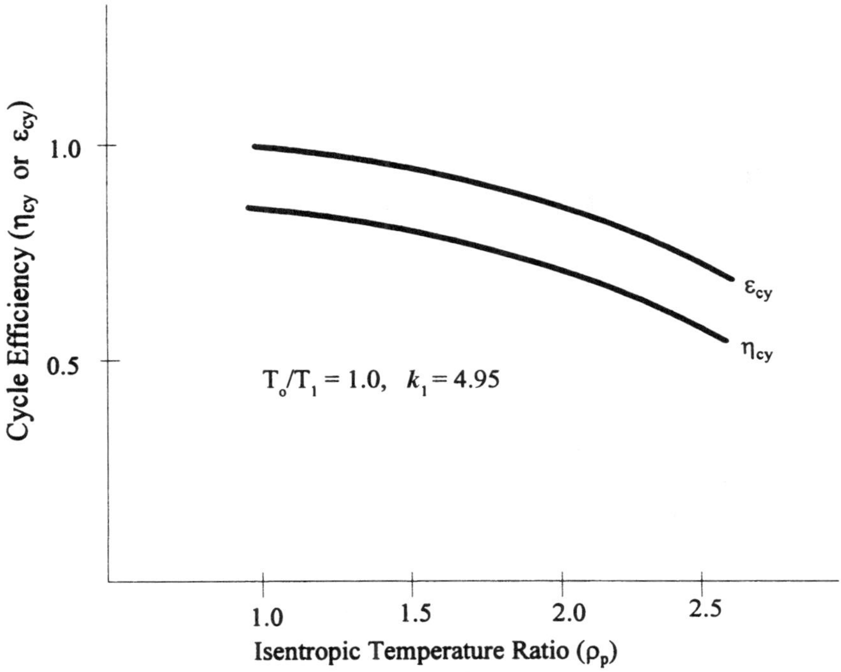

Figure 7-8 Cycle efficiency for a regenerative Brayton cycle.

7-5 THE OTTO AND DIESEL CYCLES

This section is concerned principally with the Otto and Diesel cycles. These two cycles are the idealized cycles for the spark-ignition engine and the compression-ignition engine, respectively. These engines frequently are referred to as *internal combustion engines*. Similar to the gas-turbine system, internal combustion engines have replaced steam with air as the working substance. Since there is a combustion process, the working substance changes from air and fuel to products of combustion. Even though the engines operate in a mechanical cycle, the working substance does not go through a thermodynamic cycle because of intake and exhaust processes. To analyze internal combustion engine, we again utilize the air-standard assumption. These assumptions include replacement of the combustion process by a heat-transfer process from an external source, and the intake and exhaust processes by heat rejection to the environment. Also, the working substance is assumed to be air, which obeys the ideal gas equation and has constant specific heat. Like the air-standard Brayton cycle, all processes are internally reversible. There is no question that the results thus obtained will be quite different from those of an actual internal combustion engine . However, the air-standard cycle approach is a convenient means to qualitatively analyze the performance of internal combustion engines.

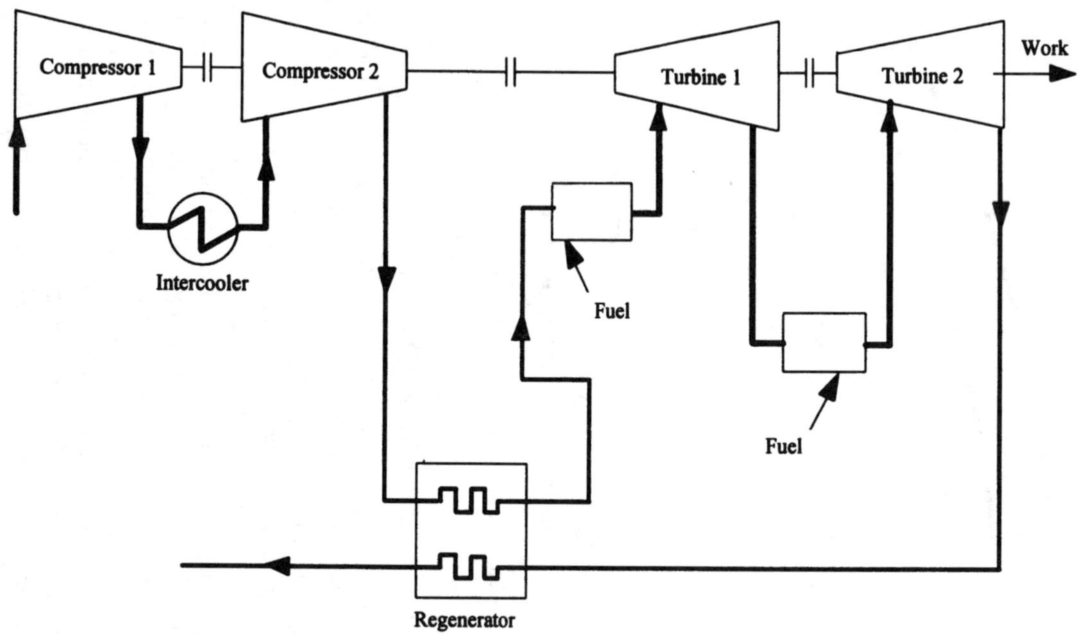

Figure 7-9 A reheat gas turbine system.

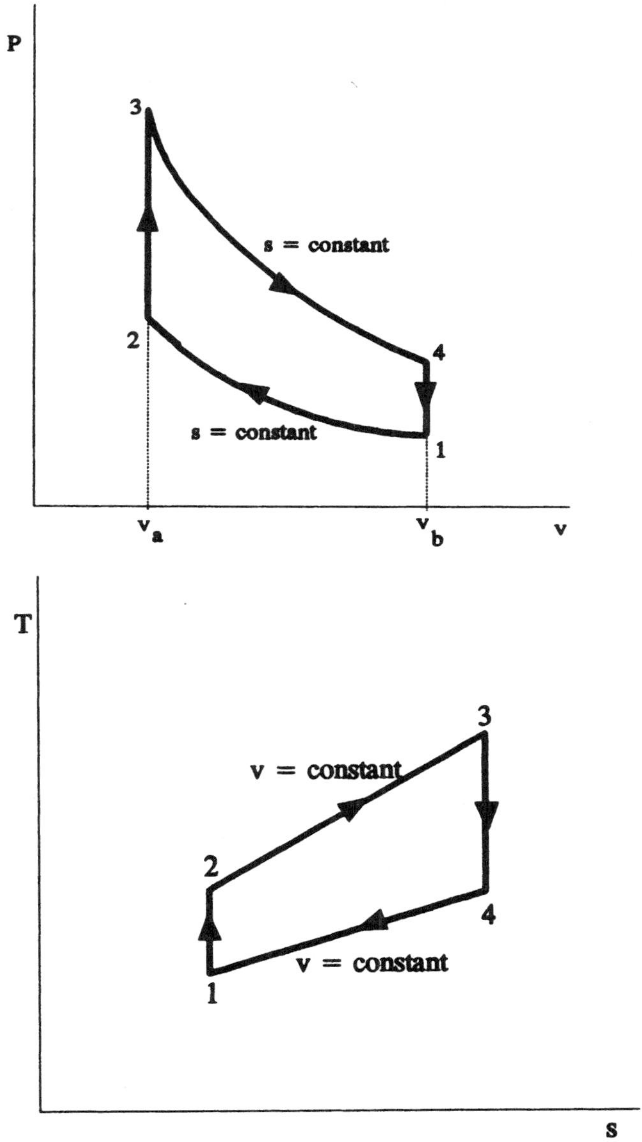

Figure 7-10 The air-standard Otto cycle.

Figure 7-10 shows the *P-v* and *T-s* diagrams for an air-standard Otto cycle. Process 1-2 is reversible and adiabatic compression. The process takes place just as the piston moves from crank-end dead center to head-end dead center. Process 2-3 is reversible and takes place at constant volume. In this process the working substance (air plus fuel) ignites and burns very rapidly. Next, the working substance expands isentropically in Process 3-4 and produces useful work. Process 4-1 is reversible and of constant volume. This idealized process is used to replace the exhaust and intake processes of the actual engine.

The first law cycle efficiency is simply the ratio of cycle net work to the heat supplied to the cycle. In equation form it is

$$\eta_{cy} = 1 - \frac{q_1}{q_h} = 1 - \frac{c_v(T_4 - T_1)}{c_v(T_3 - T_2)} = 1 - \frac{T_1}{T_2}\frac{\left(\dfrac{T_4}{T_1} - 1\right)}{\left(\dfrac{T_3}{T_2} - 1\right)}$$

Since $T_4/T_1 = T_3/T_2$, the above equation becomes

$$\eta_{cy} = 1 - \frac{T_1}{T_2} \tag{7-45}$$

Let the compression ratio r_v be defined as

$$r_v = \frac{V_1}{V_2} = \frac{V_4}{V_3} = \frac{V_b}{V_a} \tag{7-46}$$

Since

$$\frac{T_2}{T_1} = \frac{T_3}{T_4} = \left(\frac{V_b}{V_a}\right)^{k-1} \tag{7-47}$$

the compression ratio r_v becomes

$$r_v^{\,k-1} = \frac{T_2}{T_1} = \frac{T_3}{T_4} \tag{7-48}$$

Thus the first law efficiency for an Otto cycle is

$$\eta_{cy} = 1 - \frac{1}{r_v^{\,k-1}} \tag{7-49}$$

This expression indicates that the efficiency of the air-standard Otto cycle is a function of the compression ratio r_v. As the compression ratio increases, so does the cycle efficiency. However, there is an upper limit beyond which undesirable detonation of the fuel will take place. In practice, the compression ratio for a spark-ignition engine is determined by the fuel antiknock characteristics and is usually between 6 and 12.

The second law cycle efficiency is simply

$$\varepsilon_{cy} = \frac{w_{cy}}{\int_2^3 \left(1 - \dfrac{T_0}{T}\right) c_v\, dT} \tag{7-50}$$

The numerator w_{cy} is the net work from the cycle. It is given by

$$w_{cy} = q_h - q_1 = c_v(T_3 - T_2) - c_v(T_4 - T_1)$$

Rearranging these terms will yield

$$w_{cy} = c_v T_1 \left(\frac{T_2}{T_1} - 1 \right) \left(\frac{T_3}{T_2} - 1 \right)$$

or

$$w_{cy} = c_v T_1 \left(r_v^{k-1} - 1 \right) \left(\frac{k_1}{r_v^{k-1}} - 1 \right) \tag{7-51}$$

where

$$k_1 = \frac{T_3}{T_1} \tag{7-52}$$

The denominator in Eq. (7-50) represents the heat availability transfer in Process 2-3. An integration will give

$$\int_2^3 \left(1 - \frac{T_0}{T} \right) c_v \, dT = c_v (T_3 - T_2) - c_v T_0 \ln \frac{T_3}{T_2}$$

or

$$\int_2^3 \left(1 - \frac{T_0}{T} \right) c_v \, dT = c_v T_1 \left[\left(k_1 - r_v^{k-1} \right) - \frac{T_0}{T_1} \ln \frac{k_1}{r_v^{k-1}} \right] \tag{7-53}$$

Substituting Eqs. (7-51) and (7-53) into Eq. (7-50), we have the second law efficiency for the Otto cycle as

$$\varepsilon_{cy} = \frac{\left(r_v^{k-1} - 1 \right) \left(\dfrac{k_1}{r_v^{k-1}} - 1 \right)}{k_1 - r_v^{k-1} - \dfrac{T_0}{T_1} \ln \dfrac{k_1}{r_v^{k-1}}} \tag{7-54}$$

It is of interest to note that the second law cycle efficiency is a function of three variables. These are the compression ratio r_v, normalized maximum temperature (T_3/T_1), and normalized reference temperature (T_0/T_1). Figure 7-11 shows the influence of these variables on the first law and second law cycle efficiencies.

Figure 7-12 shows the *P-v* and *T-s* diagrams for an air-standard Diesel cycle. The compression and expansion processes are assumed to be isentropic, as they were in the Otto cycle. The combustion process is approximated by a reversible and constant-pressure process. The fuel injection rate is maintained so that the pressure will remain unchanged even in the expansion process 2-3. Because only air is being compressed, the compression ratio usually is higher than that for an Otto cycle. The heat rejection is a reversible and constant-volume process, which approximates the exhaust and intake processes of the actual Diesel engine.

The efficiency for an air-standard Diesel cycle is given by $\eta_{cy} = 1 - q_1/q_h$. Since heat is added in a constant-pressure process and rejected in a constant-

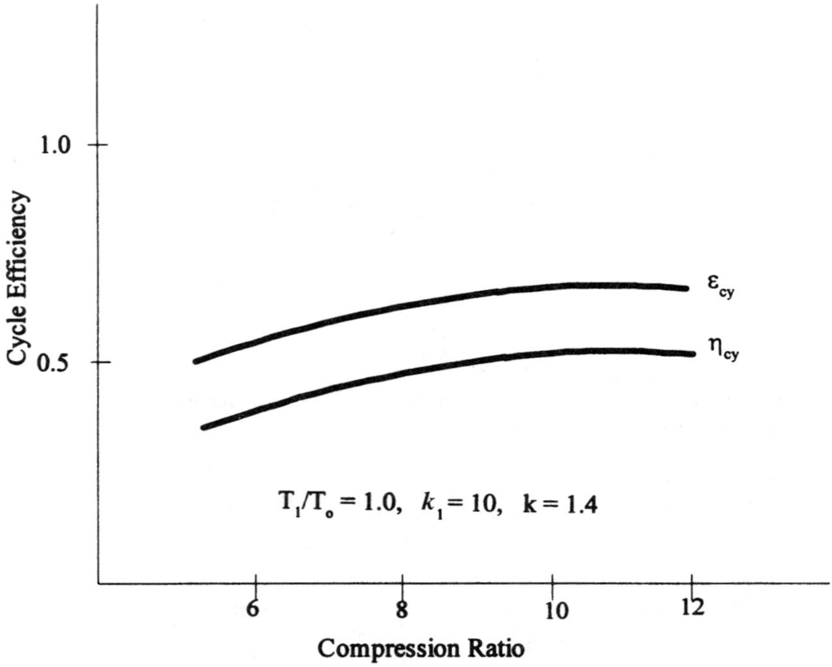

Figure 7-11 Cycle efficiency for an Otto cycle.

volume process, the above equation can be simplified and expressed as

$$\eta_{cy} = 1 - \frac{c_v(T_4 - T_1)}{c_p(T_3 - T_2)} = 1 - \frac{1}{k}\frac{T_1}{T_2}\frac{\left(\dfrac{T_4}{T_1} - 1\right)}{\left(\dfrac{T_3}{T_2} - 1\right)} \tag{7-55}$$

Next, we determine these temperature ratios and express them in a convenient form. For the isentropic process 1-2,

$$\frac{T_2}{T_1} = \left(\frac{V_1}{V_2}\right)^{k-1} = r_v^{\,k-1} \tag{7-56}$$

and for the constant pressure process 2-3,

$$\frac{T_3}{T_2} = \frac{V_3}{V_2} = \alpha \tag{7-57}$$

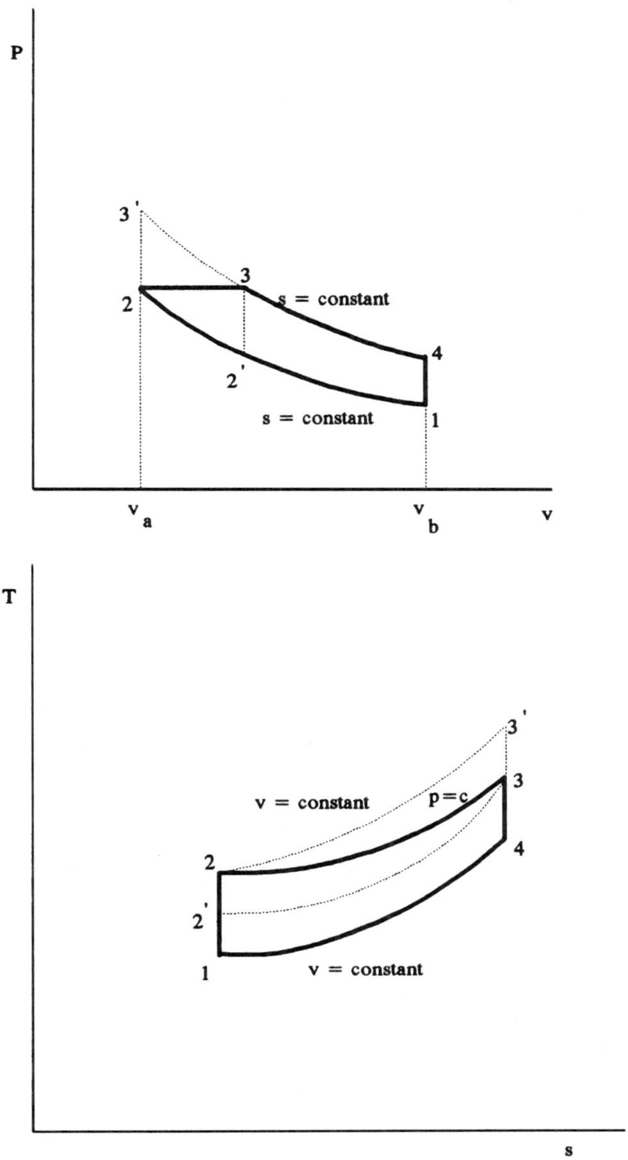

Figure 7-12 The air-standard diesel cycle.

The term α is called the *fuel cutoff ratio* or *load ratio*. To determine the temperature ratio (T_4/T_3), we obtain

$$\frac{T_4}{T_3} = \left(\frac{V_3}{V_4}\right)^{k-1} = \left(\frac{V_3 V_2}{V_2 V_4}\right)^{k-1} \tag{7-58}$$

Making use of the definitions for r_v and α, we can express Eq. (7-58) as

$$\frac{T_4}{T_3} = \left(\frac{\alpha}{r_v}\right)^{k-1} \tag{7-59}$$

Finally, combining Eqs. (7-56), (7-57), and (7-59) will determine the temperature ratio to be

$$\begin{aligned}\frac{T_4}{T_1} &= \left(\frac{T_4}{T_3}\right)\left(\frac{T_3}{T_2}\right)\left(\frac{T_2}{T_1}\right) = \left(\frac{\alpha}{r_v}\right)^{k-1}(\alpha)(r_v)^{k-1} \\ &= \alpha^k \end{aligned} \tag{7-60}$$

In terms of the compression ratio and the fuel cutoff ratio, the first law efficiency for an air-standard Diesel cycle becomes

$$\eta_{\text{cy}} = 1 - \frac{1}{r_v^{k-1}}\left[\frac{\alpha^k - 1}{k(\alpha - 1)}\right] \tag{7-61}$$

The efficiency of the air-standard Diesel cycle depends not only on the compression ratio r_v but also on the cutoff ratio α. For the same compression ratio, the Diesel cycle efficiency always is less than the Otto cycle efficiency, because the term in the square bracket of Eq. (7-61) always is greater than unity. This is evident also from the *T-s* diagram in which the enclosed area 1-2-3-4-1 is 3′ smaller than the area 1-2-3′-4-1. However, it should not be concluded that the compression-ignition engines (also referred to as diesel engines) are less efficient than the spark-ignition engines. In fact, the reverse is true. Because only air is compressed in the compression process, Diesel engines always have a much higher compression ratio than the Otto engines and therefore a higher efficiency. For a given maximum temperature as shown in Fig. 7-12, the Diesel cycle network represented by the area 1-2-3-4-1 is greater than the Otto cycle network represented by the area 1-2′-3-4-1. The second law efficiency for an air-standard Diesel cycle is

$$\varepsilon_{\text{cy}} = \frac{w_{\text{cy}}}{\int_2^3 \left(1 - \dfrac{T_0}{T}\right) c_p \, dT} \tag{7-62}$$

The numerator w_{cy} is the net work from the cycle and is given by the equation

$$w_{cy} = c_p(T_3 - T_2) - c_v(T_4 - T_1) = c_p T_1 \left[\frac{T_2}{T_1}\left(\frac{T_3}{T_2} - 1\right) - \frac{c_v}{c_p}\left(\frac{T_4}{T_1} - 1\right) \right]$$

In terms of r_v and α, the above equation becomes

$$w_{cy} = c_p T_1 \left[r_v^{k-1}(\alpha - 1) - \frac{1}{k}(\alpha^k - 1) \right] \tag{7-63}$$

The denominator in Eq. (7-62) is the heat availability input to the cycle. After an integration we have

$$\int_2^3 \left(1 - \frac{T_0}{T}\right) c_p \, dT = c_p(T_3 - T_2) - T_0 c_p \ln \frac{T_3}{T_2}$$

In a convenient form the equation becomes

$$\int_2^3 \left(1 - \frac{T_0}{T}\right) c_p \, dT = c_p T_1 \left[r_v^{k-1}(\alpha - 1) - \frac{T_0}{T_1} \ln \alpha \right] \tag{7-64}$$

Substituting Eqs. (7-63) and (7-64) into Eq. (7-62), we have the second law efficiency for the Diesel cycle as

$$\varepsilon_{cy} = \frac{r_v^{k-1}(\alpha - 1) - \dfrac{1}{k}(\alpha^k - 1)}{r_v^{k-1}(\alpha - 1) - \dfrac{T_0}{T_1} \ln \alpha} \tag{7-65}$$

This equation indicates that the second law efficiency for the Diesel cycle is a function of three variables: the compression ratio r_v, cutoff ratio α, and normalized reference temperature (T_0/T_1). Figure 7-13 shows their influences on the cycle efficiency.

It should be emphasized that the cycle efficiency calculated with the air-standard assumptions is always greater than the actual value. This is simply because the air-standard assumptions are not compatible with reality and very difficult to implement in practice. In an actual engine, the combustion may not be complete and the compression and expansion process are not isentropic. The engine operation involves an inlet process and an exhaust process, and a certain amount of work usually is required to overcome the friction in the processes. However, as previously indicated, the main value of the air-standard-cycle approach is to enable engineers to identify the important parameters and to determine qualitatively their influence on the engine performance.

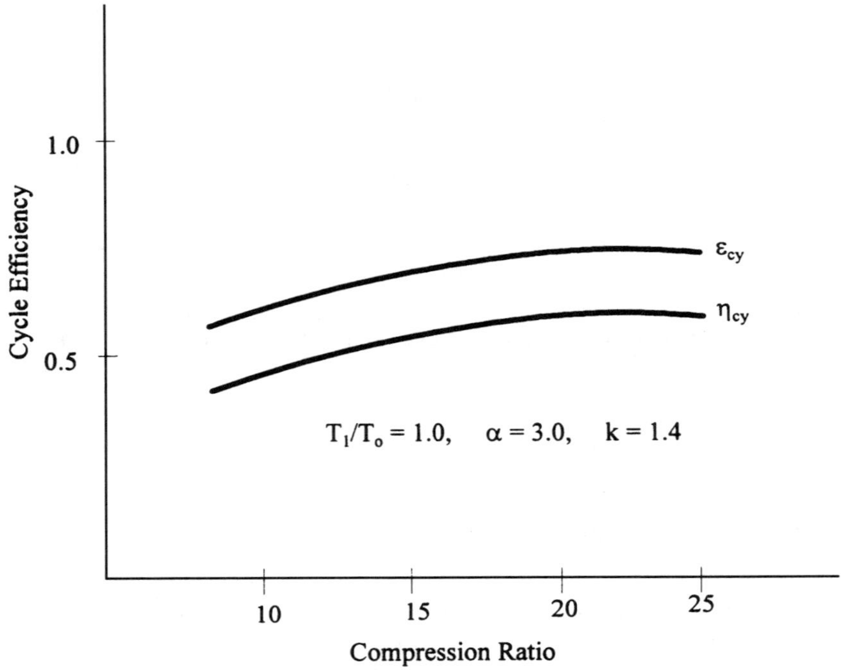

Figure 7-13 Cycle efficiency for a diesel cycle.

7-6 THE IDEAL GAS-REFRIGERATION CYCLES

In this section we present two ideal gas-refrigeration cycles. They are the reversed Carnot and reversed Brayton cycles. The assumptions used are identical to those in the air-standard cycle. All processes in the cycle are internally reversible. The main purpose is to develop the second law cycle efficiency and compare it with the coefficient of performance, which traditionally is used in engineering practice.

The reversed Carnot cycle

Figure 7-14 shows the *T-s* diagram for a reversed Carnot cycle. When we apply the availability equation to the cyclic process 1-2-3-4-1, it gives

$$a_w = -a_q \tag{7-66}$$

where

$$a_w = -\oint \delta w = -\oint \delta q$$

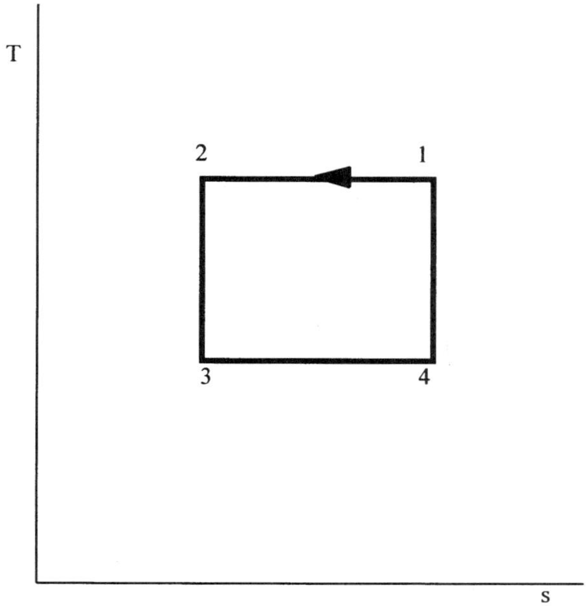

Figure 7-14 *T-s* diagram for a reversed Carnot cycle.

and

$$a_q = \oint \left(1 - \frac{T_0}{T}\right) \delta q$$

Next, we conveniently express the term a_w

$$\begin{aligned} a_w &= -\int_1^2 \delta q - \int_2^3 \delta q - \int_3^4 \delta q - \int_4^1 \delta q = -\int_1^2 \delta q - \int_3^4 \delta q \\ &= -q_{1\text{-}2} - q_{3\text{-}4} \end{aligned}$$

In terms of absolute values, we have

$$a_w = q_h - q_l \tag{7-67}$$

Similarly, we express a_q as

$$\begin{aligned} a_q &= \int_1^2 \left(1 - \frac{T_0}{T}\right) \delta q + \int_2^3 \left(1 - \frac{T_0}{T}\right) \delta q + \int_3^4 \left(1 - \frac{T_0}{T}\right) \delta q + \int_4^1 \left(1 - \frac{T_0}{T}\right) \delta q \\ &= \left(1 - \frac{T_0}{T_h}\right) q_{1\text{-}2} + \left(1 - \frac{T_0}{T_l}\right) q_{3\text{-}4} \end{aligned}$$

In terms of absolute values, the above equation becomes

$$a_q = -\left(1 - \frac{T_0}{T_h}\right) q_h + \left(1 - \frac{T_0}{T_l}\right) q_l \tag{7-68}$$

Substituting Eqs. (7-67) and (7-68) into Eq. (7-66), we have the availability equation in form of input and output. That is,

$$q_h - q_l = \left(1 - \frac{T_0}{T_h}\right) q_h - \left(1 - \frac{T_0}{T_l}\right) q_l \tag{7-69}$$

The second law efficiency of the reversed Carnot cycle when it is operating as a refrigerator is given by

$$\beta_{\text{ref}} = \frac{-\left(1 - \dfrac{T_0}{T_l}\right) q_l}{q_h - q_l} \tag{7-70}$$

As a heat pump, the second law efficiency is

$$\beta_{\text{hp}} = \frac{\left(1 - \dfrac{T_0}{T_h}\right) q_h}{q_h - q_l} \tag{7-71}$$

In terms of the coefficient of performance, Eqs. (7-70) and (7-71) can be further modified and expressed respectively as

$$\beta_{\text{ref}} = -\left(1 - \frac{T_0}{T_l}\right) \text{COP}_{\text{ref}} \tag{7-72}$$

and

$$\beta_{\text{hp}} = \left(1 - \frac{T_0}{T_h}\right) \text{COP}_{\text{hp}} \tag{7-73}$$

The minus signs in Eqs. (7-70) and (7-72) are not surprising. In refrigeration the temperature T_0 is greater than the temperature T_l, and the direction of heat flow is opposite to the direction of associated availability flow. Because of this, the term $(1 - T_0/T_l)$ in both equations is always negative, and the second law efficiency is still positive.

The reversed Brayton cycle

Figure 7-15 shows the *T-s* diagram for the reversed Brayton cycle. The differences between the reversed Carnot and Brayton cycles are in the heat addition and rejection processes. The processes in the Carnot cycle are isothermal, whereas those in the Brayton cycle are isobaric. Because of this, we write the availability equation for the cyclic Brayton process 1-2-3-4-1 as simply

$$q_h - q_l = -\int_1^2 \left(1 - \frac{T_0}{T}\right) \delta q - \int_3^4 \left(1 - \frac{T_0}{T}\right) \delta q \tag{7-74}$$

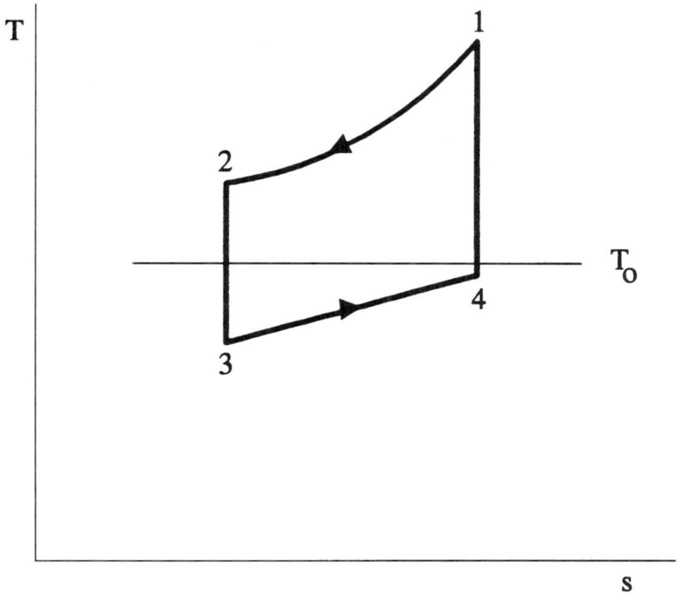

Figure 7-15 *T-s* diagram for a reversed Brayton cycle.

and the second law efficiencies for a refrigerator and a heat pump respectively as

$$\beta_{\text{ref}} = \frac{-\int_3^4 \left(1 - \frac{T_0}{T}\right) \delta q}{q_h - q_l} \tag{7-75}$$

and

$$\beta_{\text{hp}} = \frac{-\int_1^2 \left(1 - \frac{T_0}{T}\right) \delta q}{q_h - q_l} \tag{7-76}$$

It is of interest to note that Eqs. (7-75) and (7-76) can be further modified by using the mean temperatures from Process 1-2 and Process 3-4. The mean temperature in Process 1-2 is defined by

$$T_{h_m} = \frac{1}{s_2 - s_1} \int_1^2 T\, ds = \frac{T_2 - T_1}{\ln \dfrac{T_2}{T_1}} \tag{7-77}$$

Similarly, the mean temperature in Process 3-4 is

$$T_{l_m} = \frac{1}{s_4 - s_3}\int_3^4 T\,ds = \frac{T_4 - T_3}{\ln\frac{T_4}{T_3}} \tag{7-78}$$

Thus, the second law efficiencies for the reversed Brayton cycle would become

$$\beta_{\text{ref}} = -\left(1 - \frac{T_0}{T_{l_m}}\right)\frac{q_l}{q_h - q_l} \tag{7-79}$$

and

$$\beta_{\text{hp}} = \left(1 - \frac{T_0}{T_{h_m}}\right)\frac{q_h}{q_h - q_l} \tag{7-80}$$

Both equations are similar to those for the reversed Carnot cycle.

7-7 THE IDEAL VAPOR-COMPRESSION CYCLES

Figure 7-16 shows a schematic diagram for an ideal vapor-compression system. The working substance, also referred to as a *refrigerant*, will follow a path in the direction of the reversed Rankine cycle. The refrigerant first enters the com-

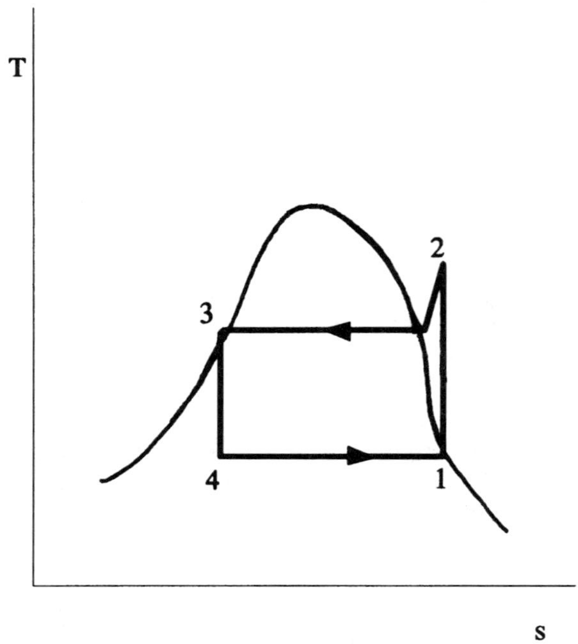

Figure 7-16 *T-s* diagram for an ideal vapor-compression cycle.

pressor as dry saturated vapor. In an ideal vapor-compression cycle, the compression process is isentropic. Next the refrigerant enters the condenser and leaves it as a saturated liquid. The pressure drop is assumed to be zero in the process. In the third process, the saturated liquid enters a liquid expander and leaves it as a mixture. The expander process is isentropic. To complete the cycle, the refrigerant enters the evaporator and absorbs the heat removed from the refrigerated space. Unless otherwise indicated, there will be zero superheating at state 1 and zero subcooling at state 3 in the ideal vapor-compression cycle. The coefficient of performance of an ideal vapor-compression cycle when it operates as a refrigerator is given by

$$\mathrm{COP}_{\mathrm{ref}} = \frac{q_l}{q_h - q_l} = \frac{h_1 - h_4}{(h_2 - h_3) - (h_1 - h_4)}$$

As a heat pump, the coefficient of performance would be given by

$$\mathrm{COP}_{\mathrm{hp}} = \frac{q_h}{q_h - q_l} = \frac{h_2 - h_3}{(h_2 - h_3) - (h_1 - h_4)}$$

To develop the second law efficiency for the ideal vapor-compression cycle, we apply the availability equation to the cyclic process 1-2-3-4-1 and have $a_w = -a_q$, or

$$(h_2 - h_1) - (h_3 - h_4) = -\int_2^3 \left(1 - \frac{T_0}{T}\right) \delta q - \int_4^1 \left(1 - \frac{T_0}{T}\right) \delta q$$

Thus, the second law efficiency for a refrigeration cycle is

$$\beta_{\mathrm{ref}} = \frac{-\int_4^1 \left(1 - \frac{T_0}{T}\right) \delta q}{(h_2 - h_1) - (h_3 - h_4)}$$

After an integration the equation becomes

$$\beta_{\mathrm{ref}} = \frac{(h_4 - h_1) - T_0(s_4 - s_1)}{(h_2 - h_1) - (h_3 - h_4)}$$

Similarly, the second law efficiency for a heat pump cycle is

$$\beta_{\mathrm{hp}} = \frac{-\int_2^3 \left(1 - \frac{T_0}{T}\right) \delta q}{(h_2 - h_1) - (h_3 - h_4)} = \frac{(h_2 - h_3) - T_0(s_2 - s_3)}{(h_2 - h_1) - (h_3 - h_4)}$$

It should be pointed out that the actual refrigeration system is quite different from the ideal vapor-compression cycle shown in Fig. 7-16. Not only is the compression process not isentropic, but there also may be some degree of subcooling of the liquid leaving the condenser and superheating of the vapor entering the compressor. For technical as well as economic reasons, a simple

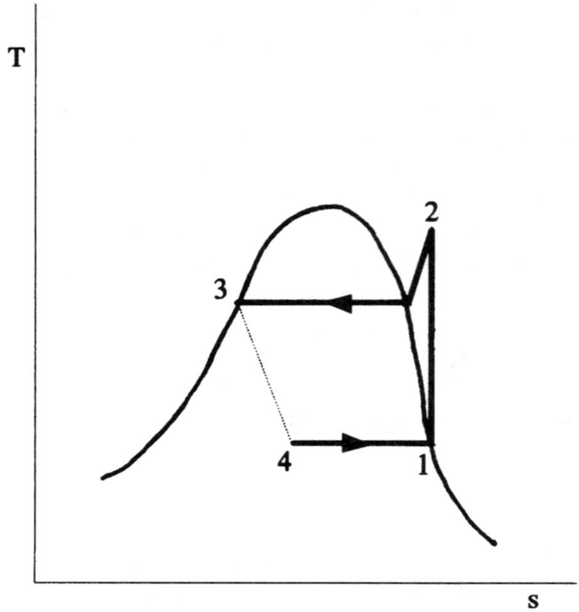

Figure 7-17 *T-s* diagram for a quasi-ideal vapor-compression cycle.

throttling device frequently is used to replace the liquid expander or turbine. Because of this, the expansion process will have essentially constant enthalpy. Figure 7-17 shows the *T-s* diagram for this cycle arrangement, in which all processes are reversible except the throttle expansion process 3-4. In this text this cycle will be referred to as the *quasi-ideal vapor-compression cycle*. The second law efficiencies for refrigerators and heat pumps can be developed in the same fashion.

7-8 PROBLEMS

7-1 A Carnot cycle operates between two thermal reservoirs. One has a temperature of 1000°F while the other has a temperature of 120°F. What is the second law cycle efficiency? Also, compare the result with the first law cycle efficiency.

7-2 It is known that the second law cycle efficiency frequently is greater than the first law cycle efficiency. Show how the ratio of cycle efficiency for the Carnot cycle varies with the temperature of a thermal reservoir, and present the results in a graphical form.

7-3 An ideal Rankine cycle operates under the following conditions:

Turbine inlet	1000 psia, 1000°F
Turbine exit	5 in. Hg. abs.

Determine the second law cycle efficiency and compare the result with the first law cycle efficiency.

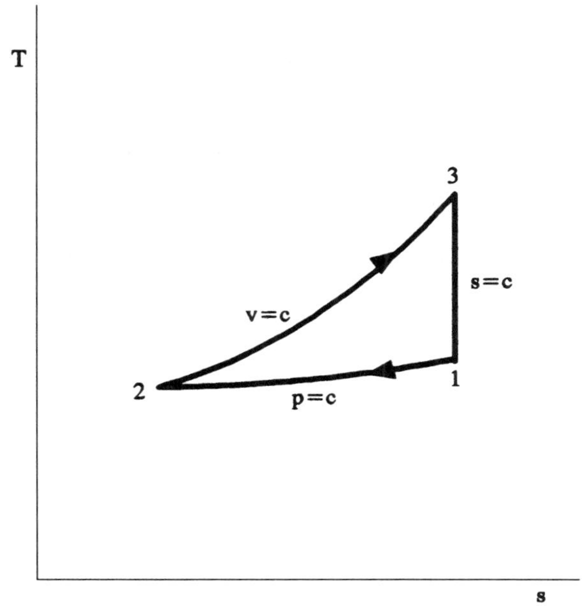

Figure 7-18 *T-s* diagram for Problem 7-6.

7-4 It is well known that the performance of an ideal Rankine cycle will be improved by reducing the turbine exhaust pressure. Repeat Problem 7-3 for various turbine exhaust pressures and present the results in a tabular form.

7-5 Repeat problem 7-3 by adding a single reheat process to the Rankine cycle. The temperature of steam leaving the reheater is the same as the turbine inlet temperature. Determine the reheat pressure at which the cycle has a maximum cycle efficiency. Does it make any difference whether the cycle efficiency is based on the concept of energy or availability?

7-6 Air undergoes a thermodynamic cycle consisting of three reversible processes, as shown in Fig. 7-18. For this cyclic process, derive an expression for the work availability a_w and the heat availability a_q.

7-7 Consider an air-standard Brayton cycle with air entering the compressor at 14.7 psia and 90°F and exhausting at 150 psia. The maximum cycle temperature is 2000°F. Calculate the first law and second law cycle efficiencies. Also calculate the net work per pound of air, using the specific heat of air 0.24 Btu/lbm-R.

7-8 Consider a regenerative Brayton cycle. Air enters the compressor at 14.7 psia and 90°F and exhausts at 60 psia. The cycle maximum temperature is 2000°F at the turbine inlet. Determine the second law cycle efficiency and compare the result with the first law cycle efficiency.

7-9 An air-standard Otto cycle has a compression ratio of 9. The pressure at the beginning of the compression stroke is 14.7 psia, and the temperature is 80°F. The heat transfer to the air is 850 Btu/lbm per cycle. Determine the first law and the second law efficiencies.

7-10 An air-standard Diesel engine intakes air at 14.7 psia and 80°F. The compression ratio of the cycle is 13, and the cutoff ratio is 2.2. Determine the first law and second law efficiencies. Also calculate the cycle net work per pound of air.

7-11 Calculate the second law efficiency of a Carnot refrigerator operating between 20°F and 100°F. Compare the result with the coefficient of performance for the same refrigerator.

7-12 A Carnot heat pump operates between two thermal reservoirs of temperatures 90°F and 20°F. What is the second law efficiency for the Carnot heat pump?

7-13 Consider an air-standard Brayton refrigeration cycle. Air enters the compressor at 14.7 psia and 10°F and leaves at 85 psia. Air at the expander inlet is 60°F. Calculate the second law efficiency for this refrigeration cycle.

7-14 A refrigerating system operates on the quasi-ideal vapor-compression cycle as defined in Fig. 7-17. The refrigerant is Freon-12, and the saturation temperatures in the evaporator and condenser are 0°F and 80°F, respectively. Calculate (a) the refrigerating effect per pound of Freon-12, (b) the coefficient of performance, (c) the second law cycle efficiency, and (d) the horsepower required per ton of refrigeration.

CHAPTER

EIGHT

ELECTRIC-POWER GENERATING SYSTEMS

The thermodynamics of electric-power generating systems has long been a classical study area for engineers. The objectives of such study traditionally have been the determination and maximization of the system efficiency. The usual approach is to apply the first law of thermodynamics and to minimize the fuel input for a given electric output. As indicated in the previous chapters, the first law is not sufficient and so the concept of availability must be utilized. In this chapter we apply the availability method to electric-power generating systems. These systems will include steam turbine, gas turbine, and combined-cycle systems. In addition to the calculation of the system second law efficiency, attention will be focused on the availability losses and destruction among the components. We will also identify the component in which the most availability loss and destruction occurs. This critical component should be the key area in any power plant–improvement project.

8-1 STEAM TURBINE SYSTEMS

The steam turbine power plant system consists of four major components: boiler, turbine-electric generator, condenser, and pump. As shown in Fig. 8-1, steam leaves the boiler at high pressure and high temperature and enters the turbine for expansion. In the turbine expansion process, the steam's thermal energy is converted into first mechanical work and then electrical energy

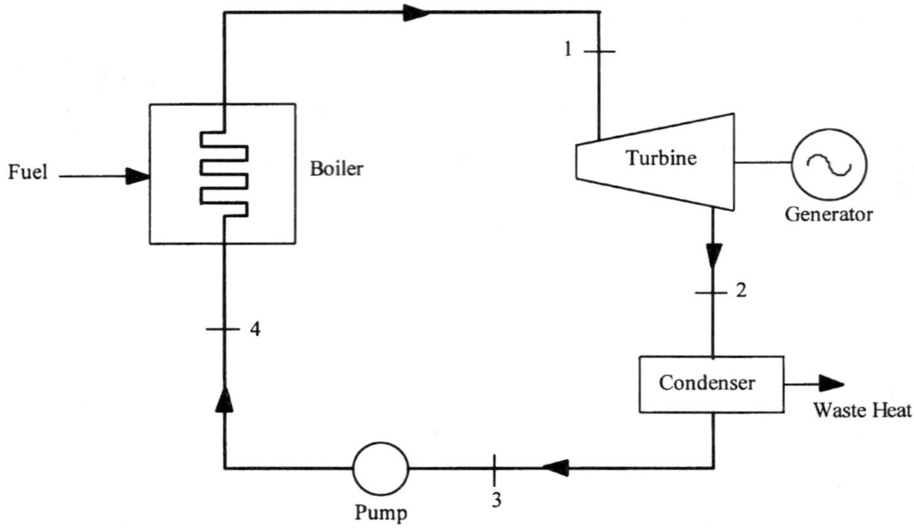

Figure 8-1 Schematic diagram for a steam turbine system.

through the electric generator. At the turbine exhaust end, steam is usually at low-pressure and saturated conditions. Then the steam passes through a condenser and exits as a saturated liquid. In the final process, the water is pumped back to the boiler and completes the steam turbine cycle. The steam turbine cycle efficiency is the ratio of the cycle net output to the cycle input. That is,

$$\eta_{\text{cy}} = \frac{w_{\text{st}} - w_p}{q_b} \tag{8-1}$$

The cycle net output is the turbine-generator output minus the pump power. The cycle input is simply the thermal energy received by the water and steam. The second law cycle efficiency is defined as the ratio of the cycle availability output to the cycle availability input. In an equation form, it is

$$\varepsilon_{\text{cy}} = \frac{w_{\text{st}} - w_p}{a_{q_b}} \tag{8-2}$$

where a_{q_b} is the availability increase in the water and steam. Both equations (8-1) and (8-2) do not take into consideration the boiler performance. As indicated in Fig. 8-1, the boiler is the equipment in which energy (or availability) is transferred to the water and steam. The first law and second law efficiencies are respectively

$$\eta_b = \frac{q_b}{m_{\text{fuel}}(\text{LHV})} \tag{8-3}$$

and

$$\varepsilon_b = \frac{a_{q_b}}{m_{\text{fuel}}(a_{\text{fuel}})} \tag{8-4}$$

where q_b = energy transferred to steam and water
a_{q_b} = availability transferred to steam and water
m_{fuel} = fuel consumption
a_{fuel} = unit fuel availability
LHV = fuel lower heating value.

The overall efficiency of the electric-power generating system including the boiler is simply the product of the cycle efficiency and the boiler efficiency, or

$$\eta_o = \eta_{\text{cy}} \times \eta_b \tag{8-5}$$

for the first law calculation and

$$\varepsilon_o = \varepsilon_{\text{cy}} \times \varepsilon_b \tag{8-6}$$

for the second law calculation.

The application of the availability method to the steam turbine power plant is quite straightforward. As an example, we consider a highly simplified system as shown in Fig. 8-1 and prepare an availability analysis with the following conditions:

Turbine inlet conditions	320 psia (sat. vapor)
Condenser pressure	1 psia
Turbine internal efficiency	85%
Pump efficiency	80%
Boiler efficiency	82%
Coal lower heating value	13,200 Btu/lbm
Coal unit availability	13,860 Btu/lbm

First we take the steam turbine as a control volume and utilize the Mollier diagram to determine the steam properties as shown in Fig. 8-2. The turbine work is

$$w_t = h_1 - h_2 = 1204.3 - 895.3 = 309 \text{ Btu/lbm}$$

and the turbine available work is the decrease of flow availability in the process. That is,

$$\begin{aligned} w_{\text{ava}} &= a_{f_1} - a_{f_2} = (h_1 - h_2) - T_o(s_1 - s_2) \\ &= (1204.3 - 895.3) - 537(1.5060 - 1.6032) \\ &= 361.2 \text{ Btu/lbm} \end{aligned}$$

Applying the availability equation, the availability destruction in the turbine process is calculated as $i = w_{\text{ava}} - w_t = 361.2 - 309 = 52.2$ Btu/lbm. Then the

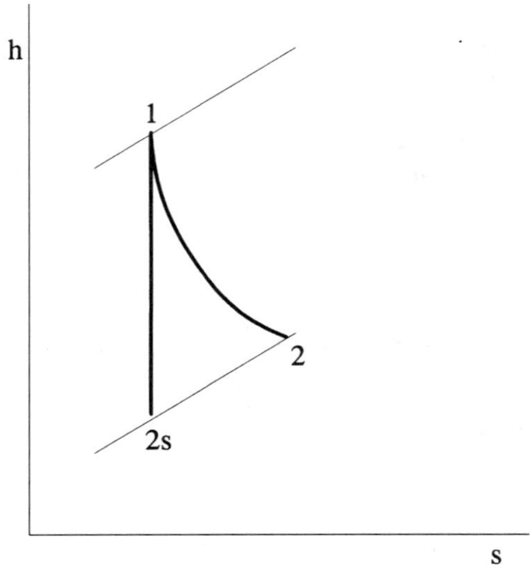

Figure 8-2 *h-s* diagram fore the turbine process.

second law turbine efficiency is

$$\varepsilon_t = \frac{w_t}{w_{\text{ava}}} = \frac{309}{361.2} = 0.86$$

Second, we take the condenser as a control volume and have $w_{\text{act}} = 0$ and

$$\begin{aligned} w_{\text{ava}} &= a_{f_2} - a_{f_3} = (h_2 - h_3) - T_o(s_2 - s_3) \\ &= (895.2 - 69.7) - 537(1.6032 - 0.1327) \\ &= 35.8 \text{ Btu/lbm} \end{aligned}$$

Applying the availability equation to the condenser, we have the availability destruction in the condenser process as $i = w_{\text{ava}} - w_{\text{act}} = 35.8 - 0 = 35.8$ Btu/lbm. For comparison, the heat loss from the condenser is calculated by

$$q_{\text{loss}} = h_2 - h_3 = 895.3 - 69.7 = 825.4 \text{ Btu/lbm}$$

Now we take the pump as a control volume and calculate the actual work and the available work in the following:

$$w_p = \frac{-v_3(P_4 - P_3)}{\eta_p} = \frac{-0.0161(320 - 1)(144)}{0.8(778)} = -1.19 \text{ Btu/lbm}$$

and

$$\begin{aligned} w_{\text{ava}} &= a_{f_3} - a_{f_4} = (h_3 - h_4) - T_o(s_3 - s_4) \\ &= (69.74 - 70.93) - 537(0.1327 - 0.1328) \\ &= -1.14 \text{ Btu/lbm} \end{aligned}$$

The availability destruction in the pump is $i = w_{ava} - w_p = -1.14 - (1.19) = 0.05$ Btu/lbm, and the pump second law efficiency is

$$\varepsilon_p = \frac{w_{ava}}{w_p} = \frac{1.14}{1.19} = 0.95$$

Finally, the water and steam in the boiler are taken as a control volume, and the heat and availability received by the steam and water are calculated. They are, respectively,

$$q_b = h_1 - h_4 = 1204.3 - 70.93 = 1133.4 \text{ Btu/lbm}$$

and

$$\begin{aligned} a_{q_b} &= (h_1 - h_4) - T_o(s_1 - s_4) \\ &= (1204.3 - 70.93) - 537(1.506 - 0.1328) \\ &= 395.96 \text{ Btu/lbm} \end{aligned}$$

The coal consumption is calculated with the boiler first law efficiency. That is,

$$m_{fuel} = \frac{q_b}{(\eta_b)(\text{LHV})} = \frac{1133.4}{0.82 \times 13200} = 0.1047 \text{ (lbm of coal/lbm of water)}$$

Similarly, the boiler second law efficiency is given by

$$\varepsilon_b = \frac{a_{q_b}}{(m_{fuel})(a_{fuel})} = \frac{395.95}{(0.1047)(13860)} = 0.2729 \text{ or } 27.29\%$$

The availability loss and destruction in the boiler are determined with the availability equation. It is the fuel availability minus the availability received by water and steam, that is,

$$i = (m_{fuel})(a_{fuel}) - a_{q_b} = (0.1047)(13860) - (395.96) = 1055.2 \text{ Btu/lbm}$$

With the above calculations, we can easily determine the steam turbine cycle efficiency.

$$\eta_{cy} = \frac{w_t - w_p}{q_b} = \frac{309 - 1.19}{1133.4} = 0.2716$$

for the first law cycle efficiency and

$$\varepsilon_{cy} = \frac{w_t - w_p}{a_{q_b}} = \frac{309 - 1.19}{395.96} = 0.7774$$

for the second law cycle efficiency. The overall system efficiency is simply the cycle efficiency multiplied by the boiler efficiency. Thus, we have the first law system efficiency as $\eta_o = \eta_{cy} \times \eta_b = 0.2716 \times 0.82 = 0.2227$, and the second law system efficiency is $\varepsilon_o = \varepsilon_{cy} \times \varepsilon_b = 0.7774 \times 0.2729 = 0.2122$.

Table 8-1 Heat and availability balance for a simple steam turbine system.

Characteristic	Heat balance (Btu/lbm)[a]	Availability balance (Btu/lbm)[a]
System input	1382.1	1451.1
Losses and destructions		
Steam turbine	0	52.20
Condenser	825.53	35.84
Pump	0	0.05
Boiler	248.76	1055.2
System output		
Steam turbine	309.0	309.0
Pump	−1.19	−1.19

*Based on one pound of steam in the cycle.

Table 8-1 presents the heat and availability balance for this simple steam turbine power plant. It is seen that the energy loss distribution by the first law analysis is quite different from the availability loss and destruction. Most availability loss and destruction take place in the boiler, and thus the boiler becomes the critical component in any power plant–improvement scheme. It is also of interest to note that the condenser availability loss is not substantial in comparison with other components in the power plant system. This conclusion is exactly opposite to that of the first law analysis.

Table 8-2 presents the second law efficiencies for all components. They are different from the first law efficiencies. It is of interest to note that the boiler has a relatively low second law efficiency. This is because the second law analysis takes into account not only the availability loss but also the availability destruction in the process.

The steam turbine system in the above example has a relatively low efficiency. To improve the system efficiency, engineers usually take the following approaches:

a. Increase the pressure and temperature of the steam from the boiler.
b. Reduce the turbine exhaust pressure.

Table 8-2 Component efficiencies for a simple steam turbine system.

Items	The first law efficiency (%)	The second law efficiency (%)
Steam turbine	85.0	86.0
Condenser	—	—
Pump	80.0	95.0
Boiler	82.0	27.29

c. Add one or two reheat processes to the system.
d. Increase the boiler feedwater temperature by using feedwater heaters.

Some discussions have been made about these approaches in Chapter 7. the following is a use of feedwater heaters in the steam turbine power plant system.

There are two basic types of feedwater heaters. The first is the open or direct-contact heater. In this heater the extraction steam is mixed directly with the incoming subcooled feedwater to produce saturated water at the steam pressure. The contact heater usually is followed by the feedwater pump by which the saturated water is transferred to the next heater or steam generator. In a power plant, the contact heater also performs as a deaerator.

The second type is the closed or surface heater. This feedwater heater is essentially a shell-and-tube heat exchanger in which feedwater passes through the tubes, and the extraction steam on the shell side transfers its energy to it. There will be no mixing between the steam and the feedwater. The extraction steam may be saturated or superheated depending on the conditions at the turbine extraction point. In the feedwater heater, steam condenses and leaves in the form of a liquid often called the *heater drain*. Frequently a drain cooler is combined with a feedwater heater in power plant design. As indicated by its name, the drain cooler is used to further reduce the temperature of the drain water below the saturation level. Figure 8-3 presents two heater drain disposal arrangements. In one arrangement the drain water is pumped back to the feedwater circuit and in another arrangement, it is cascaded to a low-pressure feedwater heater or condenser. The terms *heater terminal temperature difference* (TTD) and *drain cooler approach* (DCA) frequently are used in the calculation. They are defined as

$$\text{TTD} = T_{\text{sat}} - T_y \tag{8-7}$$

and

$$\text{DCA} = T_{\text{drain}} - T_x \tag{8-8}$$

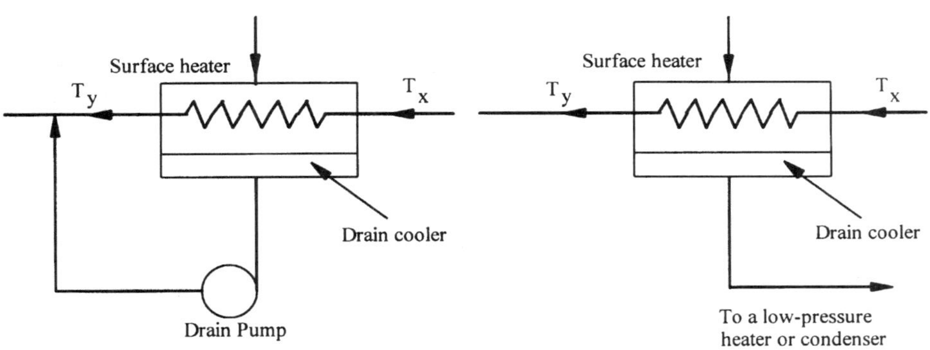

Figure 8-3 Two drain disposals.

where T_{sat} is the saturation temperature of the extraction steam, T_{drain} is the temperature of the drain leaving the heater, and T_x and T_y are the feedwater inlet and outlet temperatures, respectively. In power plant design, the terminal temperature difference ranges from −3 to 10°F, and the drain cooler approach from 10 to 20°F.

The heat transfer process in a contact heater or a surface heater is highly irreversible, and there will be substantial availability destruction in the process. (The availability calculations for heat exchangers in general have been presented in Chapter 4.) Because of this, the steam turbine system with feedwater heaters can be readily treated by the availability method. The following example is used to illustrate the application of the availability principle in a steam turbine system with two feedwater heaters.

Figure 8-4 presents a simplified steam turbine power plant. Steam enters the turbine at 1250 psia and 950°F and exhausts at 3.0 in. Hg abs. to the condenser. There are two extraction points at pressures of 100 psia and 25 psia. The high-pressure steam is used in the contact heater while the low-pressure steam is directed to the surface heater. Other conditions are specified below:

Turbine internal efficiency	90%
Pump efficiency	85%
Terminal temperature difference	10°F
Drain cooler approach	16°F
Steam flow rate at the turbine inlet	1.3×10^6 lbm/hr

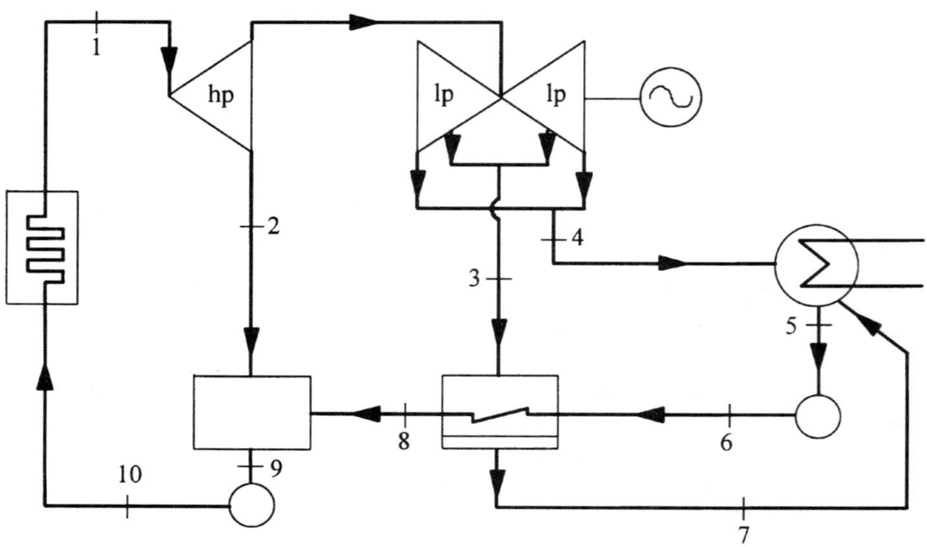

Figure 8-4 A regenerative steam turbine system.

For simplicity, the pressure drop in the steam turbine system is presently omitted from the consideration. The first law calculation usually consists of the following items:

1. the thermodynamic properties at various locations,
2. the extraction steam flows for feedwater heating,
3. the turbine output and the pump work, and
4. the heat received by the cycle and the cycle efficiency.

The detailed calculation is left as an exercise. Table 8-3 summarizes the thermodynamic properties and flow rates at various locations. Other results are expressed below:

Turbine work	594.5×10^6 Btu/hr
Pump work	6.17×10^6 Btu/hr
Heat received by the cycle	1515×10^6 Btu/hr
Cycle efficiency	38.8%

The availability analysis, as previously indicated, involves the calculation of the second law efficiency as well as the availability loss and destruction. It may be of interest to show the calculation in terms of availability input and availability output for each component. The difference between the input and output is the availability loss and destruction. The second law efficiency is defined simply as the ratio of the availability output to the availability input. First we examine the components of the steam turbine using the dead state conditions 14.7 psia and 77°F.

Table 8-3 Properties and flow rates of the power plant in Figure 8-4.

Location	Pressure (psia)	Temperature (°F)	Enthalpy (Btu/lbm)	Entropy (Btu/lbm-R)	Flow rate (10^3 lbm/hr)
1	1250	950	1468.8	1.6036	1300.0
2	100	397.4	1226.0	1.6501	127.5
3	25	240	1130.0	1.6703	130.5
4	1.474	115	970.4	1.7001	1042.0
5	1.474	115	83.0	0.1560	1172.5
6	100	115.4	83.35	0.1566	1172.5
7	25	131.4	99.3	0.1841	130.5
8	100	230.0	198.0	0.3388	1172.5
9	100	327.8	298.8	0.4744	1300.0
10	1250	328.6	303.3	0.4754	1300.0

Steam Turbine

The turbine is assumed to be well insulated, and the availability equation for the turbine process is given by

$$\dot{I} = \dot{m}_1 a_{f_1} - \dot{m}_2 a_{f_2} - \dot{m}_3 a_{f_3} - \dot{m}_4 a_{f_4} - \dot{W}_t$$

In terms of availability input and output, we have $\dot{A}_{\text{loss}} = \dot{A}_{\text{input}} - \dot{A}_{\text{output}}$, where $\dot{A}_{\text{input}} = \dot{m}_1 a_{f_1} - \dot{m}_2 a_{f_2} - \dot{m}_3 a_{f_3} - \dot{m}_4 a_{f_4}$ and $\dot{A}_{\text{output}} = \dot{W}_t$. In words, the availability input in the turbine process is the decrease of flow availability, and the availability output is simply the mechanical work produced by the turbine. For convenience, the availability input is changed to

$$\dot{A}_{\text{input}} = \dot{m}_2(a_{f_1} - a_{f_2}) + \dot{m}_3(a_{f_1} - a_{f_3}) + \dot{m}_4(a_{f_1} - a_{f_4})$$

or

$$\dot{A}_{\text{input}} = \dot{m}_2[(h_1 - h_2) - T_o(s_1 - s_2)] + \dot{m}_3[(h_1 - h_3) - T_o(s_1 - s_3)] + \dot{m}_4[(h_1 - h_4) - T_o(s_1 - s_4)]$$

With the numeric values, we have the availability input to the turbine process as

$$\begin{aligned}\dot{A}_{\text{input}} &= 127.5 \times 10^3[(1468.8 - 1226) - 537(1.6036 - 1.6501)] \\ &\quad + 130.5 \times 10^3[(1468.8 - 1130) - 537(1.6036 - 1.6703)] \\ &\quad + 1042 \times 10^3[(1468.8 - 970.4) - 537(1.6036 - 1.7001)] \\ &= 656.36 \times 10^6 \text{ Btu/hr}\end{aligned}$$

Since $\dot{A}_{\text{output}} = 594.5 \times 10^6$ Btu/hr, the availability loss and destruction is $\dot{A}_{\text{loss}} = \dot{A}_{\text{input}} - \dot{A}_{\text{output}} = (656.36 - 594.5) \times 10^6 = 61.86 \times 10^6$ Btu/hr, and the second law efficiency of the steam turbine is

$$\varepsilon = \frac{\dot{A}_{\text{output}}}{\dot{A}_{\text{input}}} = \frac{594.5 \times 10^6}{656.36 \times 10^6} = 0.9058$$

Feedwater Heaters

We take heater #1 (counted from the condenser side) as a control volume. When the heat loss from the control volume is negligible, the availability equation becomes

$$\dot{I} = \dot{m}_3 a_{f_3} + \dot{m}_6 a_{f_6} - m_7 a_{f_7} - \dot{m}_8 a_{f_8}$$

or

$$\dot{I} = \dot{m}_3(a_{f_3} - a_{f_7}) - \dot{m}_6(a_{f_8} - a_{f_6})$$

The first term on the right-hand side of the latter equation is the decrease of flow availability for the steam and thus the availability input for the process. The second term on the right-hand side is the increase of flow availability for the feedwater and so becomes the availability output for the process. Numerically, they are

$$\begin{aligned}\dot{A}_{\text{input}} &= \dot{m}_3(a_{f_3} - a_{f_7}) = \dot{m}_3[(h_3 - h_7) - T_o(s_3 - s_7)] \\ &= 130.5 \times 10^3[(1130 - 99.3) - 537(1.6703 - 0.1840)] \\ &= 30.36 \times 10^6 \text{ Btu/hr}\end{aligned}$$

and

$$\begin{aligned}\dot{A}_{\text{output}} &= \dot{m}_6(a_{f_8} - a_{f_6}) = \dot{m}_6[(h_8 - h_6) - T_o(s_8 - s_6)] \\ &= 1172.5 \times 10^3[(198 - 83.3) - 537(0.3388 - 0.1567)] \\ &= 19.83 \times 10^6 \text{ Btu/hr}\end{aligned}$$

For heater #1, the availability loss and destruction is

$$\dot{A}_{\text{loss}} = \dot{A}_{\text{input}} - \dot{A}_{\text{output}} = (30.36 - 19.83) \times 10^6 = 10.53 \times 10^6 \text{ Btu/hr}$$

and the second law efficiency is

$$\varepsilon = \frac{\dot{A}_{\text{output}}}{\dot{A}_{\text{input}}} = \frac{19.83 \times 10^6}{30.36 \times 10^6} = 0.6532$$

Similarly, for heater #2, we have the availability input and output as

$$\begin{aligned}\dot{A}_{\text{input}} &= \dot{m}_2(a_{f_2} - a_{f_9}) = \dot{m}_2[(h_2 - h_9) - T_o(s_2 - s_9)] \\ &= 127.5 \times 10^3[(1226 - 298.8) - 537(1.6501 - 0.4744)] \\ &= 37.72 \times 10^6 \text{ Btu/hr}\end{aligned}$$

and

$$\begin{aligned}\dot{A}_{\text{output}} &= \dot{m}_8(a_{f_9} - a_{f_8}) = \dot{m}_8[(h_9 - h_8) - T_o(s_9 - s_8)] \\ &= 1172.5 \times 10^3[(298.8 - 198) - 537(0.4744 - 0.3388)] \\ &= 32.81 \times 10^6 \text{ Btu/hr}\end{aligned}$$

For heater #2, the availability loss and destruction is $\dot{A}_{\text{loss}} = \dot{A}_{\text{input}} - \dot{A}_{\text{output}} = 37.72 \times 10^6 - 32.81 \times 10^6 = 4.91 \times 10^6$ Btu/hr, and the second law efficiency is

$$\varepsilon = \frac{\dot{A}_{\text{output}}}{\dot{A}_{\text{input}}} = \frac{32.81 \times 10^6}{37.72 \times 10^6} = 0.8698$$

Condenser

We take the condensing steam and the drain from heater #1 as a control volume. The control volume loses heat and also the associated heat availability. Because of this, the availability output for this process must be $\dot{A}_{\text{output}} = 0$, and the availability input is

$$\begin{aligned}
\dot{A}_{\text{input}} &= \dot{m}_4(a_{f_4} - a_{f_5}) + \dot{m}_7(a_{f_7} - a_{f_5}) \\
&= \dot{m}_4[(h_4 - h_5) - T_o(s_4 - s_5)] + \dot{m}_7[(h_7 - h_5) - T_o(s_7 - s_5)] \\
&= 1042 \times 10^3[(970.4 - 83) - 537(1.7001 - 0.1560)] \\
&\quad + 130.5 \times 10^3[(99.3 - 83) - 537(0.1841 - 0.1560)] \\
&= 60.82 \times 10^6 \text{ Btu/hr}
\end{aligned}$$

For the condenser the availability loss and destruction is numerically equal to that of the availability input, and therefore the second law efficiency is zero.

Pumps

The availability input for a pump is the mechanical work it receives, and the availability output is simply the increase of flow availability in the water. Numerically, for pump #1

$$\begin{aligned}
\dot{A}_{\text{input}} &= \dot{W}_{p_1} = \dot{m}_5(h_6 - h_5) = 1172.5 \times 10^3(83.35 - 83) \\
&= 410.4 \times 10^3 \text{ Btu/hr}
\end{aligned}$$

and

$$\begin{aligned}
\dot{A}_{\text{output}} &= \dot{m}_5(a_{f_6} - a_{f_5}) = \dot{m}_5[(h_6 - h_5) - T_o(s_6 - s_5)] \\
&= 1172.5 \times 10^3[(83.35 - 83) - 537(0.1566 - 0.1560)] \\
&= 32.6 \times 10^3 \text{ Btu/hr}
\end{aligned}$$

Similarly, for pump #2

$$\dot{A}_{\text{input}} = \dot{m}_9(h_{10} - h_9) = 1300 \times 10^3(303.3 - 298.8) = 5850 \times 10^3 \text{ Btu/hr}$$

and

$$\begin{aligned}
\dot{A}_{\text{output}} &= \dot{m}_9(a_{f_{10}} - a_{f_9}) = \dot{m}_9[(h_{10} - h_9) - T_o(s_{10} - s_9)] \\
&= 1300 \times 10^3[(303.3 - 298.8) - 537(0.4754 - 0.4744)] \\
&= 5152 \times 10^3 \text{ Btu/hr}
\end{aligned}$$

Finally, the availability loss and destruction and the second law efficiency are

$$\dot{A}_{\text{loss}} = (410.4 - 32.6) \times 10^3 = 377.8 \times 10^3 \text{ Btu/hr}$$

$$\varepsilon = \frac{32.6 \times 10^3}{410.4 \times 10^3} = 0.0794$$

for pump # 1 and

$$\dot{A}_{\text{loss}} = (5850 - 5152) \times 10^3 = 698 \times 10^3 \text{ Btu/hr}$$

$$\varepsilon = \frac{5152 \times 10^3}{5850 \times 10^3} = 0.8807$$

for pump #2.

Boiler

We take the water and steam in the boiler as a control volume. The increase of flow availability in the substance is the input to the cycle. Numerically the input is given by

$$\begin{aligned}\dot{A}_{\text{input}} &= \dot{m}_1(a_{f_1} - a_{f_{10}}) = \dot{m}_1[(h_1 - h_{10}) - T_o(s_1 - s_{10})] \\ &= 1300 \times 10^3[(1468.8 - 303.3) - 537(1.6036 - 0.4754)] \\ &= 727.55 \times 10^6 \text{ Btu/hr}\end{aligned}$$

In summary, we present the availability balance of this regenerative steam turbine cycle in Table 8-4. It is seen that the availability input to the cycle is quite close to the availability output plus the availability loss and destruction within the cycle. The difference evidently is due to the truncation error in numeric calculations. The cycle efficiency is simply the cycle output over the cycle input. For this case, the second law cycle efficiency is approximately 81 percent.

Table 8-4 Availability Balance for a Regenerative Steam Turbine Cycle.

	Availability Flow (10^6 Btu/hr)	
Cycle availability input	727.55	
Availability received by water		727.55
Availability loss and destruction	139.20	
Steam turbine		61.86
Condenser		60.82
Heater #1		10.53
Heater #2		4.91
Pump #1		0.38
Pump #2		0.70
Cycle availability output	588.24	
Turbine		594.5
Pump #1		−0.41
Pump #2		−5.85

8-2 AVAILABILITY ANALYSIS OF A COAL-FIRED POWER PLANT

Modern steam turbine power plants are much more complicated than those shown in Figs. 8-1 and 8-4. To raise the plant efficiency, engineers frequently use superheated steam instead of saturated steam. In addition, they also include reheat as well as regenerative processes in the power plant system. The regeneration process frequently is implemented by using various types of feedwater heaters. Figure 8-5 presents the flow diagram for a modern coal-fired power plant. The plant has one stage of steam reheat and seven stages of regenerative feedwater heating. The steam conditions are supercritical, with an initial pressure of 3515 psia and initial and reheat temperatures of 1000°F. The condenser pressure in this cycle arrangement is set equal to 2.5 in Hg abs. As indicated in Fig. 8-5, the feedwater pump is driven by an auxiliary steam turbine, and the air preheater is supplied with the extracted steam from the main turbine. Omitting some minor cycle losses such as gland leakages, the turbine-generator system has a net output of 726,171 kW and a turbine net heat rate (NHR) of 7823 Btu/kW · hr. The equivalent cycle efficiency is given by

$$\eta_{cy} = \frac{3413}{\text{NHR}} = \frac{3413}{7823} = 0.436$$

It should be pointed out that the above calculation is based on the assumption of 98.5 percent efficiency for the main generator. With a boiler efficiency of 90.7 percent, we have an overall system efficiency of $\eta_o = \eta_b \times \eta_{cy} = 0.907 \times 0.436 = 0.395$.

The following is the availability analysis for the same coal-fired power plant system. First, we calculate the steam and water unit availabilities at various locations. The results are shown in Fig. 8-6. The environmental state conditions are 14.7 psia and 77°F. For convenience, we present the availability analyses for each major component and then for the entire power plant system.

Steam Generator

The coal consumption is 475,000 lb/hr by the first law calculation. When the coal unit availability is assumed to be 13,860 Btu/lb, the fuel availability supplied to the steam generator is given by

$$\dot{A}_{\text{fuel}} = \dot{m}_{\text{fuel}} a_{\text{fuel}} = 475{,}000 \times 13{,}860 = 6583.5 \times 10^6 \text{ Btu/hr}$$

Since the steam is used to heat a portion of the air supply, the availability of the steam must be taken into consideration. It is given by

$$\dot{A}_{\text{steam}} = \dot{m}_{\text{steam}} \times \Delta a_{f_{\text{steam}}} = 103{,}400(416.04 - 57.94) = 37.03 \times 10^6 \text{ Btu/hr}$$

Then the availability input to the steam generator (including the air preheater) is

$$\dot{A}_{\text{input}} = \dot{A}_{\text{fuel}} + \dot{A}_{\text{steam}} = 6583.5 \times 10^6 + 37.03 \times 10^6 = 6620.5 \times 10^6 \text{ Btu/hr}$$

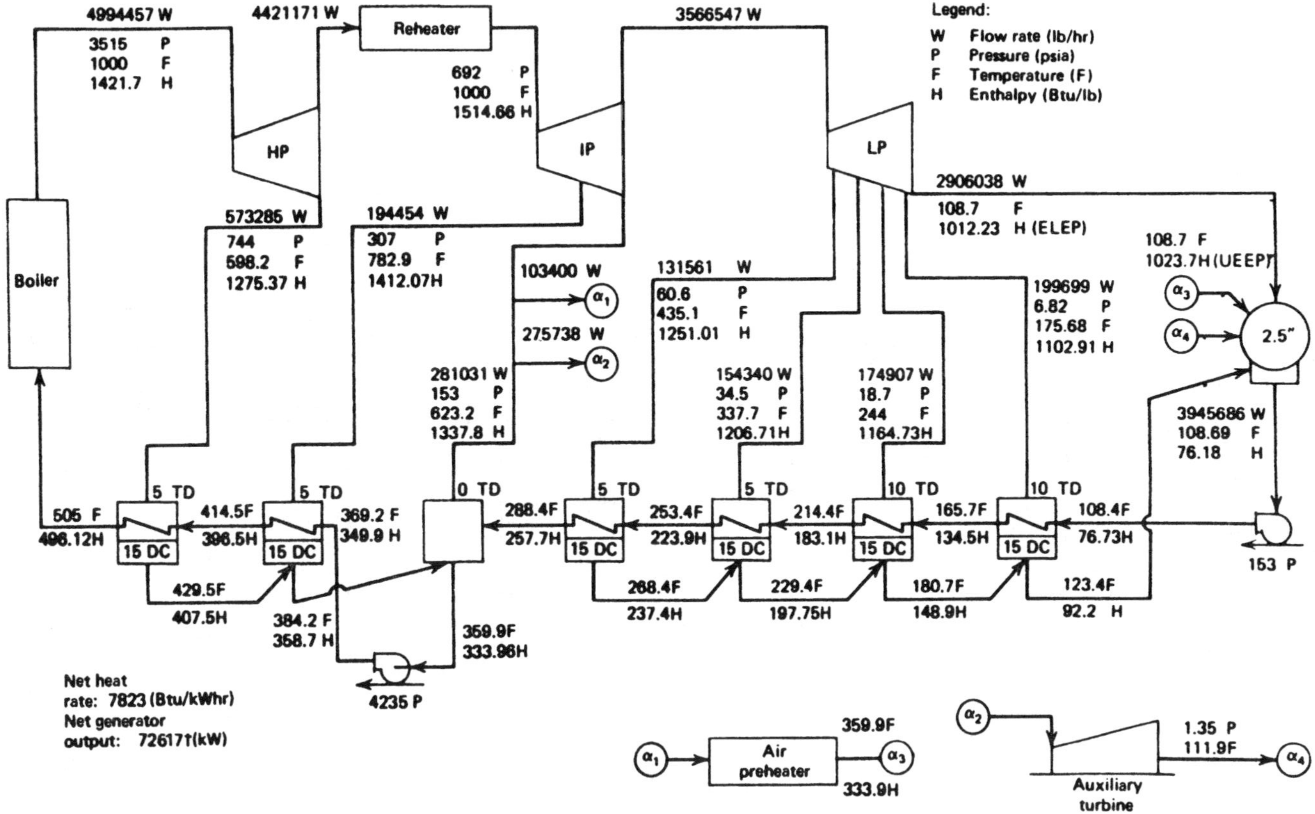

Figure 8-5 Heat balance diagram for a conventional power plant.

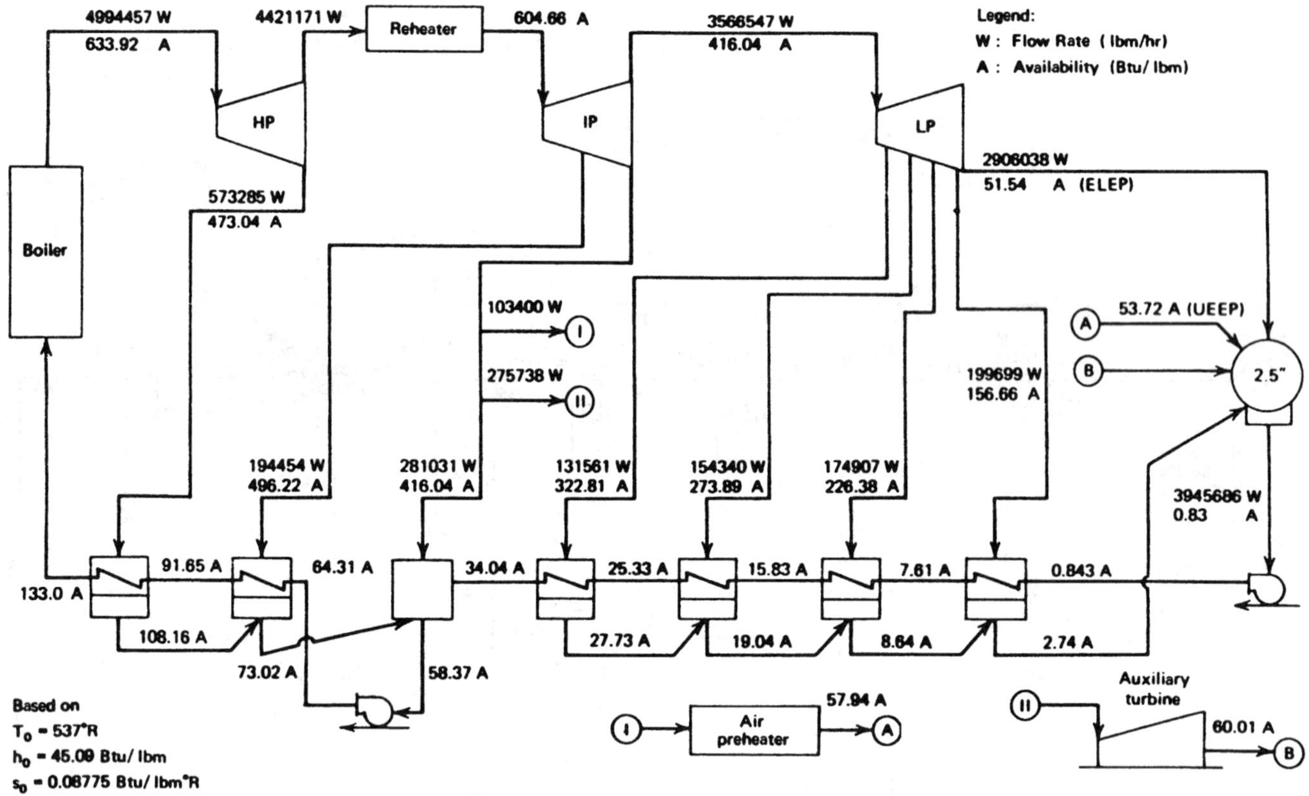

Figure 8-6 Steam availability of a conventional power plant.

The availability output of steam generator is the availability gained by water in the boiler as well as by the reheater. Numerically, it is

$$\begin{aligned}\dot{A}_{\text{output}} &= 4{,}994{,}456(633.92 - 133.0) + 4{,}421{,}171(604.66 - 473.04)\\ &= 3083.7 \times 10^6 \text{ Btu/hr}\end{aligned}$$

The availability loss and destruction in the steam generator are, by definition, the difference between the availability input and output. The availability loss and destruction for this case is $\dot{A}_{\text{loss}} = 3536.8 \times 10^6$ Btu/hr. The second law efficiency of the steam generator is calculated and given by $\varepsilon = 3083.7 \times 10^6/6620.5 \times 10^6 = 0.4658$.

Steam Turbine

The turbine consists of three turbine cylinders (high pressure, intermediate pressure, and low pressure). For each turbine cylinder, we can determine the availability input, output, and loss and destruction. For instance, the availability input for the high-pressure turbine is

$$\dot{A}_{\text{input}} = \Sigma \dot{m}_i a_i - \Sigma \dot{m}_e a_e$$

The subscripts i and e denote turbine inlet and exit, respectively. Using the numeric values in Fig. 8-6, we have

$$\begin{aligned}\dot{A}_{\text{input}} &= (4{,}994{,}456)(633.92) - (573{,}285)(473.04) - (4{,}421{,}171)(473.04)\\ &= 803.5 \times 10^6 \text{ Btu/hr}\end{aligned}$$

The availability output $\dot{A}_{\text{output}}$ for the high-pressure turbine is simply the turbine work. That is,

$$\begin{aligned}\dot{A}_{\text{output}} &= \Sigma \dot{m}_i h_i - \Sigma \dot{m}_e h_e\\ &= (4{,}994{,}457)(1421.7) - (573{,}285)(1275.4) - (4{,}421{,}171)(1275.4)\\ &= 730.8 \times 10^6 \text{ Btu/hr}\end{aligned}$$

The process availability loss and destruction $\dot{A}_{\text{loss}}$ is defined as the difference between A_{input} and A_{output}. For the high-pressure turbine, the availability loss and destruction is

$$\dot{A}_{\text{loss}} = \dot{A}_{\text{input}} - \dot{A}_{\text{output}} = 803.5 \times 10^6 - 730.8 \times 10^6 = 72.7 \times 10^6 \text{ Btu/hr}$$

Finally, we have the second law efficiency for the high pressure turbine as $\varepsilon = 730.8 \times 10^6/803.5 \times 10^6 = 0.9095$.

We complete calculations for the intermediate-pressure and low-pressure turbine similarly. The results are presented in Table 8-5.

Table 8-5 Second law analysis of major components in power plant.

Item description of component	$\dot{A}_{input}$ (10^6 Btu/hr)	$\dot{A}_{output}$ (10^6 Btu/hr)	$\dot{A}_{loss}$ (10^6 Btu/hr)	Effectiveness (%)
1. Steam generator	6620.5	3083.7	3536.8	46.58
2. HP turbine	803.5	730.8	72.7	90.95
3. IP turbine	817.8	767.5	50.30	93.85
4. LP turbine	1172.1	1021.6	150.50	87.16
5. Electric generator	2519.9	2482.1	37.80	98.50
6. Feedwater train	770.58	719.881	50.70	93.40
a. Heater 1	31.30	24.53	6.77	78.40
b. Heater 2	41.06	32.45	8.61	79.03
c. Heater 3	40.48	37.49	2.99	92.61
d. Heater 4	38.84	34.36	4.48	88.47
e. Heater 5	303.40	291.50	11.90	96.08
f. Heater 6	106.30	93.05	13.25	87.54
g. Heater 7	209.20	206.50	2.70	98.71
7. Condenser	177.19	—	177.19	—
8. Air preheater	37.03	—	—	—
9. Auxiliary turbine	98.14	79.52	18.62	81.03
10. Feedwater pump	79.61	73.17	6.44	91.91

Note: The calculations were made using a coal consumption of 475,000 lbm/hr and coal unit availability of 13,860 Btu/lbm.

Electric Generator

The electric generator has an efficiency of 98.5 percent. The output of the electric generator is equal to the product of its input and efficiency. That is,

$$A_{output} = A_{input} \times \eta_{gen} = (730.8 + 767.5 + 1021.6) \times 10^6 \times 0.985$$
$$= 2482.1 \times 10^6 \text{ Btu/hr}$$

The availability loss and destruction are simply the generator input minus the generator output:

$$A_{loss} = A_{input} - A_{output} = (2519.9 - 2482.1) \times 10^6 = 37.8 \times 10^6 \text{ Btu/hr}$$

Finally, the second law efficiency is $\varepsilon = A_{output}/A_{input} = 2482.1 \times 10^6/2519.9 \times 10^6 = 0.985$.

Feedwater Heater

Figure 8-7 presents a schematic diagram of a typical feedwater heater. Heat is transferred from the condensing steam to the feedwater. In the heat exchanger process, energy (heat) is conserved. Examining the process in terms of thermodynamic availability, however, the conservation principle is not valid. It is expected that the availability gained by the feedwater is somewhat less than that given up by the steam (or hot substance in general). The difference is defined as

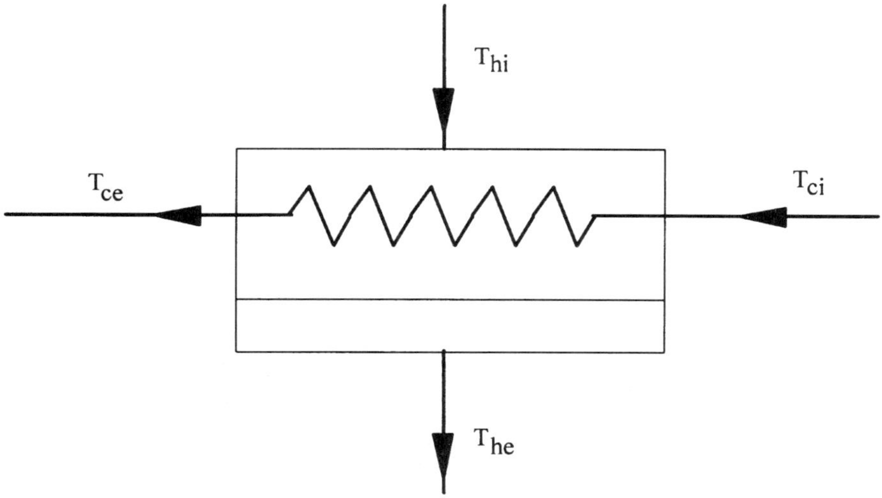

Figure 8-7 A schematic diagram for a feedwater heater.

the availability loss and destruction of the heat exchanger. The ratio of these two is the second law efficiency of heat exchanger.

The availability input to the heat exchanger, $\dot{A}_{\text{input}}$, is the availability reduction in the hot substances (such as steam and hot water). The availability output $\dot{A}_{\text{output}}$ is the availability gain of the feedwater. The expressions for both terms are

$$\dot{A}_{\text{input}} = \Sigma \dot{m}_{h_i} a_{h_i} - \Sigma \dot{m}_{h_e} a_{h_e}$$

and

$$\dot{A}_{\text{output}} = \Sigma \dot{m}_{c_e} a_{c_e} - \Sigma \dot{m}_{c_i} a_{c_i}$$

Using feedwater heater #7 as an example, we have

$$\dot{A}_{\text{input}} = 573{,}285(473.04 - 108.16) = 209.2 \times 10^6 \text{ Btu/hr}$$

and

$$\dot{A}_{\text{output}} = 4{,}994{,}457(133.0 - 91.65) = 206.5 \times 10^6 \text{ Btu/hr}$$

From these terms, we get $\dot{A}_{\text{loss}} = 209.2 \times 10^6 - 206.5 \times 10^6 = 2.7 \times 10^6$ Btu/hr and $\varepsilon = 206.5 \times 10^6 / 209.2 \times 10^6 = 0.9871$.

The results for other feedwater heaters are presented in Table 8-5.

Condenser

The calculation for the condenser is similar to that for the feedwater heaters. The hot substances in a condenser are the condensing steam and the hot water returned from the heaters and other sources. The reduction of availability in these substances is equal to the total loss of availability in the condensers. This approximation is acceptable because the availability gained by the cooling water is not utilized and eventually wasted. The numerical results are given in Table 8-5.

Miscellaneous Equipment

The equipment in this group includes the feedwater pumps, the auxiliary steam turbine, and the air preheater. The pumping process is the reverse of the turbine process, and the equations for pumps are identical to those for turbines. In the availability balance, the feedwater pumps have a second law efficiency approximately equal to 92 percent and an availability loss and destruction of 6.44×10^6 Btu/hr. The condensate pumps are relatively small and usually are omitted in the cycle analysis. As for the air preheater, the availability input is simply the product of the steam flow rate and the unit availability decrease. That is, $\dot{A}_{\text{input}} = 103{,}400(416.04 - 57.94) = 37.03 \times 10^6$ Btu/hr.

In this cycle analysis the availability loss and destruction for the air preheater is treated as a part of steam generator loss. Because of this, a separate calculation for the air preheater is not included. As for the auxiliary steam turbine, the procedure is identical to that of the main steam turbine. The numerical results are presented in Table 8-5.

Steam Turbine Cycle

Before the entire power plant system is considered, it may be of interest to have an availability balance of the steam turbine cycle. The turbine cycle consists of three turbine cylinders, seven feedwater heaters, condensers, pumps, an auxiliary turbine, and other associated equipment. The availability input to the turbine cycle is the availability gained by the water and steam in the steam generator. Numerically, it must be equal to the output of steam generator. As indicated in the previous calculation, it is $\dot{A}_{\text{input}} = 3083.7 \times 10^6$ Btu/hr. The output of the steam turbine cycle is by definition equal to the turbine work plus the steam availability used for the air preheater. That is, $\dot{A}_{\text{output}} = 2482.1 \times 10^6 + 37.03 \times 10^6 = 2519.1 \times 10^6$ Btu/hr. Therefore, the second law efficiency and availability loss and destruction of the turbine cycle must be $\varepsilon_{\text{cy}} = 2519.1 \times 10^6/3083.7 \times 10^6 = 0.817$ and $\dot{A}_{\text{loss}} = 3083.7 \times 10^6 - 2519.1 \times 10^6 = 564.6 \times 10^6$ Btu/hr.

The availability loss and destruction in the steam turbine cycle occurs mainly at locations such as turbines, pumps, and heat exchangers (condensers and feedwater heaters). As previously indicated, the availability loss and destruction is due to either internal friction or heat transfer through a finite temperature difference.

The Entire Power Plant System

Finally, the entire coal-fired power plant is taken as a system. In this case the plant output is the turbine net output 2482.1×10^6 Btu/hr, if we neglect the plant auxiliary power and other losses. The plant input is simply the fuel availability, which is equal to 6583.5×10^6 Btu/hr. The ratio, which is the second law efficiency, is

$$\varepsilon_o = \frac{2482.1 \times 10^6}{6583.5 \times 10^6} = 0.377$$

Table 8-6 presents an availability balance for the entire power plant system. It includes the availability input and output to the plant. Also, the availability loss and destruction for each major component is identified. It is seen that the steam generator is the least efficient and that approximately 54 percent of the availability input is either lost or destroyed in this equipment. This conclusion is completely in contrast to the relatively high first law efficiency of the steam generator.

8-3 GAS TURBINE SYSTEMS

The gas turbine system is another electric-power generating system. Because of its relatively simple arrangement and low capital cost, the gas turbine system often is used for peak power and emergency power generation. Figure 8-8 presents a schematic diagram for a gas turbine system. It is seen that the system consists of a compressor, burner, and turbogenerator. Air is first compressed and then mixed with fuel and burned in the burner. After leaving the burner, the products of combustion enter the turbine (sometimes referred to as the

Table 8-6 Availability balance for a coal-fired power plant.

Characteristic	Availability flow rate (10^6 Btu/hr)	Percent of availability input
Plant input	6583.5	—
Loss and destruction		
Boiler	3536.8	53.72
Turbine	273.5	4.16
Electric generator	37.8	0.58
Heater	50.7	0.77
Condenser	177.19	2.69
Miscellaneous	25.06	0.38
Plant output	2482.1	37.70

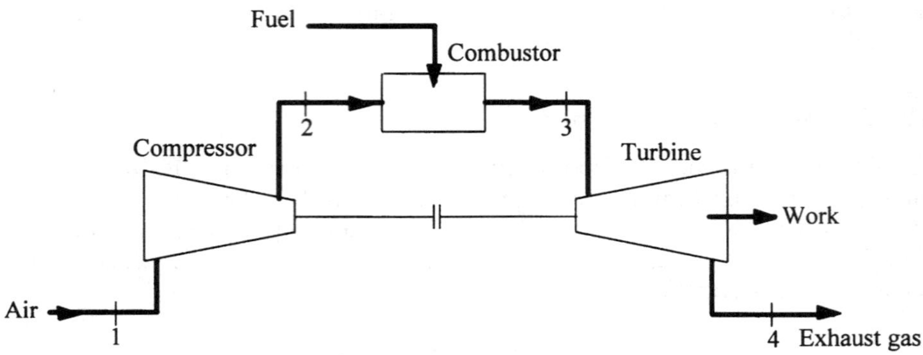

Figure 8-8 A schematic diagram for a simple gas turbine system.

expander) and produce mechanical work. Some of the turbine output is used to drive the compressor, and the remainder is delivered to meet the outside power demand. Because of the combustion process, there is a change in the composition of the working substance. For simplicity, the combustion process sometimes is approximated by a heat transfer process in the cycle analysis. In other words, the air simply receives heat energy from an external source and has its temperature increased to a desired level. Like the steam turbine system, the gas turbine system has a cycle efficiency defined by

$$\eta_{cy} = \frac{w_{gt} - w_c}{q_b} \tag{8-9}$$

When the burner efficiency is defined as

$$\eta_b = \frac{q_b}{m_{fuel}(LHV)} \tag{8-10}$$

the overall system efficiency becomes

$$\eta_o = \eta_{cy} \times \eta_b \tag{8-11}$$

In terms of availability, the second law cycle and burner efficiencies are respectively

$$\varepsilon_{cy} = \frac{w_{gt} - w_c}{a_{q_b}} \tag{8-12}$$

and

$$\varepsilon_b = \frac{a_{q_b}}{(m_{fuel})(a_{fuel})} \tag{8-13}$$

Then, the second law overall system efficiency is

$$\varepsilon_o = \varepsilon_{cy} \times \varepsilon_b \tag{8-14}$$

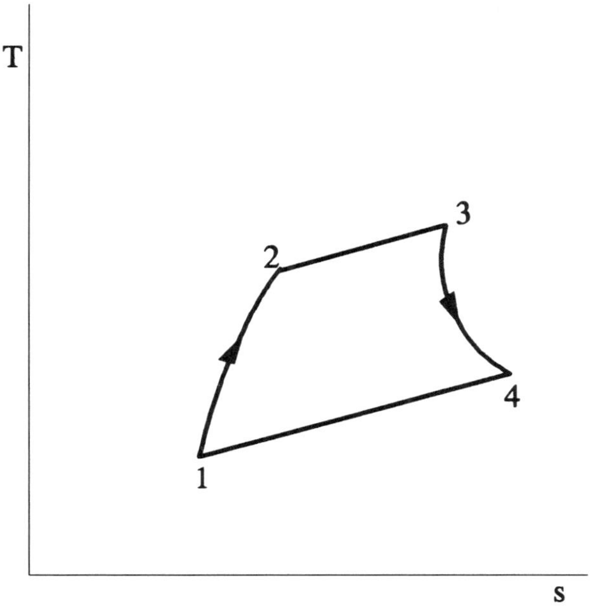

Figure 8-9 *T-s* diagram for a simple gas turbine system.

To demonstrate the use of the availability method, we examine the gas turbine system as shown in Fig. 8-8. The system has the following conditions:

Ambient air conditions	14.7 psia, 60°F
Compressor pressure ratio	10.2
Turbine inlet temperature	2000°F
Compressor efficiency	85%
Turbine efficiency	88%
Burner efficiency	100%

In this illustrative example, the combustion process is approximated by a heat transfer process, and there is no change in composition across the burner. In this availability analysis the diffusion component simply is omitted, and the conditions in the environmental state are assumed to be 14.7 psia and 60°F.

The *T-s* diagram for the gas turbine system is shown in Fig. 8-9. To calculate the compressor exit temperature T_2, we make use of the isentropic relationship and the compressor efficiency. Similarly, the turbine efficiency is utilized to find the turbine exit temperature T_4. The temperature and pressure at various locations are calculated and summarized in Table 8-7.

Next we take an availability balance for each component and then for the entire gas turbine system.

Table 8-7 Characteristics of the gas turbine system in Figure 8-9.

Location	Pressure (psia)	Temperature (°F)
1	14.7	60
2	150.0	636.4
3	150.0	2000
4	14.7	950

Compressor

The availability input is the compressor work, which is

$$A_{\text{input}} = -w_c = c_p(T_2 - T_1) = 0.25(1096.4 - 520) = 144.1 \text{ Btu/lbm}$$

The availability output is the increase of flow availability in the air stream. That is,

$$\begin{aligned} A_{\text{output}} &= a_{f_2} - a_{f_1} = c_p(T_2 - T_1) - T_o\left(c_p \ln\frac{T_2}{T_1} - R\ln\frac{P_2}{P_1}\right) \\ &= 0.25(1096.4 - 520) - 520\left(0.25\ln\frac{1096.4}{520} - \frac{53.3}{778}\ln\frac{150}{14.7}\right) \\ &= 129.9 \text{ Btu/lbm} \end{aligned}$$

The availability loss and destruction in the compressor is the difference between the availability input and output. That is, $A_{\text{loss}} = A_{\text{input}} - A_{\text{output}} = 144.1 - 129.9 = 14.2$ Btu/lbm. The second law efficiency of compressor is $\varepsilon_c = A_{\text{output}}/A_{\text{input}} = 129.9/144.1 = 0.902$.

Burner

The thermal energy received by the substance is given by

$$q_b = c_p(T_3 - T_2) = 0.25(2460 - 1096.4) = 340.9 \text{ Btu/lbm}$$

Because the burner is 100 percent efficient, the energy supplied to the burner must be equal to the energy received by the air, and so $m_{\text{fuel}}(\text{LHV}) = 340.9$ Btu/lbm.

In terms of availability, the fuel availability input to the burner is given by $A_{\text{input}} = m_{\text{fuel}}(a_{\text{fuel}})$. When the fuel unit availability is assumed to be equal to the fuel heating value, the availability input to the burner is numerically equal to the energy input. In this case, the energy input is 340.9 Btu/lbm. The availability output from the burner is given simply by

$$A_{\text{output}} = a_{f_3} - a_{f_2} = c_p(T_3 - T_2) - T_o\left(c_p \ln\frac{T_3}{T_2} - R\ln\frac{P_3}{P_2}\right)$$

$$= 0.25(2460 - 1096.4) - 520\left(0.25\ln\frac{2460}{1096.4}\right)$$

$$= 235.8 \text{ Btu/lbm}$$

Thus, the availability loss and destruction in the burner is $A_{\text{loss}} = A_{\text{input}} - A_{\text{output}} = 340.9 - 235.8 = 105.1$ Btu/lbm, and the second law efficiency of the burner is $\varepsilon_b = A_{\text{output}}/A_{\text{input}} = 235.8/340.9 = 0.692$.

Turbine

The availability input to the turbine is the availability decrease in the air stream. That is,

$$A_{\text{input}} = a_{f_3} - a_{f_4} = c_p(T_3 - T_4) - T_o\left(c_p \ln\frac{T_3}{T_4} - R\ln\frac{P_3}{P_4}\right)$$

$$= 0.25(2460 - 1410) - 520\left(0.25\ln\frac{2460}{1410} - \frac{53.3}{778}\ln\frac{150}{14.7}\right)$$

$$= 272.9 \text{ Btu/lbm}$$

The availability output is simply the turbine work, which is

$$A_{\text{output}} = c_p(T_3 - T_4) = 0.25(2460 - 1410) = 262.5 \text{ Btu/lbm}$$

Similarly, the availability loss and destruction and second law efficiency for the turbine process are respectively $A_{\text{loss}} = 10.4$ Btu/lbm and $\varepsilon_{\text{gt}} = 0.962$.

Exhaust Loss

The exhaust gas from the gas turbine system is not utilized. Because of this, the energy and availability of the exhaust gas will be treated as a loss. The exhaust energy loss is given by

$$q_{\text{exhaust}} = c_p(T_4 - T_1) = 0.25(1410 - 520) = 222.5 \text{ Btu/lbm}$$

and the exhaust availability loss is

$$a_{f_{\text{exhaust}}} = a_{f_4} - a_{f_1} = c_p(T_4 - T_1) - T_o\left(c_p \ln\frac{T_4}{T_1} - R\ln\frac{P_4}{P_1}\right)$$

$$= 0.25(1410 - 520) - 520\left(0.25\ln\frac{1410}{520}\right)$$

$$= 92.8 \text{ Btu/lbm}$$

Table 8-8 Energy and availability balance for a simple gas turbine system.

Characteristic	Energy balance (Btu/lbm)	Availability balance (Btu/lbm)
System input	340.9	340.9
Losses		
Compressor	0	14.23
Burner	0	105.05
Turbine	0	10.40
Exhaust	222.5	92.82
System output		
Compressor	−144.1	−144.1
Turbine	262.5	262.5

The Entire Generating System

The availability input to the gas turbine system is the fuel availability supplied to the burner, and the availability output is simply the turbine work minus the compressor work. Thus, the overall second law efficiency of gas turbine system is

$$\varepsilon_o = \frac{262.5 - 144.1}{340.9} = 0.347$$

The first law efficiency is the ratio of the net work to the energy supply. Because the availability supply is assumed to be numerically equal to the energy supply, the first law efficiency should have the same value as the second law efficiency in this calculation.

All numerical results are summarized in Tables 8-8 and 8-9. It is seen that the performance picture given by the availability analysis is quite different from that given by the energy analysis. The burner has a first law efficiency of 100 percent, but, in fact, it is the least efficient component in the availability analysis. In the burner, approximately one-third of the fuel availability is either lost or destroyed. It is of interest to compare the loss distribution. In the first law analysis the loss takes place in the form of the affluent, while in the availability analysis the loss and destruction are spread over all the components.

Table 8-9 Component efficiencies in a simple gas turbine system.

Component	The first law efficiency (%)	The second law Eefficiency (%)
Compressor	85.0	90.2
Burner	100.0	69.2
Gas turbine	88.0	96.2

The availability loss at the exhaust end is relatively small compared to the energy loss in the first law analysis. It should be emphasized that the availability analysis presented in this section is greatly simplified by the assumptions below:

1. The combustion process is replaced by a simple heat transfer process.
2. The fuel availability is equal to the fuel heating value.
3. No consideration is given to the diffusion availability.
4. The specific heat for air is constant.

This approach is reasonable and easy to implement. For a quick result, this approach should be adopted.

8-4 AVAILABILITY ANALYSIS OF A GAS TURBINE POWER PLANT

This section concerns a detailed availability analysis of a gas turbine power plant. The assumptions used in the last section are removed. The calculations will become lengthy, but the results will be more accurate. Let us consider the gas turbine power plant in Fig. 8-8. Air enters the compressor at 14.7 psia and 90°F. The compressor has a pressure ratio of 10 and an internal efficiency of 82 percent. Methane gas is supplied to the burner to maintain the turbine inlet temperature at 2600°R. The turbine has an internal efficiency of 82 percent and the net output of 50,000 kW. The objective of the present analysis is to prepare the availability balance for the gas turbine power plant and the second law overall plant efficiency.

In the first step we determine the temperatures at various system locations. This can be done by using the internal efficiency of the compressor and the turbine. For simplicity, the pressure drops at the burner and at the system inlet and exhaust ends are omitted from consideration. Table 8-10 is a summary of these calculations. In the second step we determine the air-fuel ratio by which a turbine inlet temperature of 2600°R can be maintained. This is done by applying

Table 8-10 Properties of the gas turbine power plant in Figure 8-8.

Locations	Pressure (psia)	Temperature (°R)
1	14.7	550
2	147.0	1175
3	147.0	2600
4	14.7	1572

the principle of conservation for mass and energy. That is,

$$\begin{aligned}&CH_4 + 2\chi O_2 + (2\chi)(3.76)N_2\\&\quad \rightarrow CO_2 + 2H_2O + (2\chi - 2)O_2 + (2\chi)(3.76)N_2\end{aligned}$$

where χ is related to the excess air used in the combustion. Because the burner is assumed to be adiabatic, the value of χ is calculated by the first law equation:

$$\Sigma n_r \bar{h}_r = \Sigma n_p \bar{h}_p$$

Assuming the methane gas has a temperature of 77°F, we substitute the numeric values into the above equation and have

$$\begin{aligned}&(-32{,}210) + 2\chi(8432 - 3725) + 2\chi(3.76)(8239 - 3729)\\&\quad = (-169{,}300 + 29{,}187 - 4030) + 2(-104{,}040 + 23{,}840 - 4258)\\&\qquad + (2\chi - 2)(20{,}311 - 3725) + 2\chi(3.76)(19{,}416 - 3729)\end{aligned}$$

or $\chi = 2.91$. Thus, the air-fuel ratio is given by $[2 \times 2.91(1 + 3.76)]/1.0 = 27.7$ moles of air per mole of fuel.

The composition of the combustion products is important. It is determined as shown in Table 8-11.

In the third step we calculate the total availability at various system locations. The total availability consists of a thermomechanical component and a diffusion component. The thermomechanical component $\bar{a}_f^{\mathrm{th}}$ is a function of temperature and pressure, whereas the diffusion component $\bar{a}_f^{\mathrm{di}}$ is related to the composition of the substance. They are given by the equations

$$\bar{a}_f^{\mathrm{th}} = \sum_{k=1}^{n} x_k\left[\bar{h}_k(T) - \bar{h}_k(T_o) - T_o(\bar{s}'_k(T) - \bar{s}'_k(T_o))\right] + \bar{R}T_o \ln\frac{P}{P_0}$$

and

$$\bar{a}_f^{\mathrm{di}} = \bar{R}T_o \sum_{k=1}^{n} x_k \ln\frac{x_k}{x_k^0}$$

Table 8-11 Composition of the combustion products for the gas turbine power plant in Figure 8-8.

Items	Number of moles	Molar fraction
CO_2	1.0	0.0348
H_2O	2.0	0.0697
O_2	3.83	0.1334
N_2	21.88	0.7621
Total	28.71	1.00

Taking the numerical calculation for location 3 as an example, we have the thermomechanical availability as

$$\begin{aligned}\bar{a}_{f_3}^{\text{th}} &= 0.0348[29{,}187 - 4030 - 537(69.245 - 51.032)] \\ &\quad + 0.0697[23{,}840 - 4258 - 537(59.414 - 45.079)] \\ &\quad + 0.1334[20{,}311 - 3725 - 537(61.287 - 48.982)] \\ &\quad + 0.7621[19{,}416 - 3729 - 537(57.436 - 45.743)] \\ &\quad + 1.986(537)\ln(10) \\ &= 12{,}320 \text{ Btu/mole of gas}\end{aligned}$$

and the diffusion availability as

$$\begin{aligned}\bar{a}_{f_3}^{\text{di}} &= (1.986)(537)\left[0.0348\ln\frac{0.0348}{0.0003} + 0.0697\ln\frac{0.0697}{0.0303} + 0.1334\ln\frac{0.1334}{0.2035}\right. \\ &\quad \left. + 0.7621\ln\frac{0.7621}{0.7567}\right] \\ &= 184 \text{ Btu/mole of gas}\end{aligned}$$

Based on one mole of gas turbine fuel, the total availability at location 3 must be the number of moles of combustion products times the unit availability. That is,

$$A_{f_3} = \bar{n}_p\left(\bar{a}_{f_3}^{\text{th}} + a_{f_3}^{\text{di}}\right) = 28.7(12{,}320 + 184) = 358{,}865 \text{ Btu/mole of fuel}$$

Similarly, we calculate the total availability at other system locations. The results are summarized in Table 8-12. The air composition at locations 1 and 2 is assumed to be identical to that in the dead state. Because of this, the diffusion availability at these states must be zero, as indicated above.

In the fourth step we examine all the components (compressor, burner, and turbine) in terms of availability input, output, and loss and destruction. Again, the calculation is based on one mole of fuel. For the compressor the availability

Table 8-12 Total availability at various system locations of the power plant in Figure 8-8.

Locations	$\bar{a}_f^{\text{th}}$ (Btu/lbmole)	$\bar{a}_f^{\text{di}}$ (Btu/lbmole)	Number of lbmoles	A_f (Btu/lbmole of fuel)
1	2.71	0	27.7	75
2	4023	0	27.7	111,437
3	12,320	184	28.7	358,865
4	3487	184	28.7	105,358

input is the compressor work, which is $A_{\text{input}} = \bar{n}_r \Sigma x_r (\bar{h}_2 - \bar{h}_1)_r$. Substituting the numeric values into the equation, we have

$$A_{\text{input}} = 27.7[0.210(8432 - 3816) + 0.7899(8239 - 3820)]$$
$$= 123{,}542 \text{ Btu/mole of fuel}$$

The availability output for the compressor is the increase of the total availability in the air flow. Numerically, it is given by $A_{\text{output}} = A_{f_2} - A_{f_1} = 111{,}437 - 75 = 111{,}362$ Btu/mole of fuel. Thus, the availability loss and destruction for the compressor is $A_{\text{loss}} = A_{\text{input}} - A_{\text{output}} = 123{,}542 - 111{,}362 = 12{,}180$ Btu/mole of fuel.

For the burner, the availability input is the fuel availability plus the availability at location 2.

$$A_{\text{input}} = \bar{a}_{\text{fuel}} + A_{f_2} = (198.42 \times 1800) + 111{,}437 = 468{,}593 \text{ Btu/mole of fuel}$$

The availability output is simply the availability at location 3. As previously indicated, the numeric value of the output is 358,865 Btu/mole. The availability loss and destruction for the burner is therefore $A_{\text{loss}} = A_{\text{input}} - A_{\text{output}} = 468{,}593 - 358{,}865 = 109{,}728$ Btu/mole of fuel.

For the turbine, the availability input is the reduction of total availability in the gas flow.

$$A_{\text{input}} = A_{f_3} - A_{f_4} = 358{,}865 - 105{,}358 = 253{,}507 \text{ Btu/mole of fuel}$$

The availability output is simply the turbine work, which is

$$A_{\text{output}} = \bar{n}_p \Sigma x_p \left(\bar{h}_3 - \bar{h}_4\right)_p$$

Substituting the numeric values into the above equation, we have

$$\begin{aligned} A_{\text{output}} &= 28.7[0.0348(29{,}187 - 15{,}478) \\ &\quad + 0.0697(23{,}840 - 13{,}231) \\ &\quad + 0.1334(20{,}311 - 11{,}604) \\ &\quad + 0.7621(19{,}416 - 11{,}197)] \\ &= 248{,}025 \text{ Btu/mole of fuel} \end{aligned}$$

Thus, the availability loss and destruction for the turbine is $A_{\text{loss}} = A_{\text{input}} - A_{\text{output}} = 253{,}507 - 248{,}025 = 5482$ Btu/mole of fuel.

Table 8-13 is a summary of the results for all component calculations. Also included is the second law efficiency for all components.

Table 8-13 Characteristics of components of the power plant in Figure 8-8.

Item	A_{input} (Btu/lbmole)	A_{output} (Btu/lbmole)	A_{loss} (Btu/lbmole)	ε (%)
Compressor	123,542	111,362	12,180	90.14
Burner	468,593	358,865	109,728	76.58
Turbine	253,507	248,025	5,482	97.84

In the final step we determine the second law efficiency for the gas turbine power plant. The equation is

$$\varepsilon_o = \frac{\overline{w}_t - \overline{w}_c}{A_{\text{fuel}} + A_{\text{air}}}$$

With the numeric values, we have

$$\varepsilon_o = \frac{248{,}025 - 123{,}542}{(198.42 \times 1800) + 75} = \frac{124{,}483}{357{,}231} = 0.3485$$

The availability balance for the gas turbine power plant is presented in Table 8-14. It is seen that 30 percent of the plant availability input is consumed and destroyed in the burner even though there is no heat loss from the burner. Thus, the burner is the critical component in any improvement plan for a gas turbine system. The availability exhaust loss also is substantial. In this example, it constitutes 29 percent of the plant availability input. Use of the availability method for the system analysis is very much needed.

Compared with the quick analysis in the last section, the present analysis would provide more accurate results. It is especially true in the calculations of availability loss and destruction for the burner and the plant exhaust loss. In general, the quick analysis underestimates the plant net output and the plant fuel availability. However, the difference of these two analyses in second law efficiency is relatively small. A detailed comparison is left as an exercise.

8-5 COMBINED-CYCLE SYSTEMS

Improving the cycle efficiency is an important objective in any cycle analysis. One convenient approach is to combine two different cycles to form a new electric-power generating cycle. One of the most popular schemes is the combination of a gas turbine cycle with a steam turbine cycle, as shown in Fig. 8-10. It is seen that the hot exhaust gas from the gas turbine is utilized to generate the

Table 8-14 Availability balance for a gas turbine power plant.

Characteristic	Availability flow (Btu/mole of fuel)	Percent of availability input
Plant input	357,231	—
Loss and destruction		
Compressor	12,180	0.41
Burner	109,728	30.72
Turbine	5,482	1.53
Exhaust gas	105,358	29.49
Plant net output	124,483	34.85

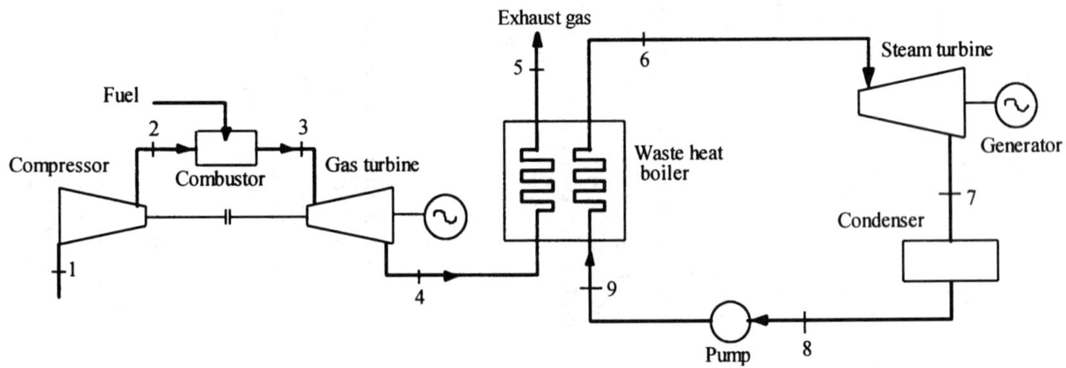

Figure 8-10 Schematic diagram for a combined-cycle system.

steam, which in turn is used to drive the steam turbine. In this combined cycle system, combustion of the fuel takes place at only one point in the cycle, namely in the burner of the gas turbine, and the cycle work is produced at two different places. The cycle efficiency of the combined-cycle system is

$$\eta_{cy} = \frac{w_{gt} + w_{st}}{q_b} \tag{8-15}$$

Let η_{gt} = gas turbine cycle efficiency, and η_{st} = steam turbine cycle efficiency. Thus the gas turbine work and steam turbine work can be expressed respectively by

$$w_{gt} = \eta_{gt} \times q_b \tag{8-16}$$

and

$$w_{st} = (1 - \eta_{gt}) \times q_b \times \eta_{st} \tag{8-17}$$

Substituting these two terms into Eq. (8-15), the cycle efficiency of the combined-cycle system becomes

$$\eta_{cy} = \eta_{st} + (1 - \eta_{st})\eta_{gt} \tag{8-18}$$

This equation shows that the thermal efficiency of the combined cycle is greater than the steam turbine cycle efficiency by an amount equal to $(1 - \eta_{st})\eta_{gt}$. With $\eta_{st} = 0.33$ and $\eta_{gt} = 0.26$ as typical values, the combined-cycle efficiency will be approximately 50 percent. This cycle efficiency represents an optimistic estimate. When a detailed design of combined-cycle system is made, the cycle efficiency is usually in the range of 38 to 42 percent.

The waste heat steam generator is the component that couples the gas and steam parts of the system. The turbine exhaust gas enters the bottom of the heat exchanger, moves upward, and releases its energy. At the end of the process, the turbine exhaust gas will leave the plant through a short stack. Figure 8-11 shows the temperature variation for both hot gas and water. Water enters the steam generator in the form of compressed liquid. As the water receives heat from the hot exhaust gas, it becomes saturated, it evaporates and it eventually gets superheated. The temperature difference between these two streams varies throughout the waste heat steam generator. The minimum value $(T_x - T_s)$ frequently is defined as the *boiler pinch point*. The pinch point selection is important and can greatly affect the physical size of the waste heat steam generator. For economic reasons, the pinch point usually ranges between 40 to 80°F. To determine the amount of steam generated in the steam generator, we take the evaporator and superheater as a control volume and apply the first law to it. That is,

$$\dot{m}_g c_p (T_{hi} - T_x) = \dot{m}_s (h_{ce} - h_{sf})$$

or

$$\frac{\dot{m}_s}{\dot{m}_g} = \frac{c_p (T_{hi} - T_x)}{h_{ce} - h_{sf}} \tag{8-19}$$

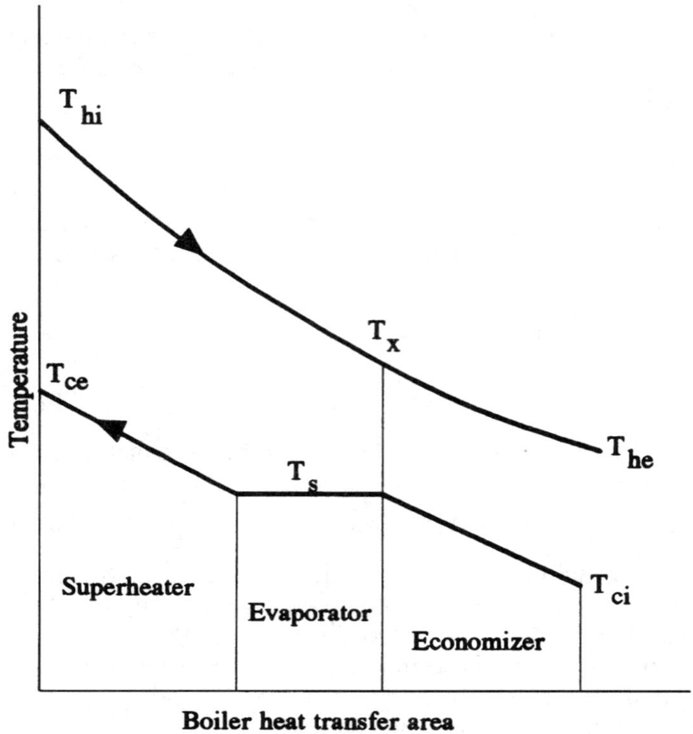

Figure 8-11 Temperature variation in a waste-heat steam generator.

This equation gives the amount of steam generated by a unit mass of exhaust gas. In calculation, the pinch point is first selected; the temperature T_x is simply the pinch point plus the steam saturation temperature. The temperature of the gas leaving the plant is also important and estimated by

$$\dot{m}_g c_p (T_x - T_{\text{he}}) = \dot{m}_s (h_{\text{sf}} - h_{\text{ci}})$$

or

$$T_{\text{he}} = T_x - \frac{1}{c_p} \frac{\dot{m}_s}{\dot{m}_g} (h_{\text{sf}} - h_{\text{ci}}) \qquad (8\text{-}20)$$

To avoid corrosion from moisture formation in the economizer and stack, the minimum gas temperature in the waste heat steam generator is always designed higher than the dew point temperature of combustion products.

There is another approach in combining the gas turbine and steam turbine systems. Figure 8-12 shows the schematic diagram of this combined-cycle system. The air supply through the compressor is used to pressurize the combustion chamber of the steam generator. The flue gases from the boiler act as the

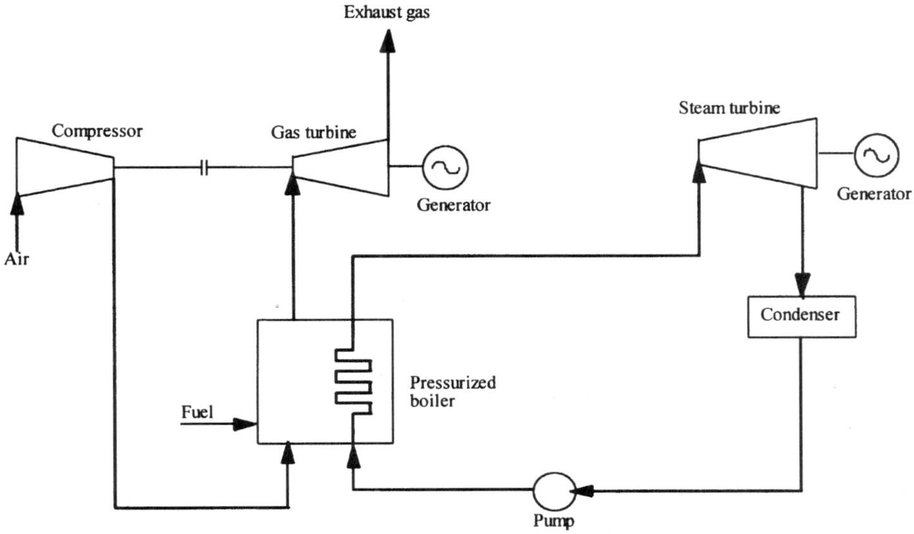

Figure 8-12 Schematic diagram for a combined-cycle system with a pressurized boiler.

working substance and expand in the gas turbine. The steam generated in the boiler goes through the steam turbine cycle as it does in the conventional steam power plant. In this combined-cycle system the combustion of fuel is effected only in the furnace of the boiler, and the useful work is produced by gas and steam turbine. The cycle efficiency can be calculated in the same manner as that for the previous combined cycle.

The application of the availability method to the combined-cycle system is straightforward. The procedure is identical to those used for the gas and steam turbine systems. As an example, a heat and availability balance is prepared for the combined-cycle system shown in Fig. 8-10. The conditions are listed below:

Gas Turbine	
Pressure ratio	9
Turbine inlet temperature	2100°F
Compressor efficiency	85%
Turbine efficiency	90%
Burner efficiency	95%
Boiler and Steam Turbine	
Steam pressure and temperature	600 psia, 750°F
Condenser pressure	3 in. Hg abs.
Boiler pinch point	60 F
Turbine efficiency	80%
Pump efficiency	75%

Table 8-15 Thermodynamic properties at various locations.

Location	Pressure (psia)	Temperature (°F)	Enthalpy (Btu/lbm)	Entropy (Btu/lbm-R)
1	14.7	77	—	—
2	132.3	629	—	—
3	132.3	2100	—	—
4	14.7	1026	—	—
5	14.7	342.3	—	—
6	600	750	1380	1.6114
7	1.47	—	1012	1.7720
8	1.47	—	83.0	0.1560
9	600	—	85.3	0.1566

Table 8-15 presents the thermodynamics properties at various locations. Assuming that the specific heat for exhaust gas is 0.25 Btu/lbm-R, we calculate 0.132 lbm of steam per pound of exhaust gas. Table 8-16 shows the energy and availability balance for the combined-cycle system. Because the fuel unit availability and heating value are assumed equal, the system input and efficiency are the same in both energy and availability analyses. Numerically, they are 387.11 Btu/lbm and 46 percent, respectively. However, the loss distributions are completely different in the energy and availability balance. The availability analysis shows that the burner and boiler are the critical components in the combined-cycle system. In these two pieces of equipment, almost 40 percent of the system availability input is either destructed or lost.

Table 8-16 Energy and availability balance for the combined-cycle system.

Characteristic	Energy balance (Btu/lbm)[a]	Availability balance (Btu/lbm)[a]
Input	387.11	387.11
Losses		
Compressor	0	14.08
Burner	19.36	134.11
Gas turbine	0	7.81
Boiler	0	20.40
Exhaust	66.33	12.43
Steam turbine	0	11.38
Condenser	122.64	8.08
Pump	0	0.04
Output		
Compressor	−138.0	−138.0
Gas turbine	268.5	268.5
Steam turbine	48.58	48.58
Pump	−0.30	−0.30

[a]Based on one pound of air passing through the gas turbine system. The steam-to-air ratio is 0.132.

Table 8-17 Component efficiencies for the combined-cycle system.

Components	The first law efficiency (%)	The second law efficiency (%)
Compressor	85	90
Burner	95	65
Gas turbine	90	97
Boiler[a]	100	77
Steam turbine	80	81
Condenser	—	—
Pump	78	86

[a]The exit loss is not taken into account.

Table 8-17 summarizes the first law and second law efficiencies of major components in the combined-cycle system. While the burner and boiler have relatively high first law efficiencies, they in fact have quite low second law efficiencies. To improve overall system performance, engineers must start with the improvement of the burner and waste heat steam generator.

At this point one can easily conclude that the production of electricity from fossil fuels is a relatively inefficient process. This low efficiency usually implies low utilization of fuel energy and that a significant amount of waste heat is produced from the power plants. To conserve fuel energy and reduce fuel cost, in recent years various cogeneration system schemes have been proposed. These cogeneration systems frequently involve the use of steam turbine, gas turbine, or combined-cycle arrangements. Topics of cogeneration will be discussed in Chapter 10.

8-6 THE BLOCK APPROACH

The system analyses in previous sections are formulated in terms of process analyses. An energy conversion system is first broken down into several processes and then analyzed in terms of these processes. This approach is quite academic and sometimes incompatible with the reality in which consulting engineers work. For example, as consulting engineers design a combined-cycle electric-power generating system, they would not have any chance to design the processes that comprise the combined system. Instead, they examine and select from the market the available components (or subsystems) such as a gas turbine, waste heat recovery boiler, and steam turbine. Under these conditions, the second law analysis of a combined-cycle system should be implemented on the performance of the actual components rather than on the hypothetical processes involved in the system.

In the block approach the second law system analysis is based on the performance data of actual components. The processes involved in any particular component are "lumped" together, and the component is evaluated in terms of its thermodynamic availability input and output. For a component or system, the general availability relationship is given by

$$\begin{pmatrix}\text{Availability input}\\ \text{to control volume}\end{pmatrix} = \begin{pmatrix}\text{Availability output}\\ \text{from control volume}\end{pmatrix} + \begin{pmatrix}\text{Availability loss}\\ \text{and destruction}\end{pmatrix} \tag{8-21}$$

The gas turbine is one of the major components in the combined-cycle system such as that in Fig. 8-13. The gas turbine usually consists of compression, combustion, and expansion processes. In the block approach, the performance data provided by equipment manufacturers are utilized to calculate the availability input and output of the gas turbine. The availability equation including loss and destruction is given by

$$A_{f_1} = A_{p_1} + A_g + A_{l_1} \tag{8-22}$$

A_{f_1} is the availability input to the gas turbine and numerically equal to the fuel consumption times the fuel unit availability if the availability of incoming air is negligible. A_{p_1} is a part of the gas turbine output, which is numerically equal to the turbine electric-power generation. A_g is another part of gas turbine output and can be calculated by multiplying the exhaust gas flow rate by the unit flow availability. That is,

$$A_g = m_g a_{f_g} \tag{8-23}$$

where

$$a_{f_g} = (h_g - h_0) - T_0(s_g - s_0) \tag{8-24}$$

At the time the gas turbine is examined and selected, sufficient performance data will be available for the calculation of availability input and output. A_{l_1} is the availability loss and destruction in the gas turbine. It is also referred to as the total irreversible of the gas turbine system. Equation (8-22) indicates that the total system irreversibility is equal to the availability input minus the availability output. In terms of physical meaning, it represents the energy not available for work production. Its presence is caused by friction, heat loss, combustion, and other irreversibility factors in the gas turbine.

To show the performance of a component or system, the concept of effectiveness (sometimes referred as the second law efficiency) frequently is utilized. Effectiveness is defined as the ratio of the increase in the availability of the desired output to the decrease in the availability of the source. For the system shown in Fig. 8-13, the effectiveness of the gas turbine unit is

$$\varepsilon_{\text{gt}} = \frac{A_{p_1} + A_g}{A_{f_1}} \tag{8-25}$$

As an electric-generating system, the gas turbine effectiveness becomes

$$\varepsilon'_{\text{gt}} = \frac{A_{p_1}}{A_{f_1}} \tag{8-26}$$

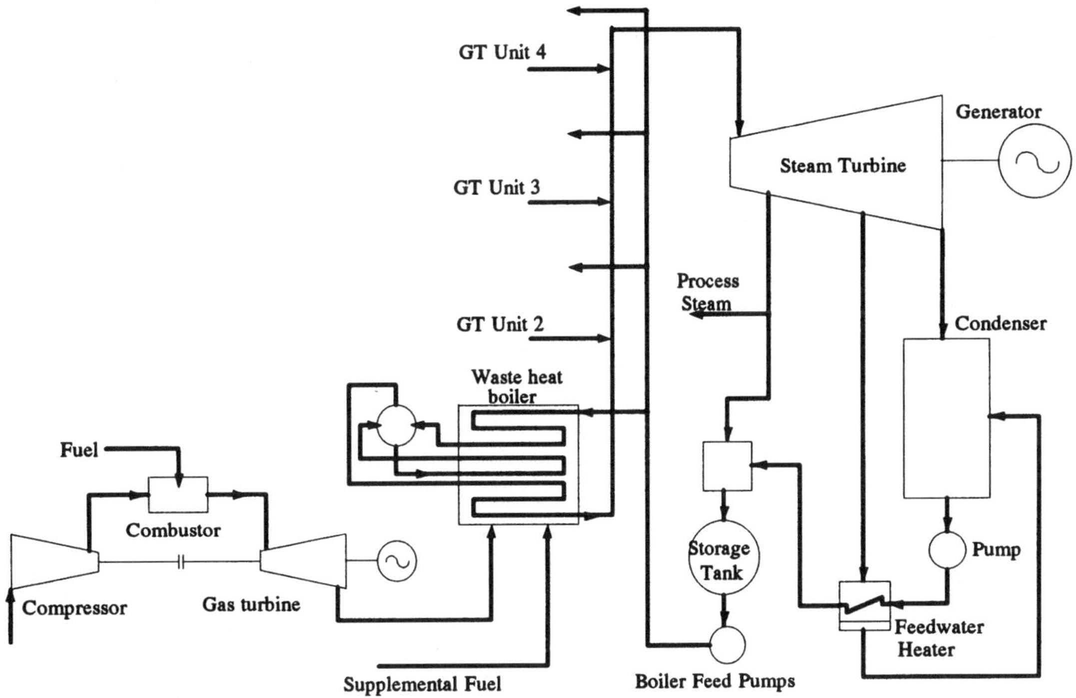

Figure 8-13 A combined gas-steam turbine system.

It is of interest to compare the block approach with the traditional approach in the gas turbine second law analysis. In the block approach, the gas turbine system is treated as an integrated unit. The turbine net output, fuel consumption, and characteristics such as heating value and unit fuel availability are the data needed to calculate the system total irreversibility and second law efficiency. The details are shown in Eqs. (8-22) and (8-25). In the traditional approach, the gas turbine system generally is "broken down" into three major processes: compression, combustion, and expansion. Because of information unavailable at the time of equipment selection, the consulting engineers usually approximate these processes by using various assumptions. In most cases they have to use a trial-and-error method to retrofit the assumptions to the equipment performance data. Because of these assumptions, the traditional approach frequently results in lengthy calculations, and the results may not be as accurate as those obtained by the block approach.

The waste heat–recovery boiler is another major component in the combined-cycle system. Boilers may be unfired or supplementally fired and may be single pressure or dual pressure. The availability equation (8-21) still is valid for the waste heat–recovery boiler. In this case, the availability input is the sum of the gas flow availability and the supplemental fuel availability. The output is the availability gained by the water and steam. Like the gas turbine, there is availability loss and destruction in the boiler. Their presence is due to combustion, heat transfer through finite temperature differences, and other factors in the boiler. In equation form, the boiler availability balance is

$$A_g + A_{f_2} = A_{\text{st}} + A_{l_2} \tag{8-27}$$

where

$$A_{\text{st}} = \dot{m}_{\text{st}} a_{f_{\text{st}}} \tag{8-28}$$

and

$$a_{f_{\text{st}}} = (h_e - h_i) - T_0(s_e - s_i) \tag{8-29}$$

The subscripts i and e denote boiler inlet and exit. When consulting engineers evaluate and select the waste-heat recovery boiler, they would have generally enough boiler data to determine the steam flow rate ($\dot{m}_{\text{st}}$) and unit availability ($a_{f_{\text{st}}}$). The steam availability (A_{st}), when combined with the other two terms (A_g and A_{f_2}), will determine the boiler total irreversibility A_{l_2}, as shown in Eq. (8-27). The boiler effectiveness is defined as the ratio of availability output to availability input. In equation form, the boiler effectiveness is

$$\varepsilon_b = \frac{A_{\text{st}}}{A_g + A_{f_2}} \tag{8-30}$$

The steam turbine is the third major component in the combined-cycle system. This component consists of a turbine, a condenser, feedwater heaters, and their auxiliary equipment. The availability input to the steam turbine is the availability received by the steam and water in the boiler. The turbine output is the sum of the turbine electric power and the availability of process steam. The availability balance for the steam turbine component is expressed as

$$A_{\mathrm{st}} = A_{p_2} + A_{\mathrm{ps}} + A_{l_3} \tag{8-31}$$

and the steam turbine effectiveness is

$$\varepsilon_{\mathrm{st}} = \frac{A_{p_2} + A_{\mathrm{ps}}}{A_{\mathrm{st}}} \tag{8-32}$$

Figure 8-14 shows a block diagram for the combined-cycle system. Each block represents a component. For each component there is availability input, output, and loss and destruction. All of these availability terms can be determined with the performance information either provided by the equipment manufacturer or determined from simple calculations.

The overall effectiveness of the system, ε_o, gives the ratio of the total availability output, electricity and process steam, to the total fuel availability input.

$$\varepsilon_o = \frac{A_{p_1} + A_{p_2} + A_{\mathrm{ps}}}{A_{f_1} + A_{f_2}} \tag{8-33}$$

The overall system effectiveness can be expressed also in terms of component performance parameters. Combining previous equations leads to the expression

$$\varepsilon_o = \frac{\varepsilon'_{\mathrm{gt}}(1 - \varepsilon_{\mathrm{st}}\varepsilon_b) + \varepsilon_{\mathrm{st}}\varepsilon_b(\varepsilon_{\mathrm{gt}} + Z)}{1 + Z} \tag{8-34}$$

where Z, the supplemental firing ratio is defined as

$$Z = \frac{A_{f_2}}{A_{f_1}} \tag{8-35}$$

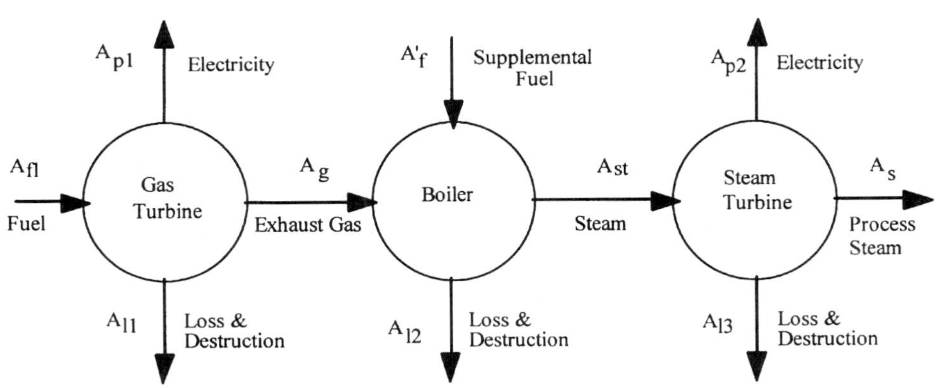

Figure 8-14 A block diagram for the cogeneration system.

If there is no supplemental firing for the waste-heat boiler, Z will be equal to zero, and Eq. (8-34) will be further simplified. Equation (8-34) is used to provide the relationship between the component performance and overall system performance. Using this equation, one can analyze how each component impacts the overall system and pinpoint the critical component. The derivation of Eq. (8-34) is left as an exercise for the reader.

It should be emphasized that the second law analysis by the block approach is straightforward. As shown by the equations in this section, only equipment performance data are needed. Compared with the traditional approach, which is characterized by individual process analyses, the block approach requires fewer calculations and no assumptions to approximate various processes. This approach is particularly convenient to the consulting engineers at the time of equipment selection.

As an example, consider the gas turbine unit of the combined-cycle system shown in Fig. 8-13. The manufacturer provides the following unit data:

Net power	73.6 MW
Fuel consumption	741.6 MBtu/hr
Exhaust gas conditions	16 inches water gauge, 1010°F
Exhaust gas flow	2.256×10^6 lbm/hr

We determine the turbine total irreversibility and the second law efficiency by starting with the availability equation $A_{l_1} = A_{f_1} - A_{p_1} - A_g$. Assuming the availability at the compressor inlet is negligible, the availability input to the gas turbine unit is numerically equal to the fuel availability, or $A_{f_1} = 741.6$ MBtu/hr = 217,287 kW. The availability output is the sum of the net power from the turbine unit and the availability associated with the exhaust gas. They are respectively $A_{p_1} = 73{,}600$ kW and

$$A_g = \dot{m}_g\left[c_p(T_g - T_0) - T_0\left(c_p \ln\frac{T_g}{T_0} - R\ln\frac{P_g}{P_0}\right)\right]$$

$$= 2.256 \times 10^6\left[0.24(1470 - 537) - 537\left(0.24\ln\frac{1470}{537} - \frac{53.3}{778}\ln\frac{15.28}{14.70}\right)\right]$$

$$= 215.59 \text{ MBtu/hr} = 63{,}167 \text{ kW}$$

Substituting the terms A_{f_1}, A_{p_1}, and A_g into the equation will provide the gas turbine total irreversibility as

$$A_{l_1} = 217{,}287 - 73{,}600 - 63{,}167 = 80{,}520 \text{ kW}$$

To determine the gas turbine second law efficiency, we can use Eq. (8-25) to obtain

$$\varepsilon_{\text{gt}} = \frac{73{,}600 + 63{,}167}{217{,}287} = 0.6294$$

As another example, we consider the waste heat–steam generator of the combined-cycle system shown in Fig. 8-13. The manufacturer provides the following data on the waste heat–recovery boiler:

Steam conditions	1250 psia, 950°F
Steam flow rate	170,000 lbm/hr
Boiler inlet temperature	326°F
Exhaust gas conditions	16 in. water gauge, 1010°F
Exhaust gas flow rate	2.256×10^6 lbm/hr

To determine the boiler total irreversibility and the second law efficiency, we first assume that there is no supplemental firing in the waste heat–recovery boiler so that the total irreversibility is equal to the availability input minus the availability output, or $A_{l_2} = A_g - A_{st}$.

The availability input A_g is 63,167 kW, as previously calculated. The availability output is the availability received by water, A_{st}, which is given by the equation

$$\begin{aligned} A_{st} &= \dot{m}_{st}[(h_e - h_i) - T_0(s_e - s_i)] \\ &= 170{,}000[(1468.6 - 296.7) - 537(1.6036 - 0.4719)] \\ &= 95.9 \text{ MBtu/hr} = 28{,}098 \text{ kW} \end{aligned}$$

Substituting A_g and A_{st} into the equation would give the boiler total irreversibility as $A_{l_2} = 63{,}167 - 28{,}098 = 35{,}069$ kW. The second law efficiency is calculated by Eq. (8-30). The result is $\varepsilon_b = 28{,}098/63{,}167 = 0.4448$.

The combined cycle in Fig. 8-13 consists of four gas turbines, each equipped with its own waste heat–recovery boiler and all coupled to a single steam turbine. As indicated, the steam turbine produces not only electric power but also process steam. The procedure to determine the steam turbine system irreversibility and second law efficiency is similar to that for the gas turbine or waste heat–recovery boiler.

The block approach is a very convenient tool to an engineer. With this approach one can easily determine the availability balance on equipment without detailed calculations. More discussion on the block approach will be presented later in the text.

8-7 PROBLEMS

8-1 Consider the steam turbine cycle with two feedwater heaters. One is a contact heater receiving the extracted steam at a pressure of 150 psia, and the other is a surface-type heater without a drain cooler using the extracted steam at 50 psia. Calculate the second law cycle efficiency and compare the result with the first law cycle efficiency for the conditions below:

Turbine inlet conditions	1250 psia, 850°F
Condenser pressure	3.5 in. Hg abs.
Turbine efficiency	0.88
Pump efficiency	0.88
Heater terminal temperature difference	10°F

Also, prepare an availability and energy balance for this steam turbine cycle.

8-2 The steam turbine system in Problem 8-1 is supplied with the steam from a boiler. The boiler fuel is a sub-bituminous coal with the following ultimate analysis:

C = 60.1	N_2 = 1.1	A (ash) = 5.0
H_2 = 4.2	S = 1.3	H_2O = 12.2
O_2 = 16.1		

The boiler has a first law efficiency of 88 percent based on a low heating value. Estimate the boiler second law efficiency and the fuel consumption for the plant net output of 6,000 kW.

8-3 A flow diagram of a 203,000 kW steam turbine system is shown in Fig. 8-15. Prepare an availability analysis and provide the answers to the following questions:
a) What is the availability loss and destruction in the H-P turbine?
b) What is the second law efficiency for the H-P turbine?
c) What is the availability loss and destruction in heater #5?
d) What is the second law efficiency for heater #5?
e) What is the heat removal rate in the condenser?
f) What is the availability loss in the condenser?
g) What is the second law efficiency for the boiler feedwater pump?

8-4 Air enters a gas turbine system at 14.7 psia and 90°F. The compressor pressure ratio is 10. Methane gas is supplied to the burner to maintain the turbine inlet temperature at 2200°F. The turbine and compressor both have an efficiency of 82 percent. For a gas turbine of net output 50,000 kW, determine (a) the air flow rate in cfm, (b) the first law system efficiency, (c) the second law system efficiency, and (d) the energy and availability balance for the gas turbine system.

8-5 A regenerator is incorporated in the gas turbine system in Problem 8-4. The regenerator has an efficiency of 80 percent. Determine the second law efficiency for this regenerative gas turbine system.

8-6 Air is compressed from 14.7 psia and 90°F to 600 psia in a two-stage compressor with intercooling between stages. Both compressor stages have an efficiency of 82 percent. In the intercooler, the air has a pressure of 93 psia and is cooled back to the initial temperature. Determine (a) the total compressor work per pound of air, (b) the energy balance for the compressor process, and (c) the availability balance for the compressor process.

8-7 A combined gas turbine and steam turbine system is arranged as shown in Fig. 8-12. Air enters the compressor at 14.7 psia and 77°F and exhausts at 100 psia. The gas enters the turbine at 1600°F and expands to a pressure close to the ambient pressure. On the steam side, steam enters the turbine at 1250 psia and 850°F and exhausts to the condenser at the pressure 4 in. Hg abs. Other conditions are listed below:

Compressor efficiency	0.85
Gas turbine efficiency	0.90
Steam turbine efficiency	0.90
Pump efficiency	0.85
Boiler heat loss	2%
Boiler fuel	Methane gas
Boiler excess air	15%

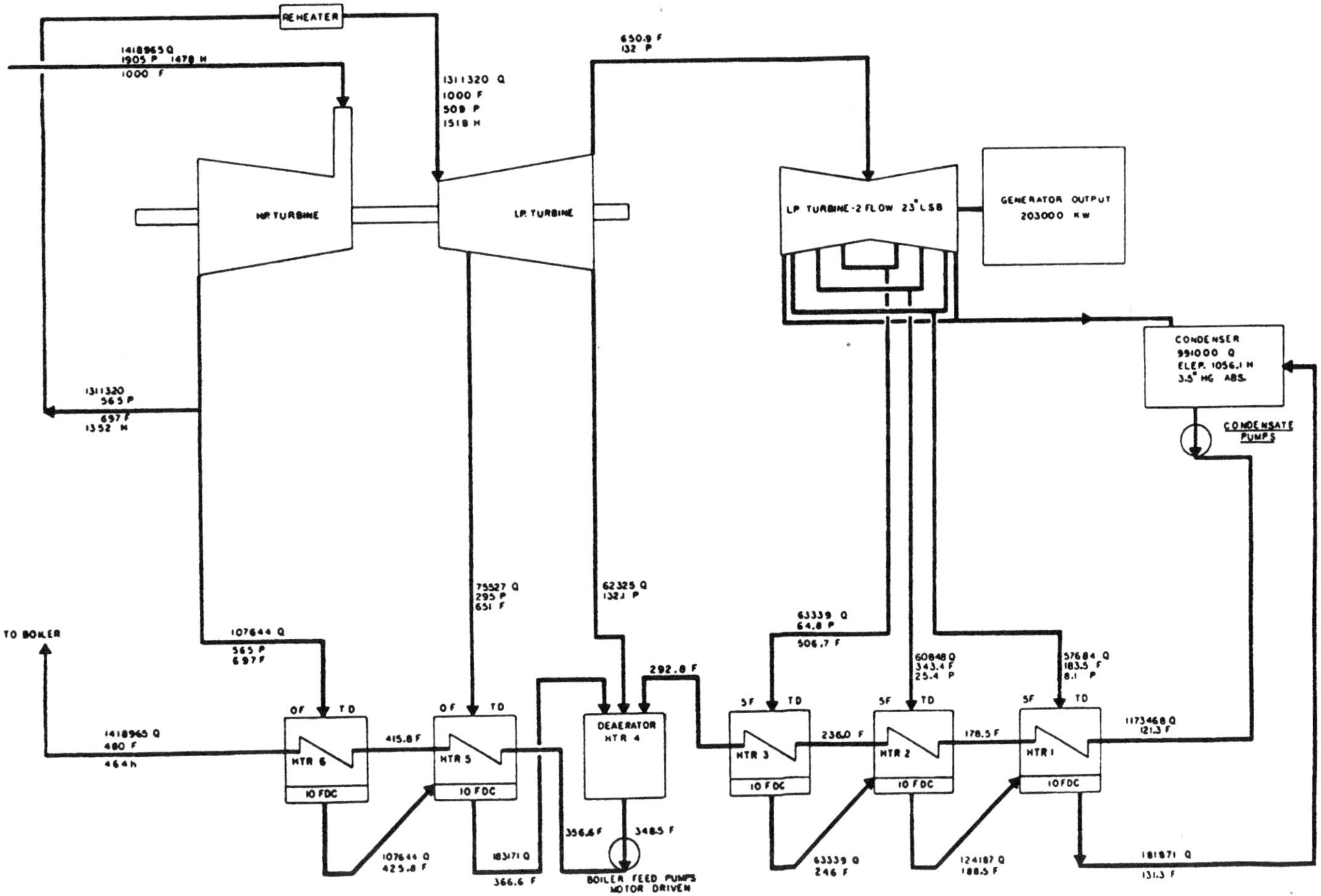

Figure 8-15 Flow diagram for a 203,000 kw turbine system.

Prepare the energy and availability balance for this combined-cycle system. Also, determine the second law efficiency for major components and the combin ed system.

8-8 A gas turbine unit has the following performance data:

Net output	86,303 kW
Fuel energy	995.5×10^6 Btu/hr
Air inlet conditions	14.7 psia, 80°F
Exhaust gas conditions	2,828,000 lbm/hr
	1010°F
	13 in. of water back pressure

Estimate the availability loss and destruction.

8-9 Consider the steam turbine in the combined-cycle system shown in Fig. 8-13. The operating data are as follows:

Turbine steam flow rate	$4 \times 170{,}000$ lbm/hr
Turbine inlet conditions	1250 psia, 950°F
High-pressure extracted steam	
Flow rate	64,793 lbm/hr
Pressure	95 psia
Enthalpy	1220.8 Btu/lbm
Low-pressure extracted steam	
Flow rate	68,939 lbm/hr
Pressure	24.5 psia
Enthalpy	1126.0 Btu/lbm
Process steam	
Flow rate	100,000 lbm/hr
Leaving conditions	95 psia, 381.6°F
Return conditions	95 psia saturated water
Condenser inlet	
Pressure	3 in Hg abs.
Enthalpy	967.0 Btu/lbm

Calculate the second law efficiency for the steam turbine.

8-10 A jet engine is tested on a test stand, and the following measurements are obtained:

Jet engine inlet	
Cross-section area	1.8 ft^2
Velocity	300 ft/s
Pressure	14.0 psia
Air flow rate	30 lbm/s
Jet engine exit	
Cross-section area	1 ft^2
Velocity	1300 ft/s
Pressure	18 psia
Gas flow rate	30.6 lbm/s
Jet engine test fuel	
Type	CH_4
Fuel flow rate	0.6 lbm/s
Pressure	150 psia
Temperature	160°F
Air and gas composition	

	Air	Gas
N_2	0.7967	0.7618
O_2	0.1877	0.0916
CO_2	0.0003	0.0441
H_2O	0.0153	0.1025

Estimate the fuel availability and the flow availability at the engine inlet and exit.

8-11 Repeat Problem 8-4 with the following additional condition:

Pressure losses	
Compressor inlet	1%
Combustion chamber	3%
Turbine outlet	2%
Energy losses	
Combustion chamber	2%
Mechanical coupling	1%

8-12 The combined-cycle system shown in Fig. 8-10 consists mainly of a gas turbine, a waste heat–recovery boiler, and a steam turbine. The conditions of this system are presented below:

Gas turbine	
Pressure ratio	10
Turbine inlet temperature	2400°F
Compressor efficiency	85%
Turbine efficiency	90%
Burner efficiency	95%
Boiler and steam turbine	
Steam pressure	900 psia
Steam temperature	900°F
Condense pressure	3 in. Hg abs.
Boiler pinch point	60 F
Turbine efficiency	80%
Pump efficiency	78%

The gas turbine utilizes the methane gas. Estimate the second law system efficiency. Also prepare the availability balance for the combined-cycle system.

CHAPTER

NINE

HEAT-PUMP AND REFRIGERATION SYSTEMS

This chapter presents availability analyses of heat-pump and refrigeration systems. The systems analyzed here are similar to actual systems. There will be pressure drops and heat losses at various locations of the system. Because the process is no longer internally reversible, as assumed in Chapter 7, there will be some availability destructions in every component of the system. For example, the compressor process is treated as nonisentropic. In this chapter availability balances will be made for several heat-pump and refrigeration systems and then compared with the corresponding energy balances. In addition, calculations will be done for the second law system efficiency.

9-1 EFFICIENCY OF HEATING PROCESS

To begin this chapter, it may be of interest to demonstrate that the heating process is generally inefficient and that the need for an alternative system such as the heat pump is real. Let us consider a hot-water system as shown in Fig. 9-1. The hot water system is fueled by electric energy. The first law equation for this system is

$$\dot{Q}_{cv} + \dot{m}_1 h_1 = \dot{W} + \dot{m}_2 h_2$$

or

$$-\dot{W} = \dot{m}_1(h_2 - h_1) - \dot{Q}_{cv}$$

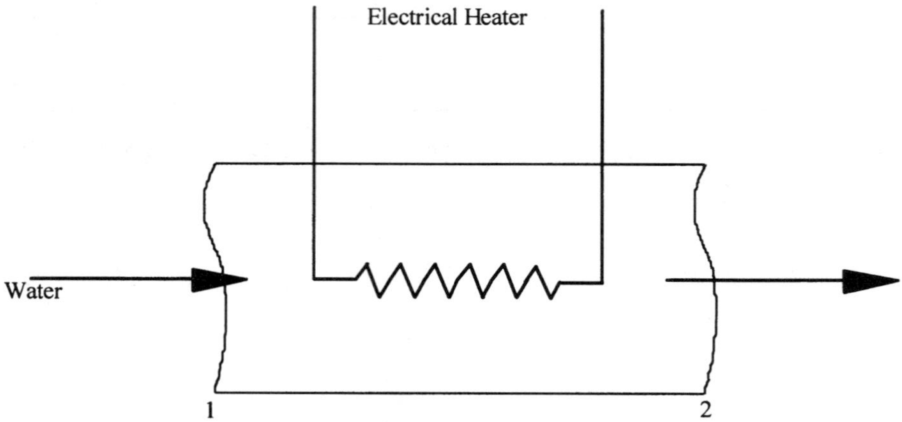

Figure 9-1 A schematic diagram for a water heater.

In terms of absolute number, we can rewrite the above equation as

$$\dot{W}_{e_1} = \dot{m}_1(h_2 - h_1) + \dot{Q}_{\text{loss}}$$

where $\dot{W}_{e_1}$ is the electric energy supplied to the system and $\dot{Q}_{\text{loss}}$ is the heat loss from the system. Thus, the first law efficiency for this heating process is simply

$$\eta = \frac{\dot{m}_1(h_2 - h_1)}{\dot{W}_{e_1}} \tag{9-1}$$

The availability equation for this hot-water system is

$$\dot{A}_q + \dot{A}_w + \dot{m}_1 a_{f_1} - \dot{m}_2 a_{f_2} = \dot{I}$$

Rearranging this equation gives

$$\dot{W}_{e_1} = \dot{I} - \dot{A}_q + \dot{m}_1(a_{f_2} - a_{f_1})$$

The second law efficiency is defined as a ratio of availability output to availability input. In this case, it is

$$\varepsilon = \frac{\dot{m}_1(a_{f_2} - a_{f_1})}{\dot{W}_{e_1}} \tag{9-2}$$

Expressing the flow availability in terms of enthalpy and entropy, we can rewrite the second law efficiency for the hot-water system as

$$\varepsilon = \frac{\dot{m}_1(h_2 - h_1) - \dot{m}_1 T_0(s_2 - s_1)}{\dot{W}_{e_1}}$$

By rearranging this equation, we have

$$\varepsilon = \frac{\dot{m}_1(h_2 - h_1)\left[1 - \dfrac{T_0(s_2 - s_1)}{h_2 - h_1}\right]}{\dot{W}_{e_1}}$$

or

$$\varepsilon = \eta\left[1 - \frac{T_0(s_2 - s_1)}{h_2 - h_1}\right] \tag{9-3}$$

Equation (9-3) can be further simplified and expressed in terms of water temperature. We recognize

$$s_2 - s_1 = c_v \ln\frac{T_2}{T_1}$$

and

$$h_2 - h_1 = (u_2 - u_1) + v_1(P_2 - P_1) = c_v(T_2 - T_1)$$

if the pressure drop is negligible. Substituting these two expressions into Eq. (9-3), we have the second law efficiency for a hot-water system as

$$\varepsilon = \eta\left[1 - \frac{T_0 \ln\dfrac{T_2}{T_1}}{T_2 - T_1}\right] \tag{9-4}$$

When the second law efficiency for the production and distribution of the electricity is

$$\varepsilon_p = \frac{\dot{W}_{e_1}}{\dot{A}_{\text{fuel}}}$$

where $\dot{A}_{\text{fuel}}$ is the fuel availability required for power plant operation, the overall second law efficiency for the process from power-plant fuel to hot water will become

$$\varepsilon_o = \varepsilon_p \times \varepsilon = \varepsilon_p \times \eta\left[1 - \frac{T_0 \ln\dfrac{T_2}{T_1}}{T_2 - T_1}\right] \tag{9-5}$$

Using the typical values $\varepsilon_p = 36\%$, $\eta = 100\%$, $T_1 = T_0 = 77°\text{F}$, and $T_2 = 150°\text{F}$, we find $\varepsilon_o = 2.25\%$. The hot-water system is evidently very wasteful. Even though the hot-water system has no heat loss, the overall second law efficiency still is no better than 3%. Let us consider a more general system for heat energy delivery. The schematic diagram in Fig. 9-2 indicates that the system receives

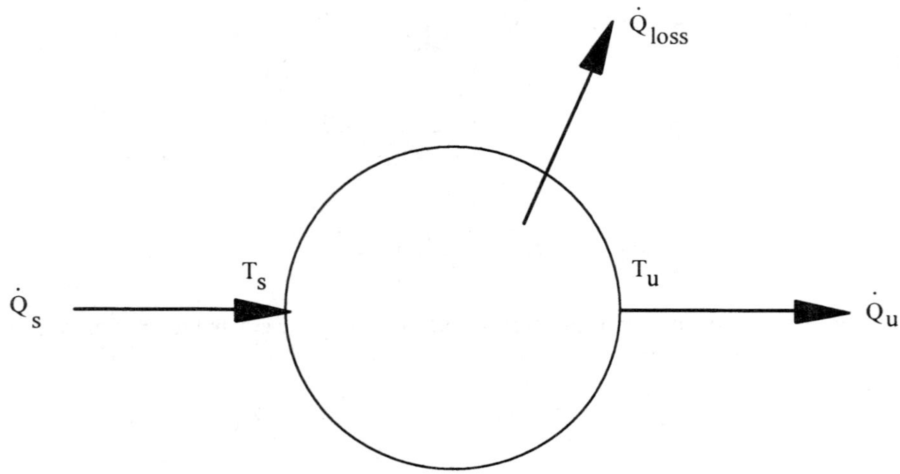

Figure 9-2 A schematic diagram for a heat-delivery system.

heat energy from a source at temperature T_s and provides a user with heat energy at temperature T_u. As expected, the system also loses heat to its surroundings. For this heat energy–delivery system, the first law equation is $\dot{Q}_s = \dot{Q}_{\text{loss}} + \dot{Q}_u$, and the first law efficiency is given by $\eta = \dot{Q}_u/\dot{Q}_s$.

The availability equation for the heat energy delivery system is simply $\dot{A}_q - \dot{I} = 0$. The term $\dot{A}_q$ consists of three parts: The first is related to the heat supply from the source, the second is related to the heat delivered to the user, and the third is related to the heat loss from the system. In equation form, it is

$$\dot{A}_{q_s} + \dot{A}_{q_u} + \dot{A}_{q_{\text{loss}}} - \dot{I} = 0$$

or

$$\left(1 - \frac{T_0}{T_s}\right)\dot{Q}_s - \left(1 - \frac{T_0}{T_u}\right)\dot{Q}_u = \dot{I} + \left(1 - \frac{T_0}{T_l}\right)\dot{Q}_l$$

The second law efficiency is a ratio of availability output to availability input for the system. In this case, the second law efficiency is given by

$$\varepsilon = \frac{\left(1 - \frac{T_0}{T_u}\right)\dot{Q}_u}{\left(1 - \frac{T_0}{T_s}\right)\dot{Q}_s}$$

In terms of the first law efficiency, the above equation becomes

$$\varepsilon = \eta \frac{1 - \dfrac{T_0}{T_u}}{1 - \dfrac{T_0}{T_s}} \tag{9-6}$$

Numerical calculations using Eq. (9-6) show the importance of matching the source temperature to the user's temperature. Table 9-1 indicates that as the source temperature increases, the second law efficiency will decrease. In other words, better fuel utilization will be realized if the source temperature is kept not much higher than the temperature at which the heat energy is provided.

9-2 HEAT PUMP SYSTEMS

The heat-pump system generally consists of four components: compressor, condenser, expansion valve, and evaporator. As shown in Fig. 9-3, the working substance, such as Freon-12, leaves the evaporator and enters the compressor. In the piping 1-2, the substance generally experiences a pressure drop and a heat gain from the environment. Next, in the compression process, mechanical work is supplied and converted into thermal energy. At the compressor discharge, the substance is at high pressure and its flow availability is increased. Then the substance passes through the condenser and leaves as saturated or subcooled liquid. In the condenser process, heat is transferred away from the heat-pump system. Associated with this heat transfer, availability also is transferred from the system. Also, there usually is a pressure drop across the condenser because of friction. When the substance goes through the expansion valve, it experiences a substantial pressure reduction while its enthalpy remains unchanged. In the final process, which takes place in the evaporator, the substance receives heat from an external source and is changed back into vapor at either a saturated or slightly superheated state. The second law efficiency for the heat pump cycle is, as defined in Chapter 7, a ratio of the cycle output to the cycle input. That is,

$$\varepsilon_{\text{cy}} = \frac{a_{q_{3\text{-}4}}}{w_c} = \frac{a_{f_3} - a_{f_4}}{w_c}$$

Table 9-1 Variation of second law efficiency[a]

Source temperature T_s (°F)	500	1000	1500	2000
Second law efficiency ε (%)	27.2	18.9	16.5	15.3

[a] Based on $T_u = 150°\text{F}$, $T_0 = 77°\text{F}$, and $\eta = 100\%$

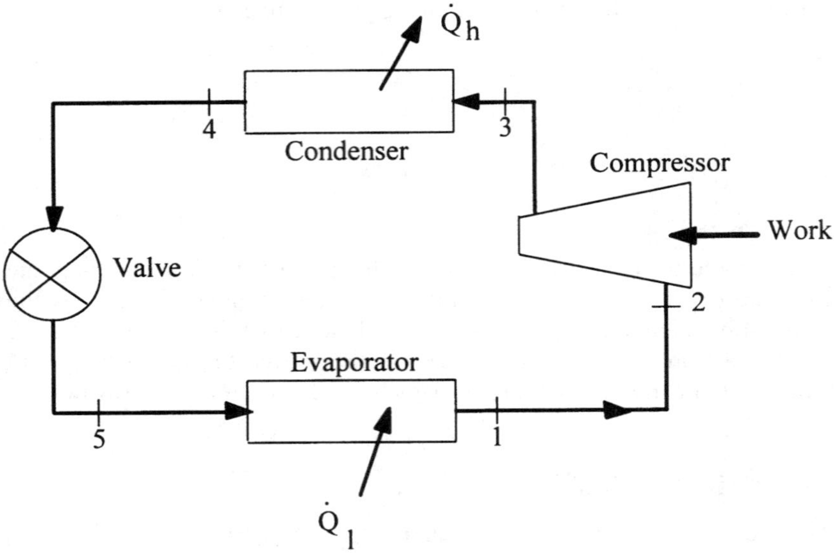

Figure 9-3 A heat-pump system.

It should be emphasized that the term $(a_{f_3} - a_{f_4})$ is the availability output from the heat pump and usually is greater than the availability received by the residential home, $a_{q_{\text{home}}}$. The difference between these two is the availability loss and destruction in the condenser. When the second law efficiency of condenser is

$$\varepsilon_{\text{cond}} = \frac{a_{q_{\text{home}}}}{a_{q_{3\text{-}4}}}$$

the second law efficiency for the entire heat-pump system will be

$$\varepsilon_o = \varepsilon_{\text{cy}} \times \varepsilon_{\text{cond}}$$

The application of the availability method to the heat-pump system is straightforward. As an example, we consider the heat pump shown in Fig. 9-3. The working substance is Freon-12, and its flow rate is 600 pounds per hour.

Table 9-2 Properties of the heat-pump system in Figure 9-3.

State	Temperature (F)	Pressure (psia)	Enthalpy (Btu/lbm)	Entropy (Btu/lbm-R)
1	25	39.3	79.91	0.1668
2	30	38	80.75	0.1691
3	200	220	101.92	0.1809
4	125	215	37.60	0.0754
5	38.15	50	37.60	0.0783

Table 9-2 gives the thermodynamic properties at various locations. The present objectives are to determine the second law cycle efficiency and to prepare the availability and energy balances for this particular heat-pump cycle. For convenience, we examine the heat-pump system in terms of its processes and take 32°F as the temperature in the environmental state.

Piping (Process 1-2)

We take the piping 1-2 as a control volume. The energy equation enables us to determine the heat gain as

$$\dot{Q}_{cv} = \dot{m}_1(h_2 - h_1) = 600(80.75 - 79.91) = 504 \text{ Btu/hr}$$

The availability equation for Process 1-2 is

$$\dot{A}_q + \dot{A}_w + \dot{m}_1 a_{f_1} - \dot{m}_2 a_{f_2} - \dot{I} = 0$$

Since $\dot{A}_w = 0$, we can express the availability loss and destruction as the reduction of flow availability in this process. That is,

$$\dot{I} - \dot{A}_q = \dot{m}_1(a_{f_1} - a_{f_2}) = \dot{m}_1[(h_1 - h_2) - T_0(s_1 - s_2)]$$

With the numeric values, the availability loss and destruction is calculated as $\dot{I} - \dot{A}_q = 600[(79.91 - 80.75) - 492(0.1668 - 0.1691)] = 174.96$ Btu/hr. It should be noted that the piping receives heat from the environment but rejects heat availability to the environment. In other words, the direction of heat flow is opposite to that of heat availability flow.

Compressor (Process 2-3)

The process is assumed to be adiabatic. The energy equation yields the compressor work as

$$\dot{W}_c = \dot{m}_2(h_2 - h_3) = 600(80.75 - 101.92) = -12{,}702 \text{ Btu/hr}$$

and the availability equation gives the availability destruction for the compression process as

$$\dot{I} = \dot{m}_2(a_{f_2} - a_{f_3}) - \dot{W}_c = 600[(80.75 - 101.92) - 492(0.1691 - 0.1809)] + 12{,}702 = 3483.4 \text{ Btu/hr}$$

Condenser (Process 3-4)

As Freon-12 passes through the condenser, it loses a substantial amount of energy. By the first law equation, we can determine this energy as

$$\dot{Q}_{cv} = \dot{m}_3(h_4 - h_3) = 600(37.60 - 101.92) = -38{,}592 \text{ Btu/hr}$$

Associated with this heat transfer, the substance will reduce its flow availability by

$$\dot{m}_3(a_{f_3} - a_{f_4}) = \dot{m}_3[(h_3 - h_4) - T_0(s_3 - s_4)]$$
$$= 600[(101.92 - 37.60) - 492(0.1809 - 0.0754)]$$
$$= 7448.4 \text{ Btu/hr}$$

The reduction of flow availability in Process 3-4 usually is referred to as the availability output of the heat-pump system.

Expansion Valve (Process 4-5)

Process 4-5 is a constant enthalpy process. While the heat loss is negligible, the availability destruction is substantial in the process. By the availability equation, we have the availability destruction as

$$\dot{I} = \dot{m}_4 T_0(s_5 - s_4) = 600 \times 492(0.0783 - 0.0754) = 856.08 \text{ Btu/hr}$$

Evaporator (Process 5-1)

As Freon-12 passes through the evaporator, it receives heat from the ambient air. The amount of energy involved can be determined by the first law equation.

$$\dot{Q}_{\text{cv}} = \dot{m}_5(h_1 - h_5) = 600(79.91 - 37.60) = 25{,}386 \text{ Btu/hr}$$

While Freon-12 receives heat in this process, it loses its flow availability. The reduction of flow availability is given by

$$\dot{m}_5(a_{f_5} - a_{f_1}) = \dot{m}_5[(h_5 - h_1) - T_0(S_5 - S_1)]$$
$$= 600[(37.60 - 79.91) - 492(0.0783 - 0.1668)]$$
$$= 739.2 \text{ Btu/hr}$$

In Tables 9-3 and 9-4 we summarize the results in terms of energy balance and availability balance for the heat pump.

Table 9-3 Energy balance for the heat pump of Figure 9-3.

Item	Input (Btu/hr)	Output (Btu/hr)
Piping	504	
Compressor	12,702	
Condenser		38,592
Valve		0
Evaporator	25,386	
Total	38,592	38,592

Table 9-4 Availability balance for the heat pump of Figure 9-3.

Component	Input (Btu/hr)	Output (Btu/hr)	Loss and destruction (Btu/hr)
Piping			174.9
Compressor	12,702		3483.4
Condenser		7448.4	
Valve			856.1
Evaporator			739.2
Total	12,702	7448.4	5253.6

Finally, for the heat pump, we calculate the coefficiency of performance as $\mathrm{COP_{hp}} = 38{,}592/12{,}702 = 3.04$, and the second law efficiency as $\varepsilon_{\mathrm{hp}} = 7448.4/12{,}702 = 0.5864$.

Several observations should be made at this point. In terms of energy, the heat pump simply collects the energies from the compressor and the ambient air and channels them through the condenser to the user. No consideration is given to the quality of energy in the process. Because of the nature of energy conservation, the coefficient of performance always is greater than unity. However, the performance picture is quite different in terms of availability. The availability output of the heat pump is substantially lower than the availability input. This should not be surprising, since all components of the heat-pump system are imperfect, and availability loss and destruction does occur in the various processes. For example, in the compression process, there is no heat energy loss but rather a substantial amount of availability destruction. In fact, approximately 27 percent of the system availability input is destroyed within the compressor. A similar observation can be made for the expansion valve. The energy analysis indicates no heat loss. The energy input to the valve is exactly equal to its output. However, from the availability analysis we see that the process is relatively inefficient and that more than 6 percent of the system availability input is consumed.

The second law efficiency calculated above is an indicator of fuel utilization. In this example, it shows that approximately 59 percent of the system availability input (i.e., compressor work) is supplied to the user. It must be emphasized that not all of this availability will be received by the users. Figure 9-4 shows the arrangement of the condenser. On the hot side, Freon-12 flows at the rate of 600 pounds per hour, and other conditions are identical to those in the illustrative example. On the cold side, air enters at 80°F and leaves at 100°F. To determine the condenser second law efficiency, we first calculate the air flow rate by the energy equation:

$$\dot{m}_a = \frac{\dot{Q}_{\mathrm{cv}}}{c_p(T_e - T_i)} = \frac{38{,}592}{0.24(560 - 540)} = 8040 \text{ lbm/hr}$$

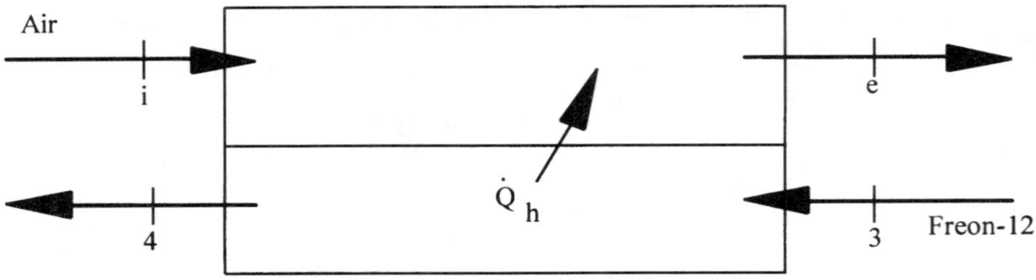

Figure 9-4 A condenser arrangement in a heat pump.

The change of flow availability in the air is given by

$$\dot{m}_a(a_{f_e} - a_{f_i})_a = \dot{m}_a[(h_e - h_i) - T_0(S_e - S_i)]_a$$
$$= 8040\left[0.24(560 - 540) - 492(0.24)\ln\frac{560}{540}\right]$$
$$= 4100.4 \text{ Btu/hr}$$

The second law efficiency of the condenser is simply the ratio of output to input in terms of availability. In this case, it is

$$\varepsilon_{\text{cond}} = \frac{\dot{m}_a(a_{f_e} - a_{f_i})_a}{\dot{m}_3(a_{f_3} - a_{f_4})}$$

With the numeric values, the condenser has the second law efficiency $\varepsilon_{\text{cond}} = 4100.4/7448.4 = 0.5505$. Finally, the overall second law efficiency of the entire heat-pump system is $\varepsilon_o = \varepsilon_{\text{cy}} \times \varepsilon_{\text{cond}} = 0.5864 \times 0.5505 = 0.3228$.

Before this section is concluded, several heat pump–related systems are introduced. The common feature of these systems is the improved utilization of fuel. In some systems, thermal energy rather than electrical energy is used to drive the heat pump and complete various industrial processes.

Figure 9-5 shows a combined system of a power plant and heat pump. In the power-plant side, steam is generated in a boiler and sent to the turbine for work production. The turbine work is exactly equal to the work input to the compressor of the heat pump. After expansion in the turbine, steam passes through the condenser and returns to the boiler through a pump. In the heat-pump side, a refrigerant such as Freon-12 will flow through the compressor, condenser, expansion valve, and evaporator and perform functions similar to those as indicated previously. In this combined system, however, the process fluid receives the heat from not only the heat-pump condenser but also the power-plant condenser. In other words, the waste heat from the power plant is used to augment the heat-pump source. This combined system will result in improved utilization of fuel.

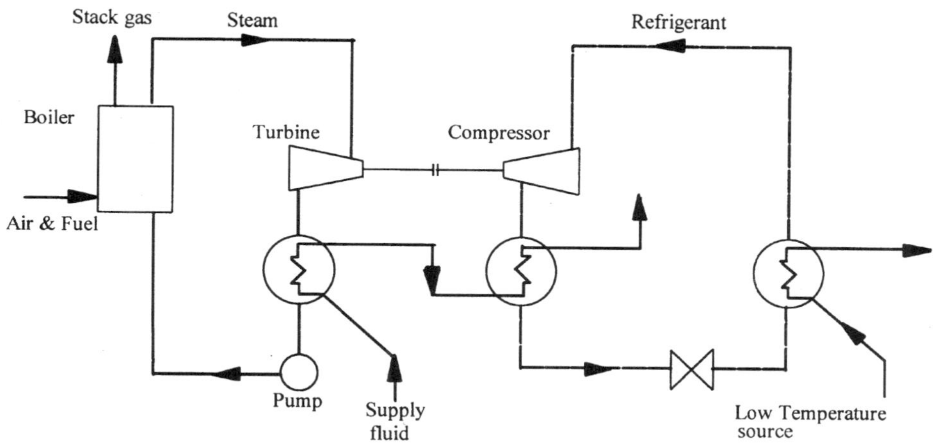

Figure 9-5 A combined system of a power plant and a heat pump.

Figure 9-6 presents the schematic diagram of a solar-energy system with a heat pump. The system consists of two parts: the heat pump and the solar heating assembly. The process fluid receives heat from solar-energy storage and transfers it to the evaporator of the heat pump. The process fluid does not mix with the refrigerant of the heat pump. In the winter, the solar collectors provide all or part of the required heat. Since the process fluid has a relatively high temperature, the heat pump is expected to have a high coefficient of performance or better fuel utilization.

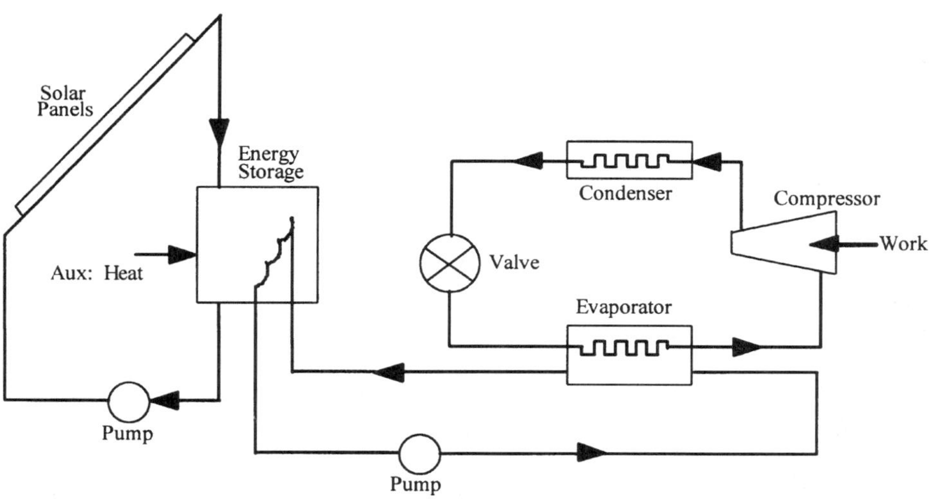

Figure 9-6 A heat pump with a solar-energy system.

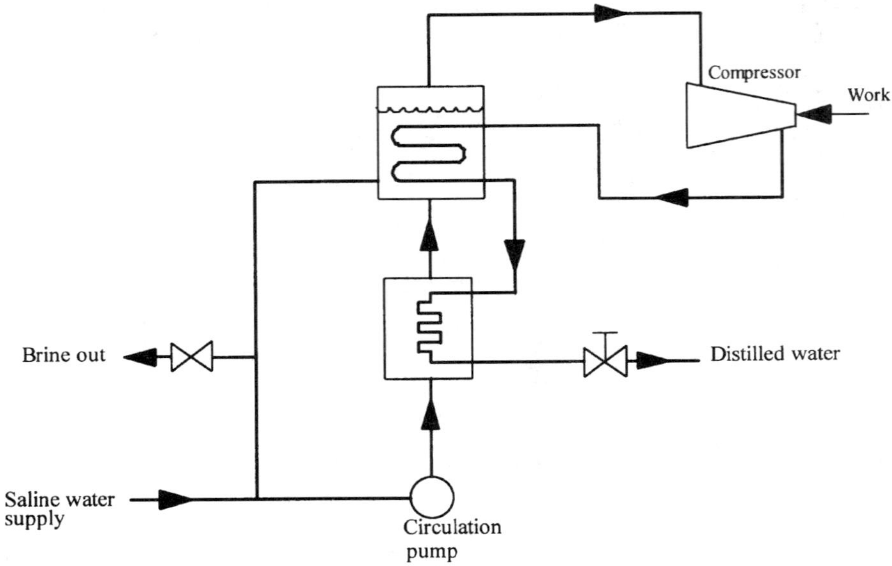

Figure 9-7 A water-distillation system with heat pump.

Figure 9-7 shows a waste-distillation system with a heat pump. Saline water enters the system through a circulating pump. Water is preheated in a heat exchanger and vaporized in the condenser of the heat pump. Generally, the pressure in the chamber is approximately equal to 14.7 psia. As indicated in Fig. 9-7, the brine leaves the system while the saturated steam enters the compressor. After the compression process, steam will pass through the condenser and heat exchanger. The steam will provide thermal energy at each step and will eventually reach the state of subcooled liquid. Finally, the distilled water leaves the system through a control valve.

Another water-distillation system with a heat pump is shown in Fig. 9-8. In this system the working substance of the heat pump, a typical refrigerant like R-114, would not mix with the saline or distilled water. The condenser and evaporator of the heat pump are located inside a horizontal cylinder. Saline water is vaporized in the cylinder, and distilled water is produced as indicated in Fig. 9-8. Because this is no mixing of the working substance and the water, there is much more flexibility in design and operation.

9-3 GAS REFRIGERATION SYSTEMS

This section presents various gas refrigeration systems. Figure 9-9 shows a typical gas refrigeration system in which air follows the path of the reversed Brayton cycle. In the piping between locations 1 and 2, the air experiences both

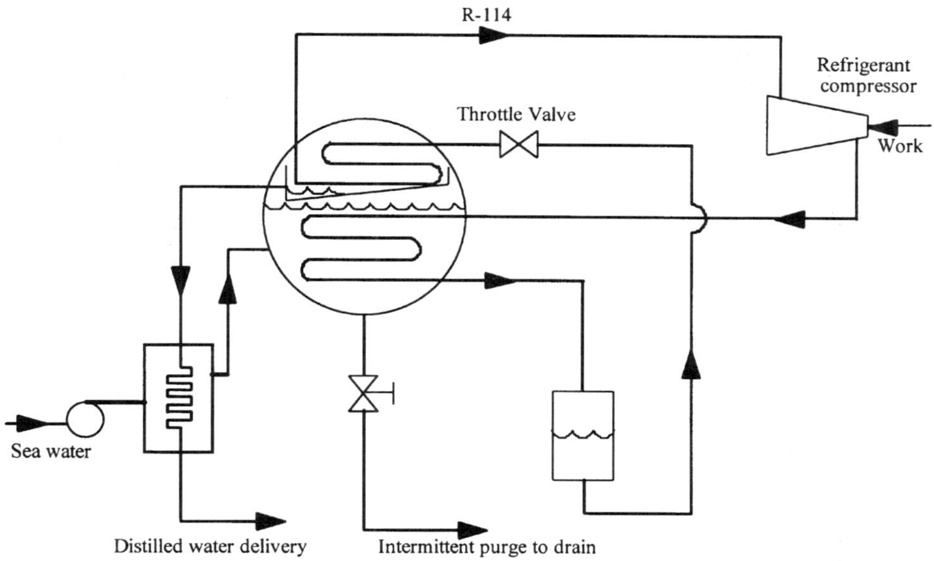

Figure 9-8 A water-distillation system with a vapor-compression heat pump.

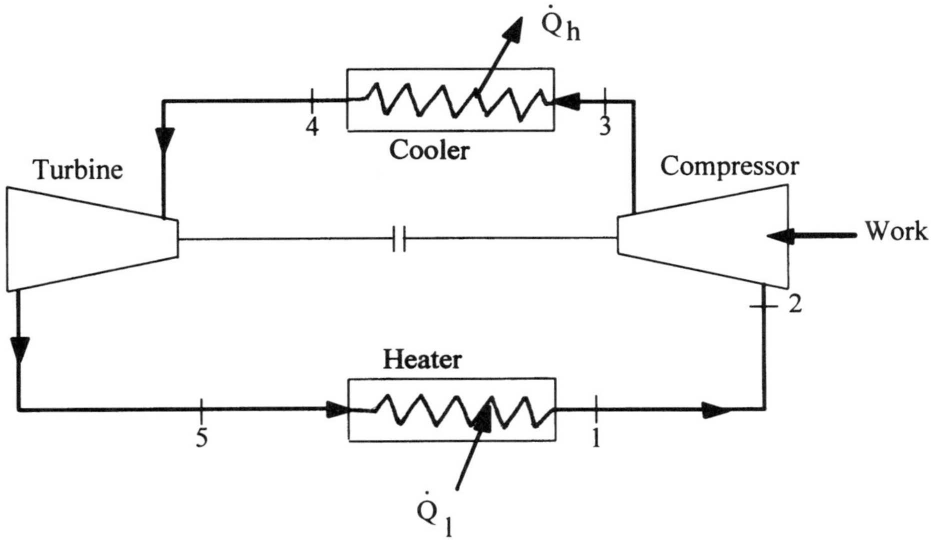

Figure 9-9 A gas refrigeration system.

a pressure drop because of friction and a heat gain because of insufficient insulation. Next, the air passes through a compressor and then a heat exchanger. In the heat exchanger, the air is cooled as a result of heat transfer to the surroundings. In the actual refrigeration system, a pressure drop is expected across the heat exchanger. After the cooling process from 3-4, the air enters an expander and then another heat exchanger. In the second heat exchanger, the air receives heat from the refrigerated space and completes the cycle. Unlike the air-standard refrigeration cycle discussed in Chapter 7, the actual system has nonisentropic processes for both the compressor and the expander. Also, a pressure drop is expected to occur throughout the gas refrigeration system.

The following is an energy and availability analysis of a typical gas refrigeration system. The air in the system has the conditions for various locations as given in Table 9-5. The refrigeration system will be examined in terms of its processes.

Process 1-2 (Piping)

We take the piping section 1-2 as a control volume. The heat gain by the control volume is determined by the energy equation.

$$q_{1\text{-}2} = h_2 - h_1 = c_p(T_2 - T_1) = 0.24(2.0 - 0) = 0.48 \text{ Btu/lbm}$$

While the air receives heat in the piping, it loses flow availability. The reduction of flow availability is due mainly to the availability loss and destruction in the pipe. The availability equation for a steady-state and steady-flow process is

$$a_q + a_w + a_{f_1} - a_{f_2} - i = 0$$

Since $a_w = 0$, the above equation becomes

$$i - a_q = a_{f_1} - a_{f_2} = c_p(T_1 - T_2) - T_0\left[c_p \ln\frac{T_1}{T_2} - R\ln\frac{P_1}{P_2}\right]$$

Table 9-5 Conditions of the air in the gas refrigeration system of Figure 9-9.

State	Temperature (°F)	Pressure (psia)
1	0	16.0
2	2	14.7
3	300	80.0
4	60	78.0
5	−110	18.0
Environmental state	0	14.7

With the numeric values, we have the availability loss and destruction as

$$i - a_q = 0.24(460 - 462) - 460\left[0.24 \ln \frac{460}{462} - \frac{53.3}{778} \ln \frac{16.0}{14.7}\right] = 2.67 \text{ Btu/lbm}$$

Process 2-3 (Compressor)

The process is assumed to be adiabatic, so the compressor work is calculated with the energy equation

$$-w_c = h_3 - h_2 = c_p(T_3 - T_2) = 0.24(760 - 462) = 71.52 \text{ Btu/lbm}$$

The availability destruction in this process is given by $i = a_{f_2} - a_{f_3} - w_c$. The above equation can be further modified and expressed as

$$i = T_0(s_3 - s_2) = T_0\left[c_p \ln \frac{T_3}{T_2} - R \ln \frac{P_3}{P_2}\right]$$

With numeric values, we have the availability destruction in the compressor

$$i = 460\left[0.24 \ln \frac{760}{462} - \frac{53.3}{778} \ln \frac{80}{14.7}\right]$$
$$= 1.55 \text{ Btu/lbm}$$

Process 3-4 (cooler)

Heat is transferred from the air to the surrounding in process 3-4. The energy equation provides

$$-q_{3\text{-}4} = h_3 - h_4 = c_p(T_3 - T_4) = 0.24(760 - 520) = 57.6 \text{ Btu/lbm}$$

Associated with this heat transfer, there is a reduction of flow availability. That is,

$$a_{f_3} - a_{f_4} = c_p(T_3 - T_4) - T_0\left[c_p \ln \frac{T_3}{T_4} - R \ln \frac{p_3}{p_4}\right]$$

Substituting the numeric values into the equation, we have

$$a_{f_3} - a_{f_4} = 0.24(760 - 520) - 460\left[0.24 \ln \frac{760}{520} - \frac{53.3}{778} \ln \frac{80}{78}\right] = 16.49 \text{ Btu/lbm}$$

Process 4-5 (Expander)

The work by the expander is given by

$$w_t = h_4 - h_5 = c_p(T_4 - T_5) = 0.24(520 - 350) = 40.8 \text{ Btu/lbm}$$

The availability destruction in the process is

$$i = T_0(s_5 - s_4)$$
$$= T_0\left[c_p \ln\frac{T_5}{T_4} - R\ln\frac{p_5}{p_4}\right]$$
$$= 460\left[0.24\ln\frac{350}{520} - \frac{53.3}{778}\ln\frac{18}{78}\right]$$
$$= 2.52 \text{ Btu/lbm}$$

Process 5-1 (Heater)

Heat is transferred to the air from the refrigerated space in process 5-1. The first law equation gives

$$q_{5\text{-}1} = h_1 - h_5 = c_p(T_1 - T_5) = 0.24(460 - 350) = 26.4 \text{ Btu/lbm}$$

While heat flows in the direction from the refrigerated space to the air, the associated heat availability flows in the opposite direction. The heat availability transfer is

$$a_{f_5} - a_{f_1} = c_p(T_5 - T_1) - T_0\left[c_p \ln\frac{T_5}{T_1} - R\ln\frac{P_5}{P_1}\right]$$
$$= 0.24(350 - 460) - 460\left[0.24\ln\frac{350}{460} - \frac{53.3}{778}\ln\frac{18}{16}\right]$$
$$= 7.50 \text{ Btu/lbm}$$

To summarize the above calculations, we present the energy balance and availability balance on the refrigeration system in Tables 9-6 and 9-7, respectively. Finally, we calculate the coefficient of performance for the gas refrigera-

Table 9-6 Energy balance of the refrigeration system of Figure 9-9.

Component	Input (Btu/lbm)	Output (Btu/lbm)
Piping	0.48	
Compressor	71.52	
Cooler		57.6
Expander		40.8
Heater	26.40	
Total	98.40	98.40

Table 9-7 Availability balance of the refrigeration system of Figure 9-9.

Component	Input (Btu/lbm)	Output (Btu/lbm)	Loss and destruction (Btu/lbm)
Piping			2.67
Compressor	71.52		1.55
Cooler			16.49
Expander		40.8	2.52
Heater		7.50	
Total	71.52	48.30	23.23

tion system as

$$COP_{ref} = \frac{q_{5\text{-}1}}{w_c - w_t} = \frac{26.4}{71.52 - 40.8} = 0.8594$$

and the second law efficiency as

$$\varepsilon_{ref} = \frac{a_{f_5} - a_{f_1}}{w_c - w_t} = \frac{7.50}{71.52 - 40.8} = 0.2441$$

It is of interest to note that the second law efficiency of the gas refrigeration system is much lower than the coefficient of performance. This is due mainly to the fact that in the availability analysis attention is paid not only to the quantity of the energy but also to its quality. For the same reason, the availability balance also is very different from the energy balance. In the availability analysis, the principle of conservation does not apply. The availability input to the refrigeration system is not equal to the availability output, and the difference is the availability loss and destruction. It should be emphasized that in the above example no consideration is given to the conditions of the refrigerated space. If the conditions are specified, the second law efficiencies of the cooler and of the overall refrigeration system can be calculated.

A gas refrigeration system can be modified and utilized for cooling of aircraft cabin. Figure 9-10 shows a schematic diagram for an open-cycle aircraft-cabin cooling. As indicated, air leaves the cooler at location 1. After cooling, air either enters an air turbine or passes through a control valve. Upon leaving these devices, the cool air is blown directly into the aircraft cabin. Generally, the cabin is pressurized, and for obvious reasons replacement of stale air usually is required.

To improve the performance of the gas refrigeration system, one approach is to incorporate a counterflow heat exchanger, as shown in Fig. 9-11. This heat exchanger often is called a regenerator or recuperator. A refrigeration system with a regenerator usually is used to produce very low temperatures. The system is similar to those used in air liquefaction plants or cryogenic refrigeration plants.

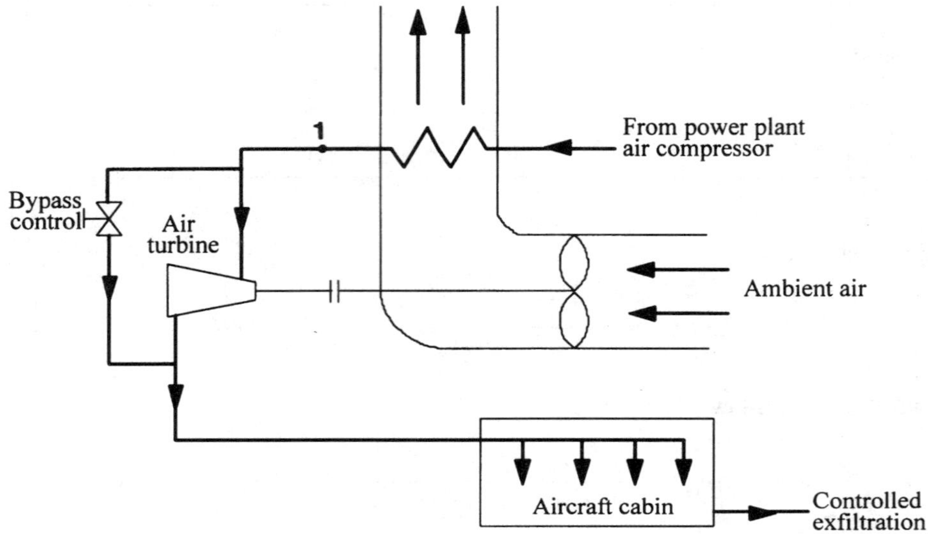

Figure 9-10 A schematic diagram for aircraft cabin cooling.

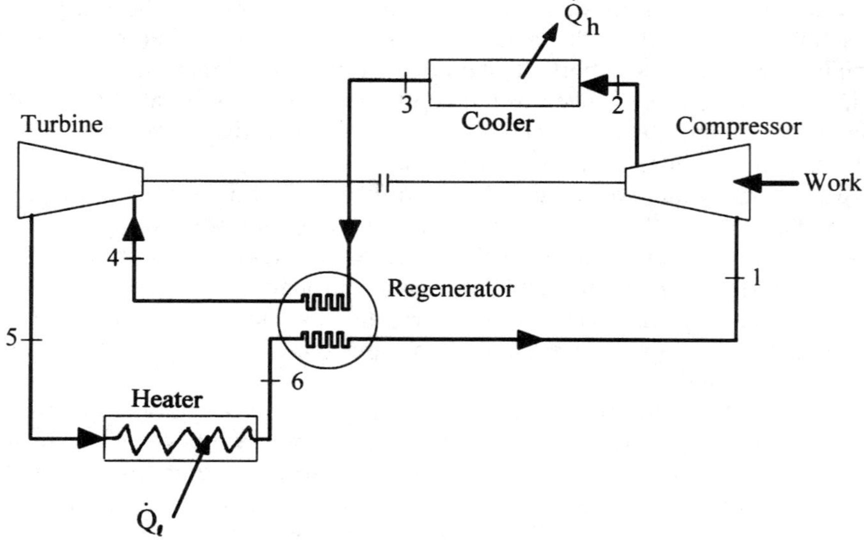

Figure 9-11 A gas refrigeration system with a regenerator.

9-4 VAPOR-COMPRESSION REFRIGERATION SYSTEMS

The ideal vapor-compression refrigeration cycle has been presented in Chapter 7. Practical refrigerating plants that operate on a simple vapor-compression cycle differ from the ideal in the following aspects:

(1). Expansion is unresisted through a simple throttling valve instead of the work-producing expander or turbine;
(2). All processes are irreversible, including heat transfer and pressure drop in various locations of the refrigeration system; and
(3). Refrigerant is subcooled at the condenser exit and slightly superheated at the compressor inlet.

The following is an energy and availability analysis for a typical vapor-compression refrigeration system, as shown in Fig. 9-12. The refrigerant is Freon-12, and the conditions are listed in Table 9-8.

We first complete energy calculations for the components. Using the absolute values, we have for the compressor, $w_c = h_2 - h_1 = 92.99 - 76.49 = 16.5$ Btu/lbm; for the condenser, $q_h = h_2 - h_3 = 92.99 - 26.37 = 66.62$ Btu/lbm; and for the evaporator, $q_1 = h_1 - h_4 = 76.49 - 26.37 = 50.12$ Btu/lbm. With these calculated values we can determine the coefficient of performance for the vapor-compression refrigeration system as

$$\mathrm{COP}_{\mathrm{ref}} = \frac{q_1}{w_c} = \frac{50.12}{16.5} = 3.04$$

The energy balance is given in Table 9-9.

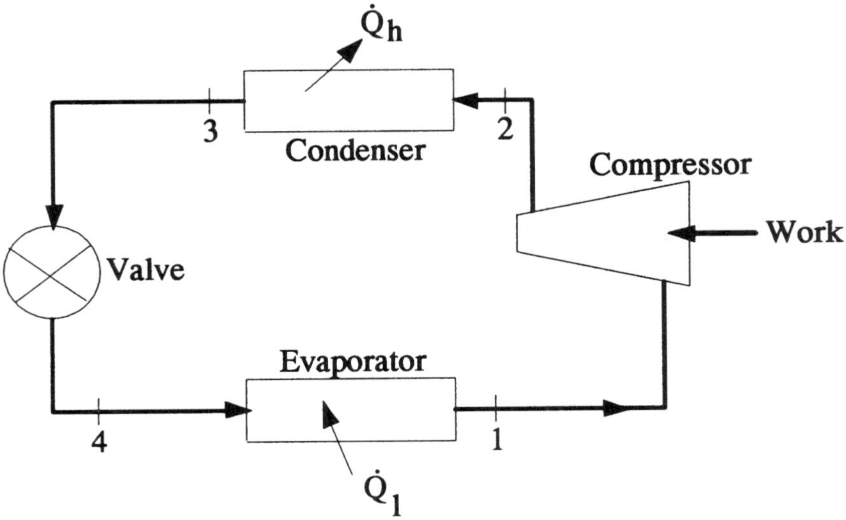

Figure 9-12 A vapor-compression refrigeration system.

Table 9-8 The conditions of the vapor-compression refrigeration system in Figure 9-12.

State	Temperature (°F)	Pressure (psia)	Enthalpy (Btu/lbm)	Entropy (Btu/lbm-R)
1	−10	15.0	76.49	0.1745
2	130	120.0	92.99	0.1748
3	80	114.5	26.37	0.0548
4	−15	17.14	26.37	0.0595
Environmental state	0	14.7	8.52	0.0193

Now, we start availability calculations for the components. The conditions of environmental state are assumed 14.7 psia and 0°F. The results are $i = T_0(s_2 - s_1) = 460(0.1748 - 0.1745) = 0.138$ Btu/lbm for the compressor; $a_{f_2} - a_{f_3} = (h_2 - h_3) - T_0(s_2 - s_3) = (92.99 - 26.37) - 460(0.1748 - 0.0548) = 11.42$ Btu/lbm for the condenser; $i = T_0(s_4 - s_3) = 460(0.0595 - 0.0548) = 2.16$ Btu/lbm for the valve; and $a_{f_4} - a_{f_1} = (h_4 - h_1) - T_0(s_4 - s_1) = (26.37 - 76.49) - 460(0.0595 - 0.1745) = 2.78$ Btu/lbm for the evaporator. With these calculated values, we determine the second law efficiency as

$$\varepsilon_{\text{ref}} = \frac{a_{f_4} - a_{f_1}}{w_c} = \frac{2.78}{16.5} = 0.1685$$

The availability balance is given in Table 9-10.

The second law efficiency for the basic vapor-compression refrigeration system is relatively low, due mainly to large compressor work. To reduce the compressor work, a multistage compression with intercooling usually is adopted. Figure 9-13 presents one scheme for two-stage compression with intercooling. The refrigerant leaving the condenser is throttled into a flash chamber from which the saturated vapor is transferred to a mixing chamber, and the saturated liquid is further throttled to the evaporator pressure. In the mixing chamber the saturated vapor from the flash chamber will be mixed with the refrigerant leaving the low-pressure compressor. In other words, the mixing chamber will act as an intercooler to the refrigerant from the low-pressure compressor. Also

Table 9-9 The energy balance of the vapor-compression refrigeration system in Figure 9-12.

Component	Input (Btu/lbm)	Output (Btu/lbm)
Compressor	16.5	
Condenser		66.62
Valve		
Evaporator	50.12	
Total	66.62	66.62

Table 9-10 The availability balance of the vapor-compression refrigeration system in Figure 9-12.

Component	Input (Btu/lbm)	Output (Btu/lbm)	Loss and destruction (Btu/lbm)
Compressor	16.5		0.138
Condenser			11.42
Valve			2.16
Evaporator		2.78	
Total	16.5	2.78	13.72

in Fig. 9-13 is the *T-s* diagram for this two-stage compression system. For simplicity, the compressor processes are assumed to be isentropic, and the pressure drops in various locations are considered negligible. While there is one refrigerant for the entire system, its flow rate obviously will vary from location to location. The flow rate for the low pressure compressor is determined by the refrigeration requirement. That is,

$$\dot{m}_1 = \frac{\dot{Q}_1}{h_1 - h_9} \tag{9-20}$$

To determine the flow rate for the high-pressure compressor, we take the flash chamber and mixing chamber together as a control volume and apply the first law equation:

$$\dot{m}_6 h_6 + \dot{m}_2 h_2 = \dot{m}_8 h_8 + \dot{m}_3 h_3 \tag{9-21}$$

Since $\dot{m}_8 = \dot{m}_2 = \dot{m}_1$ and $\dot{m}_3 = \dot{m}_6$, we can rewrite Eq. (9-21) as $\dot{m}_3(h_6 - h_3) = \dot{m}_1(h_8 - h_2)$, or

$$\dot{m}_3 = \dot{m}_1 \frac{h_8 - h_2}{h_6 - h_3} \tag{9-22}$$

The compressor work for this refrigeration system is simply the sum of the two compressor works. That is,

$$\dot{W}_c = \dot{m}_1(h_2 - h_1) + \dot{m}_3(h_4 - h_3)$$

Finally, the coefficient of performance is the ratio of refrigeration effect $\dot{Q}_1$ to the compressor work $\dot{W}_c$. The availability analysis is similar to that for the basic refrigeration system.

There are some applications in which a relatively low temperature is desired. Because of the nature of the refrigerant, the basic refrigeration system usually fails to produce the relatively low temperature. To overcome this difficulty, frequently a binary system similar to that for power generation is used. Figure 9-14 presents a typical binary system for refrigeration. The binary system, referred to also as a cascade system, comprises of two basic vapor-

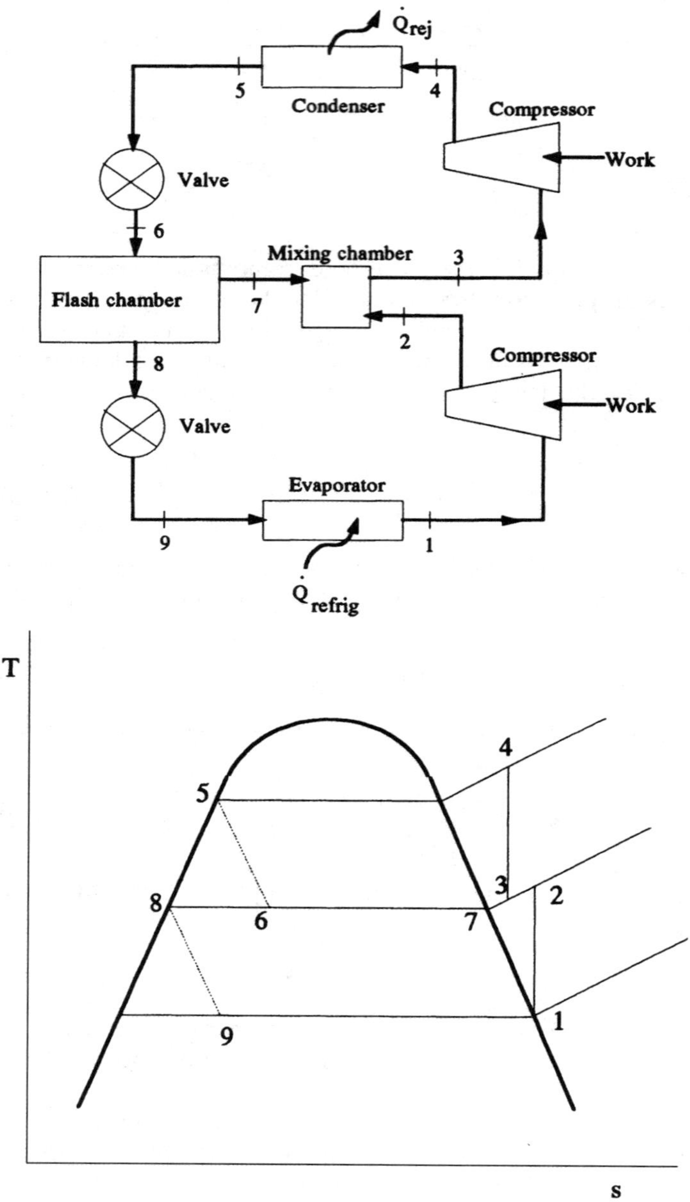

Figure 9-13 A two-stage vapor-compression refrigeration system and its *T-s* diagram.

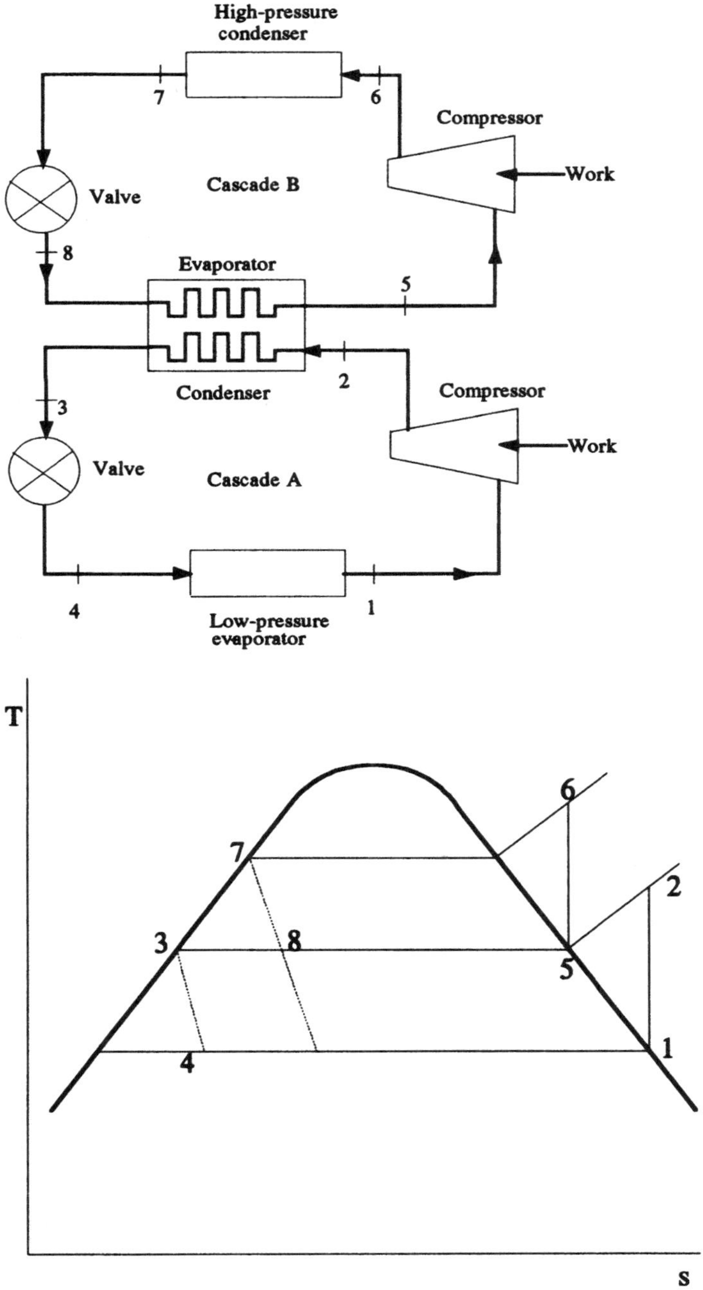

Figure 9-14 A binary refrigeration system and its *T-s* diagram.

compression systems, each having its own refrigerant. Also in Fig. 9-14 is the *T-s* diagram for a binary refrigeration system. For simplicity, the compressor processes are assumed to be isentropic, and the pressure drops in various locations are considered negligible.

In the binary system, a heat exchanger is used to couple these two basic refrigeration systems. The heat exchanger would act as an evaporator for the first system and as a condenser for the second. When the heat exchanger is taken as a control volume, the first law equation gives

$$\dot{m}_A(h_2 - h_3) = \dot{m}_B(h_5 - h_8)$$

or

$$\dot{m}_B = \dot{m}_A \frac{h_2 - h_3}{h_5 - h_8}$$

where $\dot{m}_A$ and $\dot{m}_B$ are the mass flow rate of refrigerant for systems A and B, respectively. The mass flow rate $\dot{m}_A$ is determined by refrigeration requirement. In this case it is

$$\dot{m}_A = \frac{\dot{Q}_1}{h_1 - h_4}$$

The availability calculation for the binary system is left as an exercise.

9-5 PROBLEMS

9-1 The heat pump system shown in Figure 9-3 has the following conditions:

Refrigerant	Freon-12
Refrigerant flow rate	600 lbm/hr
Compressor efficiency	80%

Temperature and pressure at various locations:

State	T (°F)	P (psia)
1	25	39.3
2	30	38
3	200	220
4	125	215
5	38.15	50

Prepare the availability balance and calculate the second law efficiency using $T_0 = 32°F$ and $P_0 = 14.7$ psia. The ambient air temperature is assumed to be 40°F.

9-2 Prepare a preliminary design, including at least a flow diagram indicating mass flow rate and thermodynamic properties at various locations, for the combined power-plant and heat-pump system as shown in Fig. 9-5. The combined system is used to provide a heat supply of 10^6 Btu/hr to an office building. The following conditions must be satisfied:

Refrigerant	Freon-12
Water temperature range	80–150°F
Ambient air temperature	40°F

Also determine the ratio of heat delivery to the fuel energy.

9-3 Prepare an availability analysis for the combined power-plant and heat-pump system as designed in Problem 9-2. The analysis should indicate at least availability input, output, and loss and destruction for all components as well as for the entire combined system.

9-4 A heat pump with Freon-12 as the refrigerant uses ground water at 50°F as an energy source. The heat pump is designed to provide the heat energy at the rate of 10^3 Btu/hr with pressures of 30 psia and 160 psia in the evaporator and condenser, respectively. For simplicity, the pressure drops in both the evaporator and condenser are neglected, and the compressor efficiency is assumed to be 80 percent. Calculate (a) the coefficient of performance and (b) the horsepower for the compressor.

9-5 For the heat pump system in Problem 9-4, calculate (a) the availability loss and destruction in the throttling valve and compressor, and (b) the change in flow availability experienced by Freon-12 in the evaporator and condenser. Also, prepare the availability balance for the entire heat-pump system.

9-6 A refrigeration system has Freon-12 as a working substance. The following are the temperatures and pressures at various locations in Figure 9-15.

State	Temperature (°F)	Pressure (psia)
1	20	18
2	220	180
3	180	175
4	110	170
5	104	168
6, 7		20
8	0	19

The heat loss from the Freon-12 in the compression process, which occurs at the location with a surface temperature of 200°F, is 2.0 Btu/lbm. Prepare the heat and availability balance. Also calculate the second law efficiency for the refrigeration system.

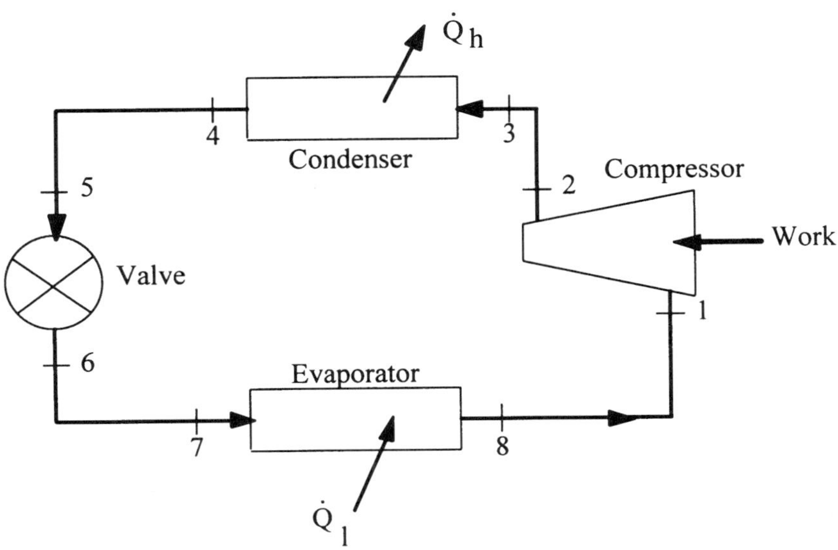

Figure 9-15 A schematic diagram for Problem 9-6.

9-7 A two-stage refrigeration system shown in Figure 9-13 utilizes Freon-12 as a working substance and has pressures of 20, 40, and 120 psia in the evaporator, flash and mixing chamber, and condenser, respectively. Assuming that both compressors have an isentropic efficiency of 85 percent and that the pressure drop is negligible except across the expansion value, calculate the following per ton of refrigeration: (a) the compressor work, (b) the Freon-12 flow rate in the condenser, (c) the coefficient of performance, (d) the availability balance for the entire refrigeration system using $T_0 = 0°F$, and (e) the second law system efficiency.

9-8 A cascade refrigeration system shown in Figure 9-14 has the following temperatures and pressures at various locations

State	Temperature (°F)		Pressure (psia)
		Fr-13	
1	−60		50
2	62		240
3	18		230
4			60
		Fr-12	
5	0		20
6	120		60
7	48		65
8			22.5

Calculate the following per ton of refrigeration: (a) the compressor work, (b) the Fr-12 and Fr-13 flow rates, (c) the coefficient of performance, (d) the availability balance for this cascade refrigeration system using $T_0 = 0°F$, and (e) the second law system efficiency.

9-9 A simple gas refrigeration system has the following conditions:

Compressor	
inlet conditions	14.7 psia, 80°F
pressure ratio	3.2
adiabatic efficiency	83%
Turbine	
inlet temperature	40°F
adiabatic efficiency	85%

The gas properties are assumed to be identical to those of the air. Determine the coefficient of performance and the second law efficiency.

9-10 A heat exchange is incorporated into a gas refrigeration system as shown in Figure 9-11. The temperatures and pressures at various locations are shown below:

Locations	Temperature (°F)	Pressure (psia)
1	60	14.7
2		200
3	60	200
4	−60	200
5		14.7
6	−60	14.7

The gas properties are assumed to be identical to those of the air, and the first law efficiency for the compressor and expander is 75 percent. Determine the coefficient of performance and the second law efficiency for this refrigeration system.

CHAPTER
TEN

COGENERATION

The production of electricity is a relatively inefficient process. The efficiency of modern power plants ranges from 32% for a nuclear power plant to 43% for advanced, sophisticated gas/steam combined-cycle plants. This means that between 57 and 68% of the fuel energy is being discharged as waste heat. Some of the waste heat such as radiation and stack loss from the boiler, is unrecoverable, but much of the waste heat can be used for industrial applications or for district heating. By utilizing waste heat, the overall conversion of fuel to usable energy can be as high as 80%.

Cogeneration is the simultaneous production of more than one form of useful energy. In industry, cogeneration has been used as a means of producing both electric and thermal energy. The cogeneration system sometimes is referred to as the total-energy system or waste heat-utilization system. Cogeneration is not a new concept. In fact, cogeneration was implemented in the United States at the end of the nineteenth century. In 1900, most industrial plants generated their own utilities, but in 1960, they generated only 21% of the electricity they consumed. This percentage continuously declined to 17% by 1968, primarily because of the availability and relatively low price of fossil fuels. The decline of cogeneration was due also to the institutional constraints and the economic distortion they produced. But in recent years the combined impact of increasing primary fuel costs and inflationary capital cost of energy systems have made energy-saving measures economically viable. Energy conservation technology, ignored in the past, again has become important in all industries.

A cogeneration system requires equipment that burns fuel to produce shaft work (including electric power) and thermal energy. Thermal energy may be used as dry heat for an industrial process or converted to steam. The equipment for that purpose is called a prime mover. Steam and gas turbines and diesel engines have been used as prime movers in industry. There are two basic arrangements for cogeneration systems. One is a topping cycle, and the other is a bottoming cycle. Figure 10-1 shows the schematic diagrams of these two basic systems. In the topping cycle, the prime mover is used to generate electric power, and the waste heat and the by-product steam from it are used for plant processes. Topping cycles are significantly more fuel-efficient than are conventional systems that generate electric energy and thermal energy separately.

The bottoming cycle is completely different from the topping cycle. In the bottoming cycle, plant processes are run by their own fuel, and the waste heat from them is utilized for electric-power generation. Bottoming cycles also save fuel when compared with conventional systems.

This chapter is intended to present a brief description of various cogeneration systems. It also includes the first and second law analyses. In addition, attention is focused to the total system, which is defined as the cogeneration unit plus standby boilers and the utility network.

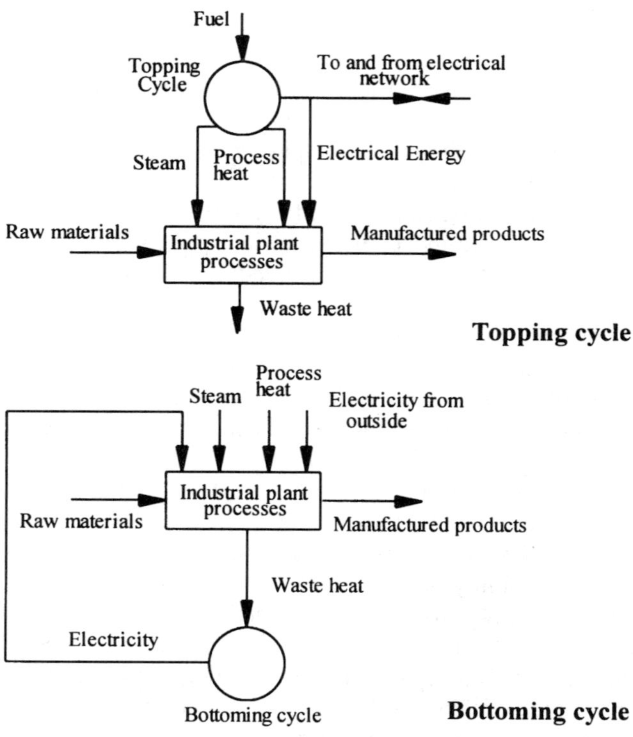

Figure 10-1 Two basic arrangements for a cogeneration system.

10-1 COGENERATION SYSTEM DESCRIPTIONS

The cogeneration systems can be classified according to the prime mover. These systems can include steam turbines, gas turbines, and diesel engine. Below is a brief description of all these cogeneration systems.

Steam Turbine

The steam turbine is the most common prime mover for cogeneration systems. There are several types of turbogenerators that can meet the widely varying demands for electric and thermal energy. One of these is the back-pressure turbine, which takes steam from the boiler and produces electric power by expanding the steam through it. Upon discharge from the turbine, the steam is directed to the steam users. While this system is simple and efficient, its major drawback is that it simultaneously tries to meet the electric power and thermal energy needs.

One convenient approach to balance electric power and process steam demands is to connect the cogeneration system with an electric power network, as shown in Fig. 10-2. In this arrangement the cogeneration system is operated according to the need of process steam and the steam flow determines the output of the turbine-generator. When the electric output exceeds the need, the surplus is exported to the electric power network. Conversely, the electric power is imported to cover any power deficiency.

Figure 10-2 presents another arrangement that would meet the electric and steam energy needs simultaneously. It involves use of pressure-reducing valve (including a desuperheater) and/or a surplus steam valve. In this case the turbine is speed-dependent and passes a quantity of steam corresponding to the electrical demand. The reducing valve makes up for any process steam deficiency, and the surplus valve takes care of any surplus steam.

When the process steam demand is relatively small compared to the electric power need, an extraction turbine may provide the best solution. Figure 10-3 shows a schematic diagram for this arrangement. The extraction turbine receives steam from the boiler and expands it to a pressure equal to that of the process steam. At this point some of the steam is extracted as process steam, and the remainder of the steam reenters the turbine through a control valve for final expansion to the condenser pressure. The turbine is speed-dependent and the pressure regulator is used to maintain the process steam at a desirable pressure level.

The extraction turbine has one advantage over the back-pressure turbine: It can easily meet the demand for both electrical and thermal energy. In turbine selection it is important that the steam inlet temperature and pressure should be as high as possible. These throttle conditions enable the turbine to generate more electricity per unit of process steam. However, high throttle steam conditions require more expensive equipment.

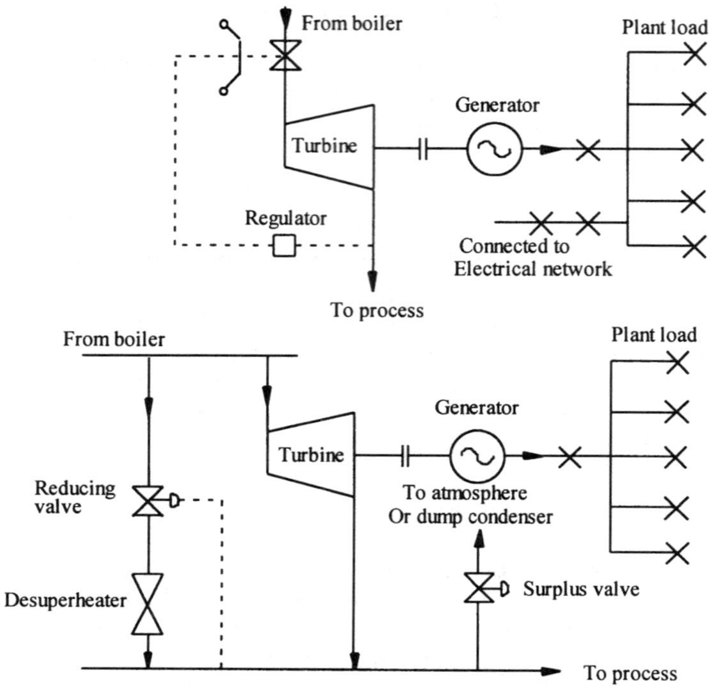

Figure 10-2 Back-pressure steam turbine in a cogeneration system.

The boiler is a key element in steam turbine cogeneration system. When the steam turbine is used as a prime mover in a topping cycle, the steam pressure and temperature are relatively high, and therefore the boiler should be evaluated and selected carefully. Based upon current energy economics, it is likely that boilers will be designed to burn coal or residual oil.

Gas Turbine

The gas turbine described in previous chapters is a convenient and compact system for electric-power generation. It usually operates at temperatures higher than those for the steam turbine and exhausts gas at about 1000°F. The thermal energy of the exhaust gas can be recovered in a variety of ways, as shown in Fig. 10-4. For instance, the exhaust gas can be used directly for heating and drying, which is important in the food industry. It can be also passed through a heat exchanger where heat is transferred to a process fluid such as air, water, or oil. It can also enter a waste heat–recovery boiler to produce steam.

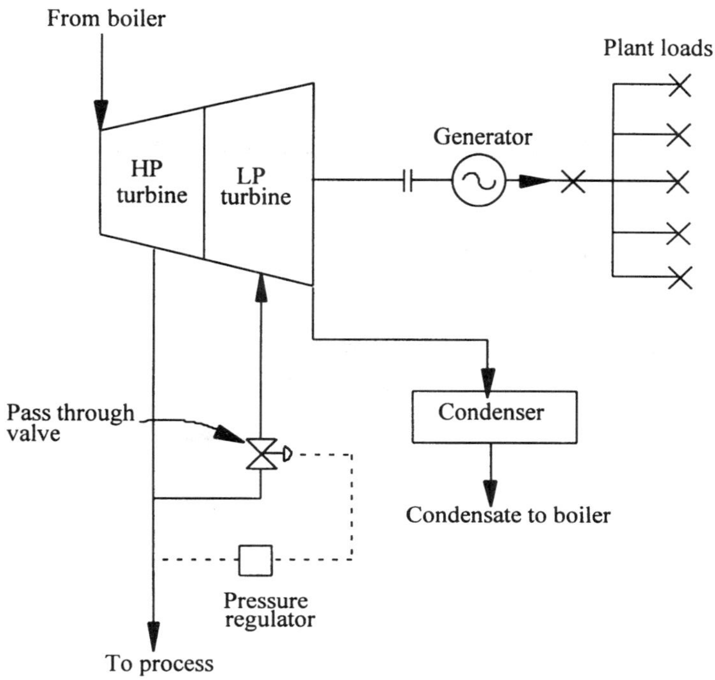

Figure 10-3 Extraction steam turbine in a cogeneration system.

The gas turbine has a drawback. Not only does it have low efficiency in electric production, but it also has a high fuel cost. While the use of gas turbines for cogeneration is expected to increase, gas turbines will not sell well until coal-derived fuels are economically available. When compared with the steam turbine, the gas turbine usually is more suitable for most cogeneration units. Because a gas turbine operates at a high temperature, it requires more maintenance. Major overhauls are required for every 10,000 to 20,000 hours of operation.

A combined gas turbine and steam turbine system has been used for electric-power generation. It has a higher efficiency than the gas turbine or steam turbine systems. When used for cogeneration, the condensing steam turbine usually is replaced by a back-pressure turbine. Figure 10-5 shows a schematic diagram for a combined-cycle cogeneration unit.

Diesel Engine

Diesel engines have relatively high thermal efficiency, usually around 39%. The waste heat rejected from a diesel engine is mainly in the form of exhaust gas and

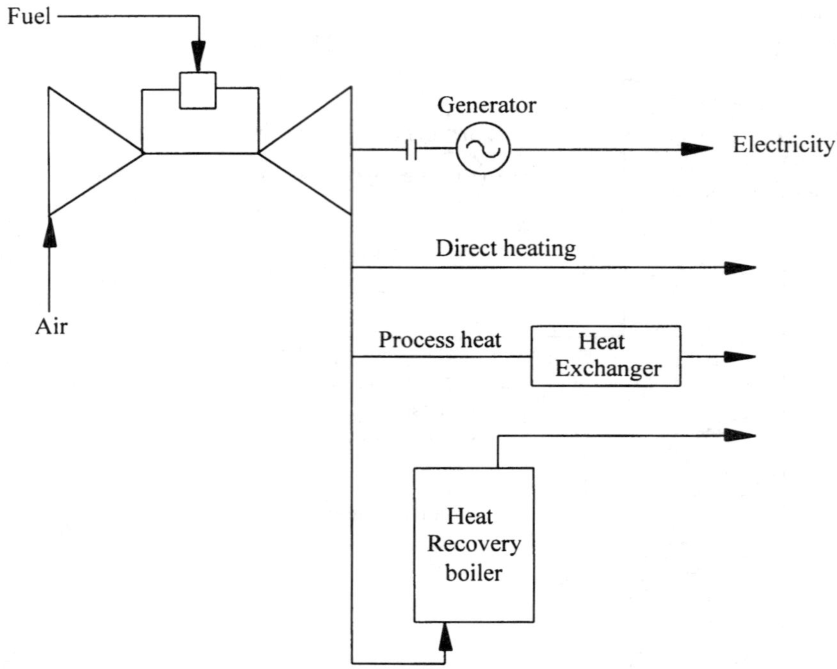

Figure 10-4 Gas turbine in a cogeneration system.

thermal energy carried away from the engine jacket water. At full load, a typical diesel engine would have the heat balance given below:

Shaft work	39.20%
Heat in exhaust gas	33.20
Heat in jacket water	13.84
Heat in lubricating oil	4.61
Radiation and other losses	9.15

It should be noted that not all waste heat can be recovered. For instance, only about 20% of the exhaust gas energy is economically usable.

Figure 10-6 presents a diesel engine cogeneration arrangement. The diesel engine drives an electric generator and exhausts the gas into a waste heat–recovery boiler for steam generation. For economic reasons, the heat removed in the engine jacket is utilized also.

The diesel engine has a small unit size and usually generates no more than 30,000 kW. A major drawback of diesels is the cost of the fuel it consumes. Unless coal-derived fuels are made more economical and feasible for use in

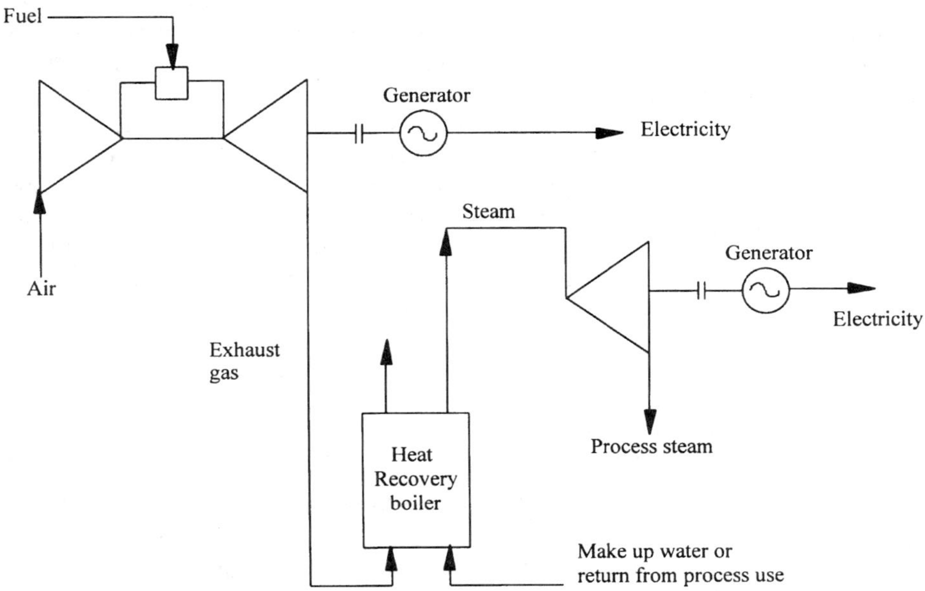

Figure 10-5 A combined-cycle cogeneration system.

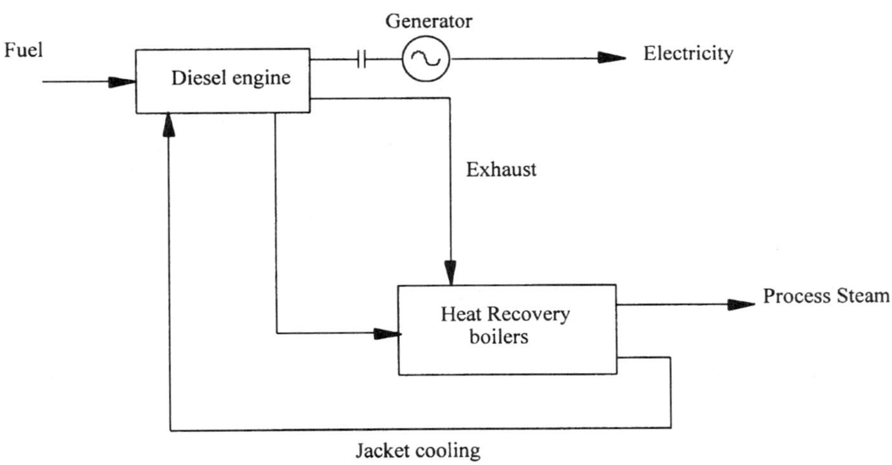

Figure 10-6 Diesel engine in a cogeneration system.

diesels, the market for diesel engines will not increase significantly in the near future.

The diesel engine cogeneration system has a relatively high electricity-to-thermal-energy ratio (E/T). Electricity generation per unit of thermal energy produced is around 400 kWh per million Btu. Therefore, to use the diesel engine as a prime mover in cogeneration, the excess of electric power must be put to productive use, either by selling it or by storing it for later use.

The selection of the prime mover for cogeneration can be made only on a system-by-system basis. Each system has its own requirements for thermal and electrical energy. The types of fuel and space available also would affect the decision. As mentioned earlier, steam and gas turbines and diesel engines have different thermal- and electrical-energy-generating capabilities. The typical ratios of electrical energy to thermal energy are

Prime Mover	E/T (kWh/MBtu)
Steam turbine	36
Gas turbine	175
Diesel engine	400

In cogeneration system design, engineers should match the energy demand with the system operating characteristics. They should consider whether the excess electrical energy can be sold. For instance, assume that a plant has thermal and electrical energy demands of 500 MBtu/hr and 40 MW, respectively. When a steam turbine is selected as a prime mover and designed to meet the plant thermal-energy need, the steam turbine would generate only 18 MW of electric power, which is 22 MW less than the plant electric-power need. However, a gas turbine would generate approximately 87 MW of electricity, producing 47 MW surplus. When a diesel engines selected and designed to meet the plant thermal-energy demand, it would generate approximately 200 MW, yielding a surplus of 160 MW.

Generally, the steam turbine has the best ratio of electric-to-thermal energy to meet the needs of various industrial cogeneration systems. Diesel engines generate the most electricity per unit of thermal energy and have the greatest surplus of electricity for most cogeneration systems. In this respect the performance of gas turbines are somewhat between those of the steam turbine and diesel engine.

10-2 PERFORMANCE CRITERIA FOR COGENERATION SYSTEMS

As described in the previous chapter, the performance criterion for a conventional power plant usually is taken as the overall system efficiency, η_o.

$$\eta_o = \frac{W_c}{F_c} \tag{10-1}$$

where W_C is the net electrical output and F_c is the fuel input to conventional power plant. This performance criterion has less relevance to a cogeneration system that provides thermal energy as well as electrical power. For cogeneration systems, one of the logical criteria is the energy utilization factor, EUF, which usually is defined as

$$\mathrm{EUF} = \frac{Q_{\mathrm{cg}} + W_{\mathrm{cg}}}{F_{\mathrm{cg}}} \tag{10-2}$$

where Q_{cg} is the thermal energy output, W_{cg} is the electrical energy output, and F_{cg} is the fuel input to a cogeneration system.

The above definition is formulated on the basis of the energy concept. In other words, electrical energy is not treated any differently from thermal energy. This, of course, is incompatible with the common-sense notion that the electrical energy is much more valuable than the thermal energy, especially those in low-temperature and low-pressure conditions. While Eq. (10-2) still is used, another term, based on the concept of thermodynamic availability, has appeared frequently in literature. It is the availability utilization factor, AUF, which is given by

$$\mathrm{AUF} = \frac{A_w + A_{\mathrm{st}}}{A_f} \tag{10-3}$$

where A_w is the availability of electrical energy output, A_{st} is the availability of thermal energy output, and A_f is the fuel availability input to a cogeneration system. Numerically, the work availability A_w is equal to the electric work itself. The fuel availability A_f is the product of fuel consumption and the fuel unit availability. As to the availability A_{st} associated with the thermal energy (either process heat or process steam), one should make use of the definition of thermodynamic availability and calculate the availability output according to the cogeneration system conditions.

Another performance criterion for a cogeneration system involves a comparison between the fuel required to meet the given loads of electricity and thermal energy in the cogeneration system with that required in separate conventional systems to meet the same loads. If the separate conventional systems consist of a conventional electric-power station with overall efficiency η_c and a boiler with the efficiency η_{b_c}, the fuel energy savings is then

$$\Delta F = \frac{Q_{\mathrm{cg}}}{\eta_{b_c}} + \frac{W_{\mathrm{cg}}}{\eta_c} - F_{\mathrm{cg}}$$

When the fuel energy savings ratio (FESR) is defined as the ratio of the saving (ΔF) to the fuel energy required in the conventional systems, we have in equation form

$$\text{FESR} = \frac{\Delta F}{\dfrac{Q_{\text{cg}}}{\eta_{b_c}} + \dfrac{W_{\text{cg}}}{\eta_c}}$$

or

$$\text{FESR} = 1 - \frac{F_{\text{cg}}}{\dfrac{Q_{\text{cg}}}{\eta_{b_c}} + \dfrac{W_{\text{cg}}}{\eta_c}} \tag{10-4}$$

The fuel availability saving ratio (FASR) can be developed in the same manner as the fuel energy savings ratio. First, the fuel availability savings ΔA_f is

$$\Delta A_f = \frac{A_{\text{st}}}{\varepsilon_{b_c}} + \frac{W_{\text{cg}}}{\varepsilon_c} - A_f$$

where ε_{b_c} is the second law boiler efficiency and ε_c is the second law power-station efficiency. Secondly, the fuel availability savings ratio is expressed as

$$\text{FASR} = \frac{\Delta A_f}{\dfrac{A_{\text{st}}}{\varepsilon_{b_c}} + \dfrac{W_{\text{cg}}}{\varepsilon_c}}$$

or

$$\text{FASR} = 1 - \frac{A_f}{\dfrac{A_{\text{st}}}{\varepsilon_{b_c}} + \dfrac{W_{\text{cg}}}{\varepsilon_c}} \tag{10-5}$$

Note that the equation for FASR is similar to that for the fuel energy savings ratio (FESR). One equation utilizes the concept of availability, whereas the other involves the concept of energy. When the fuel for a cogeneration unit is the same as that for the conventional unit, the performance parameters FESR and FASR should have the same numeric values in calculations.

In a cogeneration system, frequently an incremental heat rate for electric-power generator is used. The incremental heat rate is defined as

$$\text{IHR} = \frac{F_{\text{cg}} - Q_{\text{cg}}}{W_{\text{cg}}} \tag{10-6}$$

Equation (10-6) indicates that the fuel energy supplied to the cogeneration system is arbitrarily assumed to be composed of the sum of the thermal energy Q_{cg} and the balance $(F_{\text{cg}} - Q_{\text{cg}})$, which would go to the production of electricity. When used for a topping of process steam (or process heat), the steam turbine usually has an incremental heat rate of around 4700 Btu/kWh. For the gas

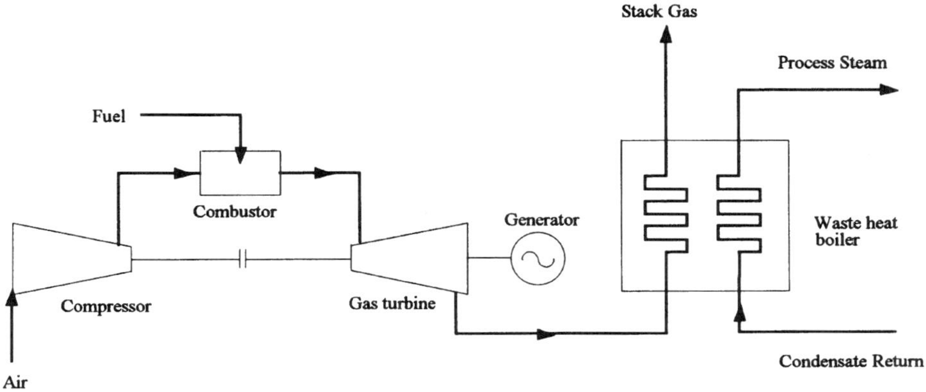

Figure 10-7 Schematic diagram for a gas turbine cogeneration system.

turbine and diesel engine, the incremental heat rate is a little higher. The range is 5500 to 6000 Btu/kWh for the gas turbine and 6000 to 7000 Btu/kWh for the diesel engine. Cogeneration systems are therefore much more efficient than conventional systems for electric-power generation.

10-3 THE FIRST LAW ANALYSIS

The cogeneration system considered in this section is shown in Fig. 10-7. This system consists of two major components: a combustion gas turbine and boiler (sometimes referred to as waste heat–recovery steam generator). These two pieces of equipment are readily available in the market. In practice, consulting engineers would not have any chance to design the processes that occur within the equipment. Instead, they examine and select the equipment directly from those available in the market. Therefore, this section will treat the gas turbine and boiler as two major components or blocks on which the cogeneration system will be constructed. This block approach has been discussed in Chapter 8.

The useful products of a cogeneration system are electrical energy W_{cg} and thermal energy Q_{cg}. In this section the thermal energy is defined as the energy in the steam and numerically equals the steam flow rate times the steam enthalpy rise in the boiler. The ratio of these two products is the heat-to-electrical-energy ratio α_{cg}, defined as

$$\alpha_{cg} = \frac{Q_{cg}}{W_{cg}} \tag{10-7}$$

The combustion gas turbine usually is described in terms of the electrical output W_{cg} and its fuel energy input F_{cg}. The thermal efficiency of a combustion gas turbine, η_{cg}, is the ratio of the electrical output to the fuel energy input. That is,

$$\eta_{cg} = \frac{W_{cg}}{F_{cg}} \tag{10-8}$$

In the block approach, these two parameters are well defined and easily determined when the combustion gas turbine and boilers are selected. The performance criteria of the cogeneration system are then expressed in terms of these parameters. Let us consider the energy utilization factor (EUF), which is given by

$$\text{EUF} = \frac{Q_{cg} + W_{cg}}{F_{cg}}$$

In terms of α_{cg} and η_{cg}, we can rewrite the above equation as

$$\text{EUF} = (1 + \alpha_{cg})\eta_{cg} \tag{10-9}$$

Using Eq. (10-4), we can express the performance criterion (FESR) in terms of α_{cg} and η_{cg}:

$$\text{FESR} = 1 - \frac{1}{\eta_{cg}\left(\dfrac{\alpha_{cg}}{\eta_{b_c}} + \dfrac{1}{\eta_c}\right)} \tag{10-10}$$

The derivations of Eqs. (10-9) and (10-10) are left as an exercise. The following is an illustrative example with numerical calculations.

Example 10-1 A gas turbine cogeneration system has been designed. The gas turbine and boiler have the performance data given below:

Gas Turbine		
	Net power output	73600 kW
	Fuel consumption	741.579 MBtu/hr
	Exhaust gas conditions	914°F, 16 in. of W.G.
	Exhaust gas flow	2,225,671 lbm/hr
Boiler		
	Exit conditions	135 psia Sat. Steam
	Inlet conditions	135 psia, 130°F
	Pinch point	100°F.

Calculate the energy utilization factor and the fuel energy savings factor for this gas turbine cogeneration system.

Solution. First we determine the steam flow rate by taking the evaporation section of boiler as a control volume. As indicated in Fig. 10-8, the energy released by the exhaust gas must be equal to the energy gained by water in the

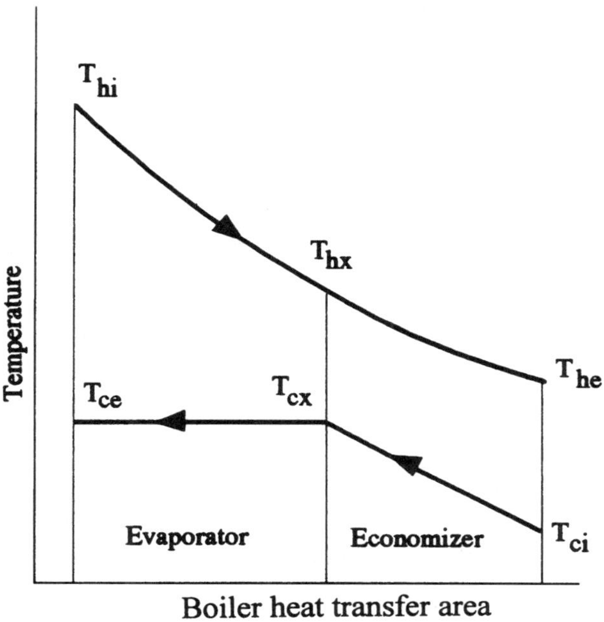

Figure 10-8 Temperature variation for gas and water in Example 8-1.

evaporator.

$$\dot{m}_g c_p (T_{\text{hi}} - T_{\text{hx}}) = \dot{m}_{\text{st}} (h_{\text{ce}} - h_{\text{ex}})$$

Since the boiler pinch point is given as 100°F, we have $T_{\text{hx}} = T_{\text{cx}}$ + pinch point = 350.3 + 100 = 450.3°F. thus the steam flow rate is

$$\dot{m}_{\text{st}} = \frac{(2{,}225{,}671)(0.25)(914 - 450.3)}{1193.2 - 322.1} = 300{,}182 \text{ lbm/hr}$$

The thermal energy supplied by the boiler is the steam flow rate times the enthalpy rise in the boiler. That is,

$$Q_{\text{cg}} = \dot{m}_{\text{st}} (h_{\text{ce}} - h_{\text{ci}}) = 300{,}182(1193.2 - 97.98) = 328.8 \times 10^6 \text{ Btu/hr}$$

The heat-to-electrical-energy ratio is therefore

$$\alpha_{\text{cg}} = \frac{Q_{\text{cg}}}{W_{\text{cg}}} = \frac{328.8 \times 10^6}{(73600)(3413)} = 1.31$$

The thermal efficiency of the combustion gas turbine is given by the equation ($\eta_{\text{cg}} = W_{\text{cg}}/F_{\text{cg}}$). Substituting numeric values into the equation gives

$$\eta_{\text{cg}} = \frac{(73600)(3413)}{741.579 \times 10^6} = 0.3387$$

By using Eq. (10-9), the energy utilization factor can be determined as EUF = (1 + 1.31)(0.3387) = 0.7824. Similarly, using Eq. (10-10), the fuel energy savings ratio is calculated as

$$\text{FESR} = 1 - \frac{1}{0.3387\left(\dfrac{1.31}{0.90} + \dfrac{1}{0.33}\right)} = 0.3418 \quad \blacksquare$$

In the above calculations, the conventional boiler and electric-generating plant are assumed to have the first law efficiencies of 90 percent and 33 percent, respectively. In this example the boiler pinch point is given as 100°F. When the boiler pinch point is reduced, more steam will be generated, and the product ratio α_{cg} will become higher. A large product ratio means a large physical size for the boiler and that the stack gas temperature is expected to be lower. In practice, it is important to maintain the stack gas temperature above the dew-point temperature of combustion products.

For a gas turbine cogeneration system it is convenient to use the energy utilization factor and the fuel energy savings ratio to describe the system performance. As indicated in Eqs. (10-9) and (10-10), these are essentially functions of the gas turbine thermal efficiency and the heat-to-electric-work ratio. Figure 10-9 presents these relationships graphically.

10-4 THE SECOND LAW ANALYSIS

In the first law analysis the thermal energy and electrical work are treated equally. This is not compatible with the understanding that the value of thermal energy is much less than the value of electrical work. The following is the second law analysis of the gas turbine cogeneration system shown in Fig. 10-7.

In terms of thermodynamic availability, the useful products of a cogeneration system are the work availability A_w and the steam thermal availability A_{st}. The product ratio β_{cg} (equivalent to α_{cg} in the first law analysis) is defined by

$$\beta_{cg} = \frac{A_{st}}{A_w} \tag{10-11}$$

Numerically, the work availability A_w is equal to the gas turbine net output W_{cg}, and the steam thermal availability A_{st} is the steam flow rate times the unit availability rise in the boiler. The second law efficiency of the combustion gas turbine is

$$\varepsilon_{cg} = \frac{A_w}{A_f} \tag{10-12}$$

where A_f is the fuel availability and numerically equal to the product of fuel consumption and fuel unit availability. Parallel to the first law analysis, the second law analysis defines an availability utilization factor by Eq. (10-3) and a

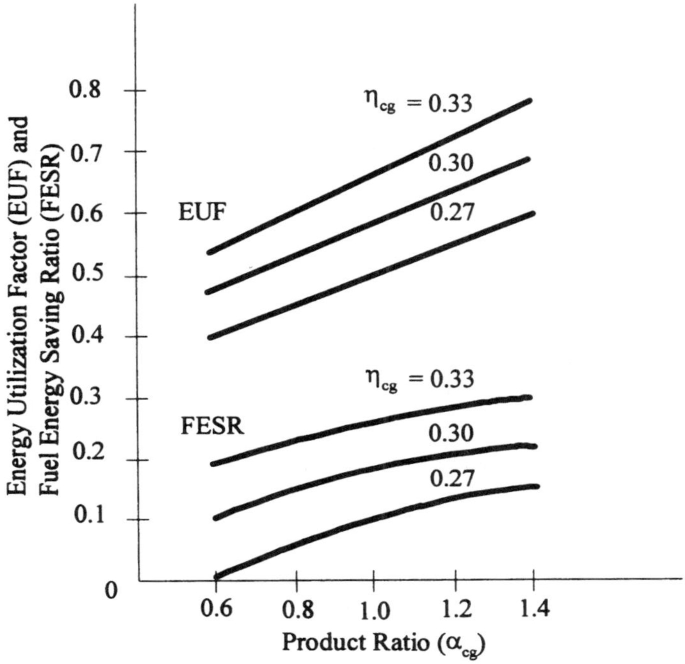

Figure 10-9 Gas turbine cogeneration system performance ($\eta_c = 0.33$ and $\eta_{b_c} = 0.9$).

fuel availability saving ratio by Eq. (10-5). For convenience, we reproduce the equations below:

$$\text{AUF} = \frac{A_w + A_{\text{st}}}{A_f}$$

and

$$\text{FASR} = 1 - \frac{A_f}{\dfrac{A_{\text{st}}}{\varepsilon_{b_c}} + \dfrac{W_{\text{cg}}}{\varepsilon_c}}$$

In terms of the product ratio β_{cg} and the second law efficiency of gas turbine ε_{cg}, we can rewrite those two equations as

$$\text{AUF} = \varepsilon_{\text{cg}}(1 + \beta_{\text{cg}}) \tag{10-13}$$

and

$$\text{FASR} = 1 - \frac{1}{\varepsilon_{\text{cg}}\left(\dfrac{\beta_{\text{cg}}}{\varepsilon_{b_c}} + \dfrac{1}{\varepsilon_c}\right)} \tag{10-14}$$

Two observations should be made at this point. For a cogeneration system, the product ratio based on the concept of thermodynamics availability, β_{cg} is always smaller than the product ratio based on the concept of energy, α_{cg}. Because of this, the availability utilization factor is much smaller than the energy utilization factor. Since the thermal energy and electrical work are of different qualities (i.e., different work-producing potential), the availability utilization factor should be more reasonable and realistic than the energy utilization factor when they are used to determine the cogeneration system performance. Secondly, the equation for the fuel availability savings (FASR) is similar to that for the fuel energy savings ratio (FESR). One utilizes the concept of availability while the other utilizes the concept of energy. When the cogeneration and conventional units use the same fuel, the parameter FASR is expected to have the same numeric value as the FESR. Figure 10-10 presents the variation between availability utilization factor and the fuel availability savings ratio under different gas turbine and boiler conditions.

In the second law analysis, the irreversibility of the cogeneration system is of interest. It represents the imperfection of the system or the departure from ideal performance. In cogeneration system design, one would like to minimize the system irreversibility. For the cogeneration system shown in Fig. 10-7, the system irreversibility is the sum of the irreversibilities for the two major

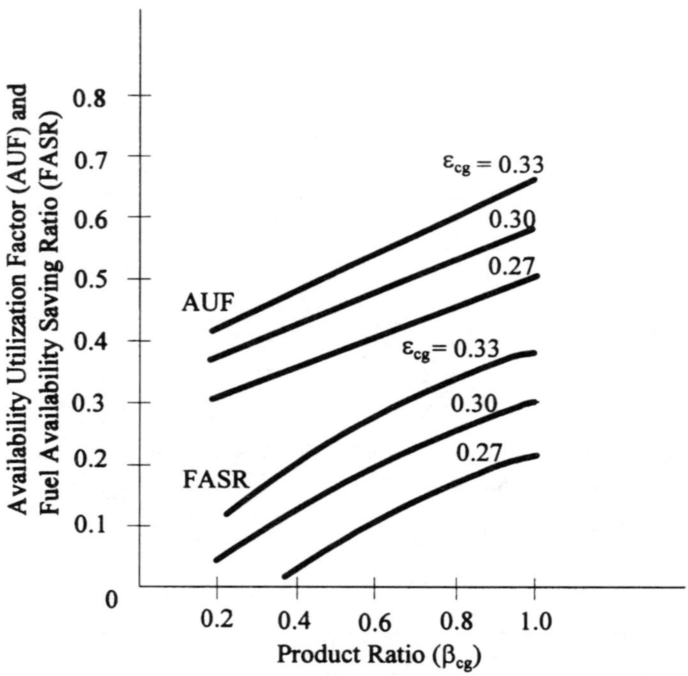

Figure 10-10 Gas turbine cogeneration system performance (with $\varepsilon_c = 0.33$ and $\varepsilon_{b_c} = 0.5$).

components, the combustion gas turbine and the boiler. That is,

$$I_{sys} = I_{gt} + I_b \tag{10-15}$$

The gas turbine irreversibility I_{gt} can be determined easily with the equipment performance data. When the availability of air entering the gas turbine system is negligible, the irreversibility must be equal to the fuel availability minus the gas turbine work output and the availability associated with the turbine exhaust gas.

$$I_{gt} = A_f - A_w - A_g \tag{10-16}$$

The boiler irreversibility I_b is defined as the boiler input minus the boiler output. In this case the boiler input is the thermodynamic availability of turbine exhaust gas, and the boiler output is the availability received by the water. In equation this is

$$I_b = A_g - A_{st} \tag{10-17}$$

Normalizing the gas turbine, boiler, and the entire system irreversibility with the fuel availability A_f, we have

$$\frac{I_{gt}}{A_f} = 1 - \varepsilon'_{cg} \tag{10-18}$$

$$\frac{I_b}{A_f} = \varepsilon'_{cg} - \varepsilon_{cg}(1 + \beta_{cg}) \tag{10-19}$$

$$\frac{I_{sys}}{A_f} = 1 - \varepsilon_{cg}(1 + \beta_{cg}) \tag{10-20}$$

where ε'_{cg} is the cogeneration efficiency for gas turbine, defined as

$$\varepsilon'_{cg} = \frac{A_w + A_g}{A_f} \tag{10-21}$$

Equation (10-20) presents the system irreversibility as a fraction of the cogeneration system fuel input. It is seen that the system irreversibility is a function of the two basic variables β_{cg} and ε_{cg}. When the gas turbine is specified, the cogeneration efficiency ε'_{cg} can be determined easily by Eq. (10-21). The term A_g is the thermodynamic availability of the exhaust gases from the turbine and numerically equal to the gas flow rate times the gas unit availability. In the gas turbine cogeneration system, the boiler is a major component. The second law boiler efficiency is defined as the ratio of steam availability produced to the availability of exhaust gas supplied. That is,

$$\varepsilon_b = \frac{A_{st}}{A_g} \tag{10-22}$$

It should be pointed out that for a given gas turbine, the boiler second law efficiency is related to the system product ratio β_{cg}. The relation shown below can be easily derived:

$$\varepsilon_b = \frac{\beta_{cg}\,\varepsilon_{cg}}{\varepsilon'_{cg} - \varepsilon_{cg}} \tag{10-23}$$

Example 10-2. For the gas turbine cogeneration system described in Example 10-1, determine the fuel availability utilization factor (AUF). Also calculate the irreversibility for the gas turbine and boiler.

Solution. First we determine the product ratio β_{cg} and the second law efficiency of gas turbine, ε_{cg}. The steam availability is given by

$$A_{st} = \dot{m}[(h_{ce} - h_{ci}) - T_0(S_{ce} - S_{ci})]$$

Substituting the numeric values into the above equation, we have $A_{st} =$ 300,182[(1193.2 − 97.98) − 537(1.579 − 0.1817)] = 103.52×10^6 Btu/hr. Then, the product ratio β_{cg} is

$$\beta_{cg} = \frac{A_{st}}{A_w} = \frac{103.52 \times 10^6}{(73600)(3413)} = 0.4121$$

The work availability A_w is numerically equal to the gas turbine output W_{cg}, and the fuel availability A_f can be approximated by the fuel energy F_{cg}. Because of this, the gas turbine has a second law efficiency numerically equal to that of the first law efficiency. Based on the calculation in Example 10-1, they are $\varepsilon_{cg} = \eta_{cg} = 0.3387$.

With β_{cg} and ε_{cg}, we calculate the availability utilization factor (AUF) as

$$\text{AUF} = \varepsilon_{cg}(1 + \beta_{cg}) = 0.3387(1 + 0.4121) = 0.4783$$

To determine the gas turbine and boiler irreversibility, we calculate the cogeneration efficiency ε'_{cg} by the equation

$$\varepsilon'_{cg} = \frac{A_w + A_g}{A_f}$$

As previously calculated, we have A_w = (73,600)(3413) = 251.2×10^6 Btu/hr and $A_f = F_{cg} = 741.579 \times 10^6$ Btu/hr. As for the availability of turbine exhaust gas, we apply the equation

$$A_g = \dot{m}_g\left[c_{p_g}(T_g - T_0) - T_0\left(c_{p_g}\ln\frac{T_g}{T_0} - R_g\ln\frac{P_g}{P_0}\right)\right]$$

Substituting the numeric values into the equation yields

$$A_g = 2{,}225{,}671\left[0.25(1374 - 537) - 537\left(0.25\ln\frac{1374}{537} - \frac{53.3}{778}\ln\frac{15.28}{14.7}\right)\right]$$

$$= 188.14 \times 10^6 \text{ Btu/hr}$$

Thus, the cogeneration efficiency is

$$\varepsilon'_{cg} = \frac{(251.2 - 188.14)10^6}{741.579 \times 10^6} = 0.5924$$

Finally, we calculate the gas turbine irreversibility by Eq. (10-18):

$$\frac{I_{gt}}{A_f} = 1 - \varepsilon'_{cg} = 1 - 0.5924 = 0.4076$$

and the boiler irreversibility by Eq. (10-19):

$$\frac{I_b}{A_f} = \varepsilon'_{cg} - \varepsilon_{cg}(1 + \beta_{cg}) = 0.5924 - 0.3387(1 + 0.4121) = 0.1141$$

The sum of these two irreversibilities must be the irreversibility for the entire cogeneration system. In this case it is given by

$$\frac{I_{sys}}{A_f} = \frac{I_{gt}}{A_f} + \frac{I_b}{A_f} = 0.4076 + 0.1141 = 0.5217$$ ■

In these calculations, all irreversibilities are normalized by the system fuel input, which is given as 741.58 × 10^6 Btu/hr.

10-5 THE TOTAL SYSTEM

The gas turbine cogeneration system is designed generally so that it may not stand alone in operation. In other words, it may need additional steam-generating facilities to meet the thermal demand and/or need the utility network to meet the electric-work demand. This section presents the performance analysis of the total system, which consists of the cogeneration unit plus the standby boilers and the utility network. Figure 10-11 indicates the schematic diagram for the total system. The following are the notations and subscripts used in this section.

Notations.

W = Electric-energy output
Q = Thermal-energy output
F = Fuel energy
η = First law efficiency

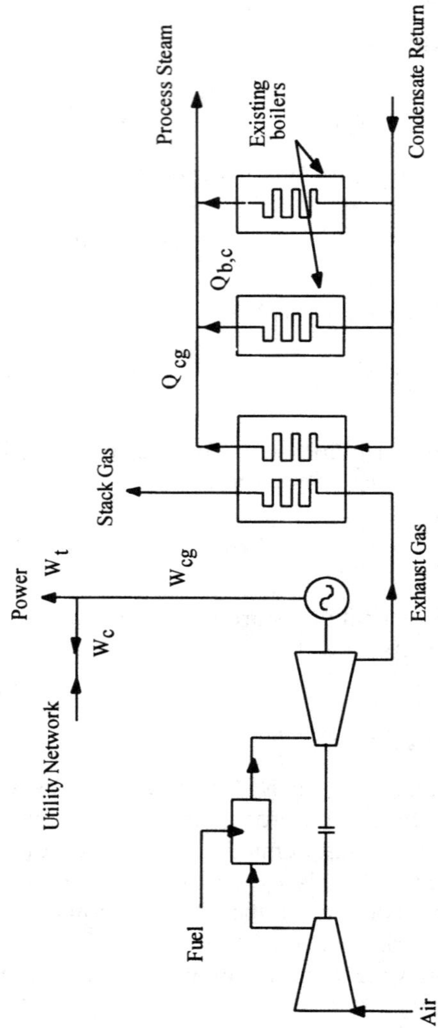

Figure 10-11 Schematic diagram for the total system.

Subscripts.

c = Conventional system (utility network)
b = Boiler
cg = Cogeneration system
t = Total system

The electric-energy output for the total system is

$$W_t = W_c + W_{cg} \tag{10-24}$$

and the thermal-energy output is

$$Q_t = Q_{b_c} + Q_{cg} \tag{10-25}$$

The fuel supplied to the total system must be the sum of the fuel supplied to the cogeneration unit, the utility network, and the standby boilers. That is

$$F_t = F_{cg} + F_c + F_{b_c} \tag{10-26}$$

or

$$F_t = \frac{W_{cg}}{\eta_{cg}} + \frac{W_c}{\eta_c} + \frac{Q_{b_c}}{\eta_{b_c}}$$

The thermal efficiency of the total system usually is defined as the ratio of the total electric-work output to the total fuel input. In equation form this is

$$\eta_t = \frac{W_c + W_{cg}}{F_t} = \frac{1}{F_t/W_c} + \frac{1}{F_t/W_{cg}} \tag{10-27}$$

To express the thermal efficiency of the total system in terms of component performance parameters, we first let

$$\overline{W}_{cg} = \frac{W_{cg}}{W_t} \tag{10-28}$$

and

$$\overline{Q}_{cg} = \frac{Q_{cg}}{Q_t} \tag{10-29}$$

Then, we derive the following two equations:

$$\frac{F_t}{W_c} = \frac{F_t}{W_{cg}}\left(\frac{\overline{W}_{cg}}{1 - \overline{W}_{cg}}\right) \tag{10-30}$$

and

$$\frac{F_t}{W_{cg}} = \frac{1}{\eta_c}\left(\frac{1 - \overline{W}_{cg}}{\overline{W}_{cg}}\right) + \frac{1}{\eta_{cg}} + \frac{\alpha_{cg}}{\eta_{b_c}}\left(\frac{1 - \overline{Q}_{cg}}{\overline{Q}_{cg}}\right) \tag{10-31}$$

By substituting Eqs. (10-30) and (10-31) into Eq. (10-27), the thermal efficiency of the total system, η_t, is expressed as

$$\frac{1}{\eta_t} = \frac{1 - \overline{W}_{cg}}{\eta_c} + \frac{\overline{W}_{cg}}{\eta_{cg}} + \frac{\alpha_{cg}\overline{W}_{cg}}{\eta_{b_c}}\left(\frac{1 - \overline{Q}_{cg}}{\overline{Q}_{cg}}\right) \tag{10-32}$$

The energy utilization factor of the total system, $(\text{EUF})_t$, is defined as the ratio of the total useful energy output to the total system fuel energy input F_t.

$$(\text{EUF})_t = \frac{W_c + W_{cg} + Q_{cg} + Q_{b_c}}{F_c + F_{cg} + F_{b_c}} \tag{10-33}$$

or

$$(\text{EUF})_t = \frac{(W_c/W_{cg}) + (Q_{cg}/W_{cg}) + (Q_{b_c}/W_{cg}) + 1}{(F_t/W_{cg})} \tag{10-34}$$

Since

$$\frac{W_c}{W_{cg}} = \frac{1 - \overline{W}_{cg}}{\overline{W}_{cg}}$$

and

$$\frac{Q_{b_c}}{W_{cg}} = \alpha_{cg}\left(\frac{1 - \overline{Q}_{cg}}{\overline{Q}_{cg}}\right)$$

Eq. (10-34) becomes

$$(\text{EUF})_t = \frac{\dfrac{1}{\overline{W}_{cg}} + \dfrac{\alpha_{cg}}{\overline{Q}_{cg}}}{\dfrac{1}{\eta_c}\left(\dfrac{1 - \overline{W}_{cg}}{\overline{W}_{cg}}\right) + \dfrac{1}{\eta_{cg}} + \dfrac{\alpha_{cg}}{\eta_{b_c}}\left(\dfrac{1 - \overline{Q}_{cg}}{\overline{Q}_{cg}}\right)} \tag{10-35}$$

The fuel energy savings ratio of the total system, FESR_t, is defined by

$$(\text{FESR})_t = \frac{F_c - F_t}{F_c}$$

or

$$(\text{FESR})_t = 1 - \frac{F_t/W_{cg}}{F_c/W_{cg}} \tag{10-36}$$

The term (F_t/W_{cg}) has been already expressed by Eq. (10-31), and the term (F_c/W_{cg}) can be derived as follows:

$$\frac{F_c}{W_{cg}} = \frac{1}{W_{cg}}\left(\frac{W_c + W_{cg}}{\eta_c} + \frac{Q_{cg} + Q_{b_c}}{\eta_{b_c}}\right) \tag{10-37}$$

or

$$\frac{F_c}{W_{cg}} = \frac{1}{\overline{W}_{cg}\eta_c} + \frac{\alpha_{cg}}{\overline{Q}_{cg}\eta_{b_c}} \tag{10-38}$$

By substituting Eqs. (10-37) and (10-38) into Eq. (10-36), the fuel energy savings ratio for the total system will become

$$(\text{FESR})_t = 1 - \frac{\dfrac{1}{\eta_c}\left(\dfrac{1-\overline{W}_{cg}}{\overline{W}_{cg}}\right) + \dfrac{1}{\eta_{cg}} + \dfrac{\alpha_{cg}}{\eta_{b_c}}\left(\dfrac{1-\overline{Q}_{cg}}{\overline{Q}_{cg}}\right)}{\dfrac{1}{\overline{W}_{cg}\eta_c} + \dfrac{\alpha_{cg}}{\overline{Q}_{cg}\eta_{b_c}}} \tag{10-39}$$

It should be pointed out that Eqs. (10-32), (10-35), and (10-39) are the expressions for thermal efficiency, the energy utilization factor, and the fuel energy savings ratio, respectively. When the terms $\overline{Q}_{cg}$ and $\overline{W}_{cg}$ are equal to unity, the total system will become a cogeneration system, and so these three equations will become identical to Eqs. (10-8), (10-9), and (10-10), respectively. For $\overline{W}_{cg}$ at unity, the total system consists of only the cogeneration unit and standby boilers. Similarly, when the term $\overline{Q}_{cg}$ is equal to unity, the total system is composed of the cogeneration unit plus the utility network. In all cases, the system performance parameters (energy utilization factor and fuel energy savings ratio) still are expressed in terms of the product ratio α_{cg} and the gas turbine system thermal efficiency η_{cg}. Figure 10-12 presents these relationships under the conditions $\overline{W}_{cg} = \overline{Q}_{cg} = 0.8$.

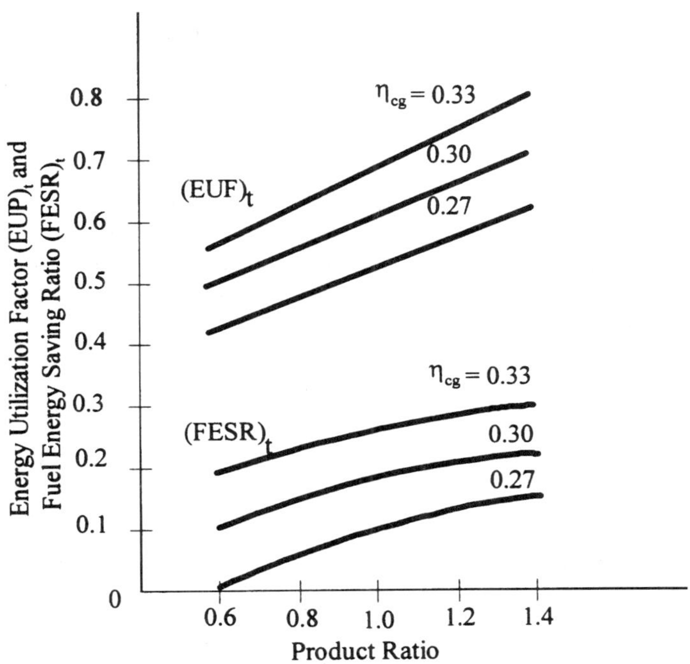

Figure 10-12 Performance parameter for the total system (with $\eta_c = 0.33$ and $\eta_{b_c} = 0.9$).

The availability analysis of the total system can be carried out in the same manner. The derivation of the expressions for the total system availability utilization factor $(\text{AUF})_t$, and fuel availability savings ratio $(\text{FASR})_t$ are left as an exercise.

10-6 STEAM COSTING

When a conventional system such as a boiler generates steam for heating purposes or for electric-power generation, there is no difficulty in estimating the cost of the steam. However, when a cogeneration system generates process steam, the costing of the process steam is not so easy. In a cogeneration system involving a back-pressure steam turbine, steam is used first for electric-power production and second as the process steam. Engineers must develop a fair means of dividing the cost between the two cogeneration products.

Different steam-costing methods have been proposed. Basically, these methods can be classified into two groups. The first group involves estimating the cost of steam and electricity on the basis of the energy content, while the second group estimates on the basis of the available energy. There is no generalized method. Some methods are particularly valuable at the time when the decision for cogeneration is made, and other methods are more suitable for an existing cogeneration unit. Each method is biased to some degree because it distributes the economic benefits of cogeneration to users of either electricity or process steam at the expense of the users of the other product. Below are brief descriptions of the three costing methods.

The Incremental Method

This steam-costing method involves the calculation of a fictitious cost for process steam from a conventional boiler. The cost for the power produced in a cogeneration unit is simply the incremental cost, that is, the cost that is incurred to produce power minus the above fictitious steam cost. In equation form the process-steam production cost is

$$c_{\text{st}} m_{\text{st}} = c_f m_{\text{fc}} + (\text{IC})_{b_c} \tag{10-40}$$

where c_{st} = steam unit cost
m_{st} = process steam flow rate
c_f = boiler unit fuel cost
m_{fc} = fuel consumption for conventional boiler
$(\text{IC})_{b_c}$ = initial cost for conventional boiler

The equation indicates that the production cost of process steam is the sum of the initial cost of a fictitious boiler (conventional type and usually a low-pressure type) and its fuel cost. Then, the unit cost for the process steam is found simply by rearranging the above equation:

$$c_{st} = \frac{1}{m_{st}}\left[c_f m_{fc} + (\mathrm{IC})_{b_c}\right] \tag{10-41}$$

The production cost of electric power in a cogeneration system is the total expenditure minus the above steam production cost. That is,

$$c_w W = \left[c_f m_f + (\mathrm{IC})_{tur} + (\mathrm{IC})_b\right] - \left[c_f m_{fc} + (\mathrm{IC})_{b_c}\right] \tag{10-42}$$

where c_w = electric energy unit cost
W = electric energy output
m_f = fuel consumption for cogeneration unit
$(\mathrm{IC})_b$ = initial cost for boiler in cogeneration unit
$(\mathrm{IC})_{tur}$ = initial cost for turbine in cogeneration unit

By rearranging Eq. (10-42) we have the electric energy unit cost as

$$c_w = \frac{1}{W}\left[c_f(m_f - m_{fc}) + (\mathrm{IC})_{tur} + \Delta(\mathrm{IC})_b\right] \tag{10-43}$$

where $\Delta(\mathrm{IC})_b = (\mathrm{IC})_b - (\mathrm{IC})_{b_c}$.

Sometimes $\Delta(\mathrm{IC})_b$ is referred to as the capital cost differential between the cogeneration boiler and the fictitious boiler. The incremental method usually is applied in situations where engineers in an industrial plant decide whether or not a cogeneration facility is needed. When the cost of electricity calculated above is lower than the purchase price, the investment in a cogeneration facility probably would be worthwhile. It should be noted that this incremental method is biased toward the decision for cogeneration, because it uses all economic benefits of cogeneration to lower the electric-power cost.

The lost-Kilowatt Method

This steam-costing method involves the calculation of kilowatts lost in the production of process steam. The lost kilowatt can be determined easily by comparing the output of an electric-power generating system and the electric output of the corresponding but hypothetical cogeneration plant. Once the reduction in electric output is determined, the market value of the electric energy is simply the cost of the process steam. This steam-costing method is suitable in situations where engineers in a power plant consider developing a steam market. When the cost of the steam is lower than the steam market price, the decision for cogeneration is probably positive.

The lost-kilowatt method is applicable to both back-pressure and extraction turbine systems. Unlike the incremental method, the lost-kilowatt method distributes al cogeneration economic benefits to the steam users; therefore, the cost of process steam by this method usually is lower than that by other methods.

The Availability Method

This method utilizes the concept of thermodynamic availability. The thermodynamic availability represents the potential of a substance to produce useful work. This concept provides a basis for the costing of process steam and electric power produced in a cogeneration system. For simplicity, we present the availability method as applied to the back-pressure steam turbine cogeneration system in Fig. 10-13. Like any energy converter, the cost of products from the cogeneration system must be equal to the total expenditure it requires. In equation form, this can be written as

$$c_w W + c_{st} A_{st} = c_f A_f + (\mathrm{IC})_{tur} + (\mathrm{IC})_b \tag{10-44}$$

where c_f = fuel availability unit cost
c_{st} = steam availability unit cost
A_f = fuel availability flow
A_{st} = steam availability flow

Other notations are identical to those defined earlier in this section. In Eq. (10-44), there are only two unknowns, the electric energy unit cost c_w and the steam availability unit cost c_{st}. To determine these unit costs, we apply the same cost method individually to the boiler and to the steam turbine. That is,

$$c_{st_b} A_{st_b} = c_f A_f + (\mathrm{IC})_b \tag{10-45}$$

for the boiler, and

$$c_w W + c_{st} A_{st} = c_{st_b} A_{st_b} + (\mathrm{IC})_{tur} \tag{10-46}$$

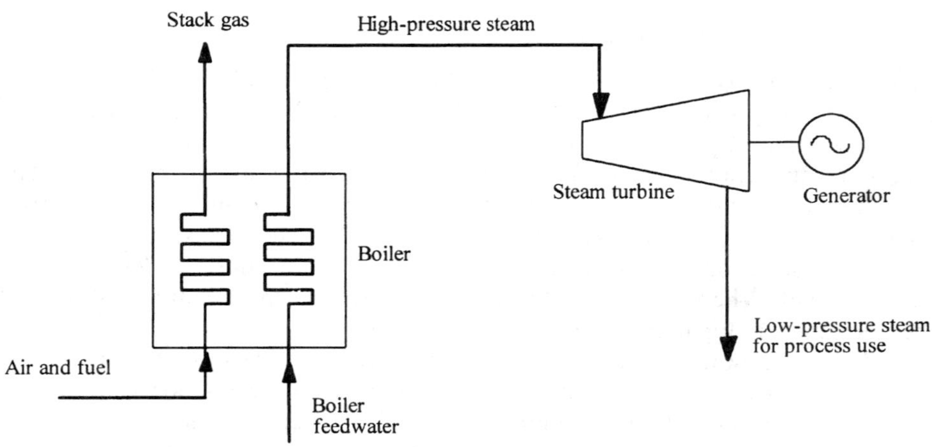

Figure 10-13 A back-pressure steam turbine for cogeneration.

for the steam turbine, where c_{st_b} is the availability unit cost for steam at the boiler outlet and A_{st_b} is the steam availability flow rate at the same location. An examination of Eqs. (10-45) and (10-46) indicates there are three unknowns in two equations. In addition to the electrical energy unit cost and the steam availability unit cost, the availability unit cost for steam at the boiler outlet, c_{st_b}, is not yet determined. Therefore, a complementary equation is needed.

Many different methods have been proposed for establishing the complementary equation. The selection of the method depends upon the situation. For instance, we can set up the complementary equation as

$$c_w = c_{st} \tag{10-47}$$

when we consider that the electric and thermal energies are equally important products in a cogeneration plant. Substituting Eq. (10-47) into Eq. (10-46) yields the unit costs for both products as

$$c_w = c_{st} = \frac{c_{st_b} A_{st_b} + (\text{IC})_{\text{tur}}}{W + A_{st}} \tag{10-48}$$

where

$$c_{st_b} = \frac{c_f A_f + (\text{IC})_b}{A_{st_b}} \tag{10-49}$$

Obviously, in this approach the electric energy output and the thermal energy output from the cogeneration unit are equally charged for all expenditures.

There may arise another situation where engineers consider the back-pressure turbine as equipment for electric-power generation. In this case the electric output must be charged for the entire cost associated with the turbine process, and the thermal energy is charged only for high-pressure steam cost. Under these conditions the complementary equation is

$$c_{st} = c_{st_b} \tag{10-50}$$

Then, we can express the electric energy unit cost as

$$c_w = \frac{1}{W}\left[c_{st}(A_{st_b} - A_{st}) + (\text{IC})_{\text{tur}}\right] \tag{10-51}$$

and the steam unit cost as

$$c_{st} = \frac{c_f A_f + (\text{IC})_b}{A_{st_b}} \tag{10-52}$$

At this point it should be emphasized that there are essentially two equations needed in the availability-energy method. The first equation is a money balance equation such as Eq. (10-44). This equation involves the use of available energy of the cogenerated products rather than the energy content. The second equation is the so-called complementary equation, which varies case by case.

There is no generalized expression for steam costings. To formulate the second equation, engineers must exercise their judgment and take into account the conditions under which the costings are made.

10-7 PROBLEMS

10-1 A simple gas turbine is used as a prime mover in a cogeneration system. The gas turbine has the following design performance:

Electric output	59,000 kW
Fuel consumption	698.8×10^6 Btu/hr
Air inlet conditions	14.7 psia, 77°F
Exhaust gas conditions	1,850,000 lbm/hr, 945°F

The turbine exhaust gas is used directly for heating, and it experiences temperature drop of 400°F in the process. Calculate (a) the incremental heat rate, (b) the availability utilization factor, and (c) the energy utilization factor.

10-2 A back-pressure steam turbine of 100,000 kW serves as a prime mover in a cogeneration system. The boiler admits the return water at a temperature of 150°F and produces the steam at 950 psia and 850°F. Steam then enters a back-pressure turbine and expands to the pressure of the process, which is 75 psia. Neglecting the effects of pumping and the pressure drops at various locations, estimate (a) the incremental heat rate, (b) the availability utilization factor, and (c) the energy utilization factor.

10-3 Estimate by the availability method the unit cost for the thermal energy from the cogeneration system described in Problem 10-1. The economic parameters are

Gas turbine capital cost	$350/kW
Fuel cost	$7.00/MBtu
Annual fixed charge rate	18%
Annual operating hours	4000 hr

Also discuss how the thermal energy unit cost changes with the selection of the complementary equation.

10-4 By the availability method, calculate the unit cost for the process steam for Problem 10-2. The economic parameters are

Boiler capital cost	$300/kW
Steam turbine capital cost	$250/kW
Boiler fuel cost	$1.10/MBtu
Annual fixed charge rate	22%
Annual operating hours	4500 hr

The two cogenerated products are assumed to be equally needed in the plant.

CHAPTER

ELEVEN

AVAILABILITY ANALYSIS OF EQUIPMENT PERFORMANCE DATA

Equipment is operated under both design and off-design conditions. The knowledge of equipment performance is essential for system simulation and optimization. This chapter presents availability analyses of several types of equipment including pumps, fans, condensers, and evaporative cooling towers, using their performance data.

11-1 PUMP PERFORMANCE

Figure 11-1 shows the typical pump performance characteristics. It is seen that the pump head and input (the power needed to drive the pump) vary with the pump capacity. In general, these performance curves are provided by the equipment vendors. Since the pump hydraulic horsepower is given by the equation

$$\text{Hydraulic hp} = \frac{(\text{WFR})(\text{DH})(\text{sp.gr})}{3960} \tag{11-1}$$

combining the hydraulic horsepower with the pump input (also called the shaft horsepower) will give the pump first law efficiency as

$$\eta_p = \frac{\text{Hydraulic hp}}{\text{Shaft hp}} \tag{11-2}$$

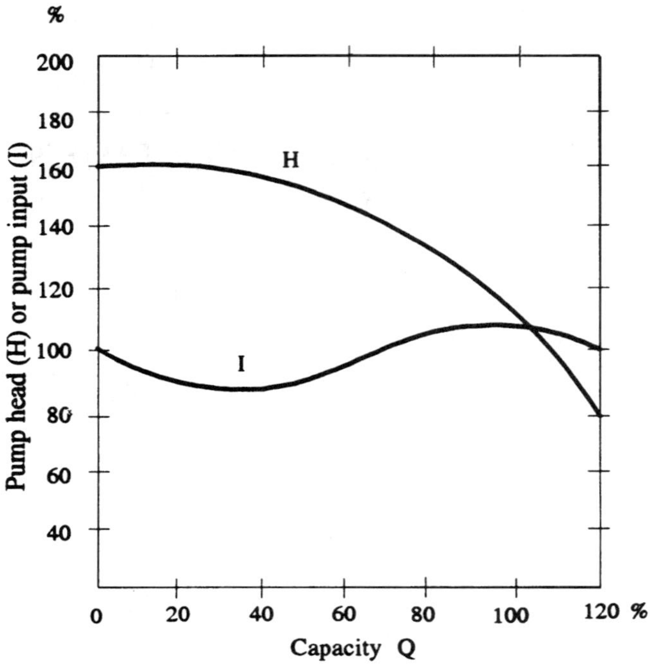

Figure 11-1 Typical pump performance curves.

where WFR is the water flow rate in gallon in minute, DH is the developed head in feet of water, and sp.gr stands for specific gravity.

Evidently, the first law efficiency also will vary with the pump capacity. The following is the application of the availability method to the pump performance data. The objective is to determine the process irreversibility and the second law efficiency. Based on a unit mass, the availability equation for the pumping process is $a_w + a_q + a_{f_i} - a_{f_e} - i = o$. Neglecting the heat loss from the pump, the process irreversibility per unit mass is given by the equation

$$i = w_p = (a_{f_e} - a_{f_i}) \tag{11-3}$$

$$= T_0 c_v \ln\left(1 + \frac{\Delta T}{T_i}\right) \tag{11-4}$$

where ΔT is the temperature rise across the pump. The equation for an estimate of temperature rise was derived in Chapter 4:

$$\Delta T = \frac{v_i(P_e - P_i)}{c_v}\left(\frac{1}{\eta_p} - 1\right)$$

Table 11-1 Pump performance data for an example pump.

Capacity (10^3 gpm)	Developed head (ft. of water)	Pump input (hp)
58	126.9	3778
87	118.8	3859
116	108.0	4020
145	90.0	4020
174	72.9	3938

The second law efficiency is simply the ratio of the availability increase in water to the pump work. Dividing Eq. (11-3) by the pump work, we have

$$\varepsilon_p = 1 - \frac{i}{w_p}$$

In terms of the temperature rise across the pump, the second law efficiency becomes

$$\varepsilon_p = 1 - \frac{T_0 c_v}{w_p} \ln\left(1 + \frac{\Delta T}{T_i}\right) \tag{11-5}$$

where w_p is the pump work per unit mass.

The following is a numerical example in which the pump performance data are utilized to determine the rate of the irreversibility and the second law efficiency. The pump performance data in Table 11-1 are given for the pump inlet conditions 14.7 psia and 80°F. Using the equation developed in this section, and assuming that the environment temperature T_0 is 77°F, we can easily determine the pump irreversibility rate and the second law efficiency. The results are presented in Table 11-2.

It is of interest to notice that the pump performance at low capacity is substantially different from that at high capacity. The irreversibility rate will increase rapidly as the pump capacity decreases. In other words, the pump

Table 11-2 The pump irreversibility rate and second law efficiency for the pump in Table 11-1.

Capacity (10^3 gpm)	Hydraulic power (hp)	η_p (%)	Temp. rise (°F)	WFR (10^6 lbm/hr)	Irreversibility rate (hp)	ε_p (%)
58	1859	49.21	0.17	29.0	1926	49.03
87	2610	67.63	0.07	43.5	1190	69.17
116	3163	78.68	0.04	58.0	906	77.45
145	3295	81.97	0.03	72.5	850	78.86
174	3203	81.34	0.02	87.0	680	82.74

irreversibility and the pump work do not change in the same direction. The higher the pump irreversibility rate, the poorer the pump performance. As expected, the second law and first law efficiencies follow a similar trends.

1-2 FAN PERFORMANCE

Fans usually perform in a fashion similar to that of the pump. When a fan develops a small pressure change, the air flow is essentially incompressible. Fans used in central power stations either are the axial-flow type or the centrifugal type. The centrifugal fan may have different curved blades, such as forward-curved, backward-curved, and straight radial blades. The different blades will result in different fan performances. Figure 11-2 presents the typical performance of a fan in terms of its capacity. It is seen that the fan design point usually is around the maximum static efficiency. In operation, the fan capacity will determine the fan static pressure head and power input. The following equations are used frequently in practice:

$$\text{Static air hp} = \frac{(\text{AFR})(\Delta p)}{6350} \tag{11-6}$$

$$\eta_f = \frac{\text{Static air hp}}{\text{Shaft hp}} \tag{11-7}$$

where AFR stands for air flow rate in cfm and Δp is developed static pressure in inches of water.

The temperature rise across the fan is the information needed in an availability analysis. The rise can be estimated by assuming an adiabatic process across the fan. In equation form, the temperature rise is

$$\Delta T = w_f / c_p \tag{11-8}$$

where w_f is the fan work per unit mass. To determine the process irreversibility and the second law efficiency, we make use of the availability equation $i = a_q + a_w + a_{f_i} - a_{f_e}$. Since the process is assumed to be adiabatic, we have $a_q = o$ and the process irreversibility

$$i = w_f - (a_{f_e} - a_{f_i}) \tag{11-9}$$

Dividing the above equation by the fan work, we have the second law efficiency

$$\varepsilon_f = 1 - \frac{i}{w_f} \tag{11-10}$$

In Eq. (11-9), w_f is the fan work per unit mass, and the term $(a_{f_e} - a_{f_i})$ is the air availability increase across the fan. This term can be expressed as

$$a_{f_e} - a_{f_i} = c_p \Delta T - T_0 c_p \ln\left(1 + \frac{\Delta T}{T_i}\right) + T_0 R \ln\left(1 + \frac{\Delta p}{P_i}\right) \tag{11-11}$$

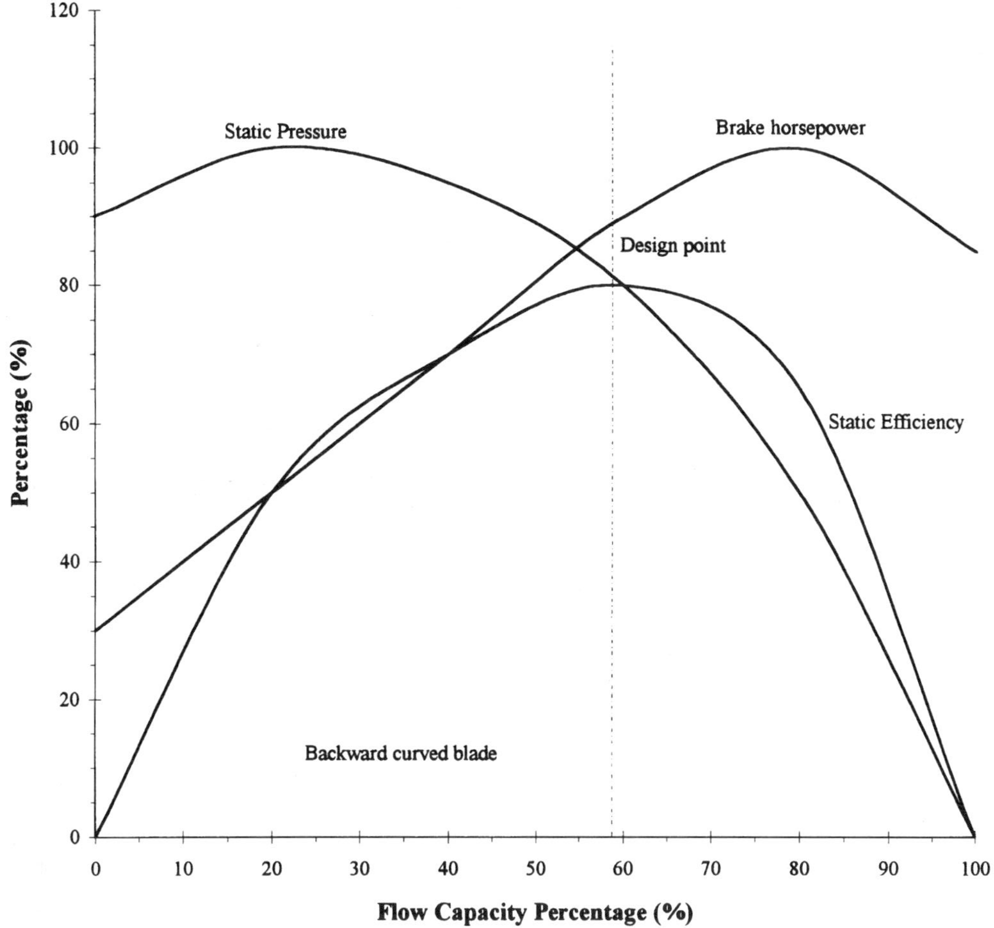

Figure 11-2 Typical fan performance characteristics.

where ΔT is the temperature rise across the fan, Δp is the pressure developed by the fan, P_i is the fan inlet pressure, and T_i is the fan inlet temperature. The following is a numerical example in which the fan performance data are utilized to determine the fan irreversibility and the second law efficiency.

Consider the boiler forced-draft fan with the performance data given in Table 11-3 at the inlet conditions 14.7 psia and 80°F. Using the equations developed in this section, we can easily determine the fan irreversibility rate and the second law efficiency. For instance, we take the performance data at the fan capacity of 200×10^3 cfm. The static air horsepower is

$$\text{Static air hp} = \frac{(300 \times 10^3)(14.1)}{6350} = 444.1 \text{ hp}$$

and the shaft power is

$$\text{Shaft hp} = \frac{444.1}{0.68} = 653.1 \text{ hp}$$

To determine the temperature rise across the fan, we assume an adiabatic process and obtain the equation $\Delta T = w_f/c_p$. Since $c_p = 0.24$ Btu/lbm-R and w_f is given by

$$w_f = \frac{\dot{w}}{\dot{m}} = \frac{(653.1)(42.42)}{(14.7 \times 144)(200 \times 10^3)/(53.3)(540)} = 1.88 \text{ Btu/lbm}$$

the temperature rise across the fan is $\Delta T = 1.88/0.24 = 7.83$°F.

To calculate the fan irreversibility, we first determine the availability increase by Eq. (11-11).

$$\begin{aligned} a_{f_e} - a_{f_i} &= 0.24(7.82) - (537)(0.24)\ln\left(1 + \frac{7.83}{540}\right) \\ &\quad + (537)(53.3/778)\ln\left(1 + \frac{0.51}{14.7}\right) \\ &= 1.27 \text{ Btu/lbm} \end{aligned}$$

Table 11-3 Performance data for an example boiler forced-draft fan.

Capacity (10^3 cfm)	Static pressure (in. of water)	Static efficiency (%)
50	14.4	28.5
100	14.4	48.0
150	14.3	60.2
200	14.1	68.0
280	13.4	74.0

The fan irreversibility rate is

$$\dot{I} = \dot{m}\left[w_f - (a_{f_e} - a_{f_i})\right] = 14709(1.88 - 1.27) = 8972.5 \text{ Btu/min} = 211.4 \text{ hp}$$

Finally, the second law efficiency at this fan capacity is

$$\varepsilon_f = \frac{a_{f_e} - a_{f_i}}{w_f} = \frac{1.27}{1.88} = 0.6755$$

The availability calculations at other capacities are summarized in Table 11-4.

It should be emphasized that an adiabatic condition has been assumed in the calculation of temperature rise. This assumption generally results in a maximum temperature rise across the fan. As an alternative, a polytropic relationship can be used. In equation form, the exit-air temperature from the fan is given by

$$T_2 = T_1\left(\frac{P_2}{P_1}\right)^{n-1/n}$$

where n is the polytropic exponent. The heat loss in the process is determined by the first law equation: $q = w_f + c_p(T_2 - T_1)$. Then the availability transfer associated with the heat loss is given by

$$a_q = \int\left(1 - \frac{T_0}{T}\right)\delta q$$

or

$$a_q = \left(1 - \frac{T_0}{T_{av}}\right)q$$

where T_{av} is the average air temperature in the process. It should be recognized also that the maximum heat loss occurs when the air temperature is assumed to be unchanged. In that case, we have the heat loss as $q = w_f$ and the associated heat availability transfer as

$$a_q = \left(1 - \frac{T_0}{T_1}\right)q$$

Table 11-4 Availability calculations for the fan in Table 11-3.

Capacity (10^3 cfm)	Shaft Hp (hp)	Temp. Rise (°F)	Air flow rate (10^3 lbm/hr)	Irreversibility Rate (hp)	ε_p (%)
50	397.8	19.13	220	277.5	30.07
100	472.4	11.33	441	242.6	48.53
150	561.1	9.0	662	223.7	60.19
200	653.1	7.83	882	211.4	67.55
280	798.5	6.83	1234	208.5	73.78

In summary, the air temperature rise across the fan cannot be predicted without uncertainty. The above calculation is at best an approximation.

11-3 CONDENSER PERFORMANCE

Condensers usually are sized to perform a heat transfer function under a given set of conditions. These conditions are called the design conditions or the conditions at the design point. In operation, the condenser may not always be at the design point. This is due mainly to either the variation of condenser heat load or a change of circulating water temperature. Because of this, attention must be paid to both design and off-design performance.

The condenser performance can be easily determined by showing that the condenser pressure is a function of condenser heat load and circulating water temperature. The derivation of the relationship is presented below.

Figure 11-3 presents a schematic diagram for condenser. The heat-transfer equation is given by

$$\dot{Q} = \text{CUA}\frac{(T_s - T_1) - (T_s - T_2)}{\ln\dfrac{T_s - T_1}{T_s - T_2}} \tag{11-12}$$

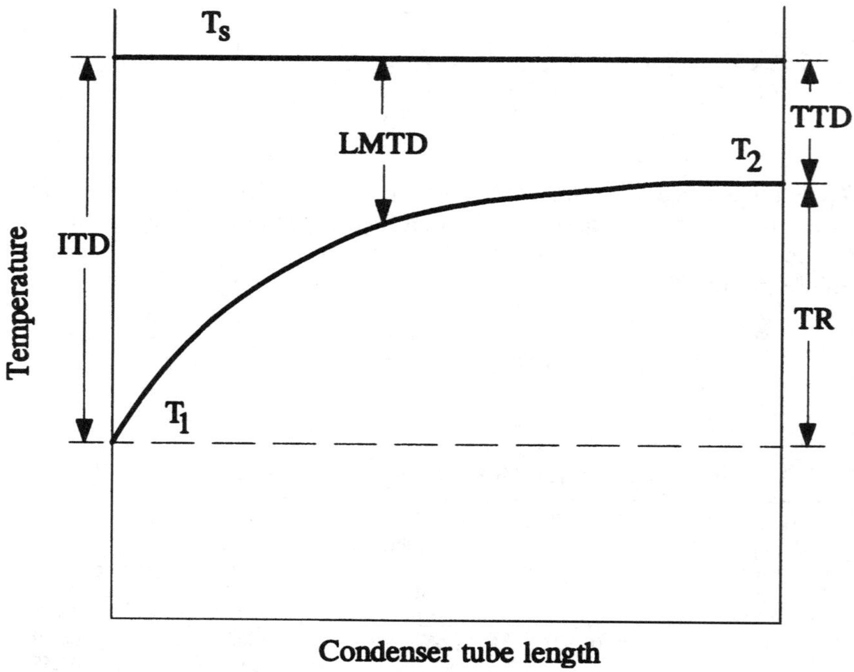

Figure 11-3 Schematic diagram for a condenser.

and the energy equation is given by

$$\dot{Q} = \dot{m}_w c_p (T_2 - T_1) \tag{11-13}$$

where C = cleanliness factor
U = overall heat transfer coefficient
A = heat-transfer surface area
$\dot{Q}$ = heat transfer rate or heat received by the cooling water
c_p = specific heat for cooling water
$\dot{m}_w$ = water flow rate
T = temperatures at various locations as indicated in Fig. 11-3.

Combining Eqs. (11-12) and (11-13), we obtain

$$\ln \frac{T_s - T_2}{T_s - T_1} = -\frac{\mathrm{CUA}}{\dot{m}_w c_p} \tag{11-14}$$

or

$$T_s = \frac{T_2 - T_1 \exp\left(-\dfrac{\mathrm{CUA}}{\dot{m}_w c_p}\right)}{1 - \exp\left(-\dfrac{\mathrm{CUA}}{\dot{m}_w c_p}\right)} \tag{11-15}$$

Substituting T_2 from Eq. (11-13) into Eq. (11-15), we get the following expression for the condenser steam temperature:

$$T_s = \frac{\dfrac{\dot{Q}}{\dot{m}_w c_p} + T_1\left[1 - \exp\left(\dfrac{-\mathrm{CUA}}{\dot{m}_w c_p}\right)\right]}{1 - \exp\left(-\dfrac{\mathrm{CUA}}{\dot{m}_w c_p}\right)} \tag{11-16}$$

or

$$T_s = c_1 \dot{Q} + T_1 \tag{11-17}$$

where

$$c_1 = \frac{1}{m_w c_p}\left[1 - \exp\left(-\frac{\mathrm{CUA}}{m_w c_p}\right)\right]^{-1} \tag{11-18}$$

It is noted in Eq. (11-17) that for a given water flow rate, the condenser temperature T_s is a function of condenser heat load $\dot{Q}$ and circulating water inlet temperature T_1. The condenser temperature has a linear relationship with the condenser heat load. In practice, frequently the condenser pressure is used instead of the condenser temperature. It must be pointed out that in the above derivation, no attention is given to the effects of noncondensible gases in the condenser and the limitation of gas removal equipment. In operation, the condenser temperature (or pressure) cannot be such that the condenser terminal temperature difference (TTD) is less than 5°F. If this should occur in calculations, the Heat Exchanger Institute recommends that the equation TTD = TR + 5 be used for an estimate of condenser temperature. Because of these, the condenser vendors frequently prepare the performance curves for their products making use of the analytical work and field experience. Figure 11-4 presents typical performance curves for a power-plant condenser.

In the following we develop the equation for the condenser irreversibility and second law efficiency in terms of the performance data. We start the availability equation for the condenser as

$$\dot{I} = \dot{A}_q + \dot{A}_w + \dot{m}_s(a_{f_i} - a_{f_e})_s - \dot{m}_w(a_{f_2} - a_{f_1})_w$$

Since $\dot{A}_q = \dot{A}_w = o$, we have

$$\dot{I} = \dot{m}_s(a_{f_i} - a_{f_e})_s - \dot{m}_w(a_{f_2} - a_{f_1})_w \tag{11-19}$$

This shows that the condenser irreversibility is simply the decrease of steam availability minus the increase of water availability. As indicated in Chapter 3, the availability change in water can be further expressed as a sum of two terms,

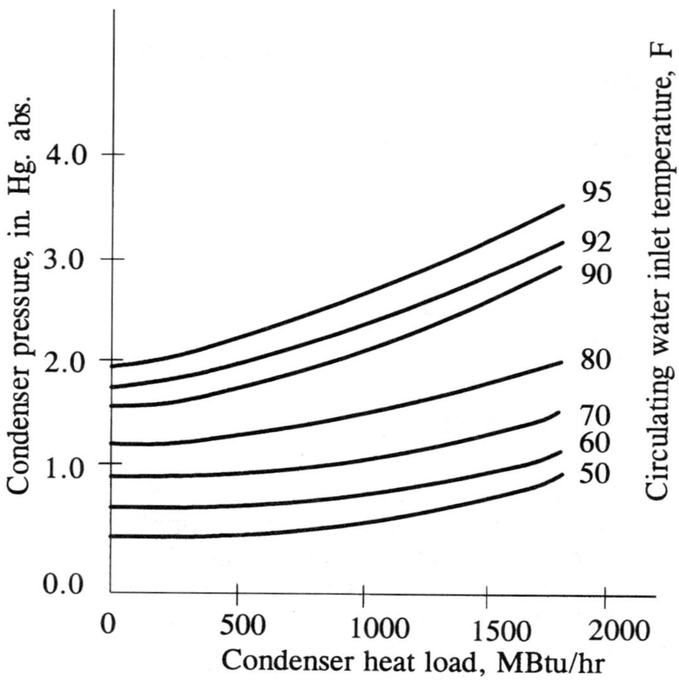

Figure 11-4 Typical condenser performance curves.

one due to the temperature and another due to the pressure. That is,

$$\dot{m}_w(a_{f_2} - a_{f_1})_w = \dot{m}_w(a_{f_2} - a_{f_1})_{w_{\Delta T}} + \dot{m}_w(a_{f_2} - a_{f_1})_{w_{\Delta p}} \tag{11-20}$$

Substituting Eq. (11-20) into Eq. (11-19), we have the condenser irreversibility

$$\dot{I} = \dot{I}_{\Delta T} + \dot{I}_{w_{\Delta p}} \tag{11-21}$$

where

$$\dot{I}_{\Delta T} = \dot{m}_s(a_{f_i} - a_{f_e})_s - \dot{m}_w(a_{f_2} - a_{f_1})_{w_{\Delta T}} \tag{11-22}$$

$$\dot{I}_{w_{\Delta p}} = \dot{m}_w(a_{f_1} - a_{f_2})_{w_{\Delta p}} \tag{11-23}$$

$\dot{I}_{\Delta T}$ is the condenser irreversibility due to heat transfer between the condensing steam and cooling water. This term would remain nonzero as long as the temperature difference between them is finite. $\dot{I}_{w_{\Delta p}}$ is the condenser irreversibility due to the friction on the water side. This term is numerically related to the pressure drop. Because of negligible pressure drop at the steam side, $\dot{I}_{s_{\Delta p}}$ is equal to zero and omitted from Eq. (11-21).

For convenience, we express $\dot{I}_{\Delta T}$ and $\dot{I}_{w_{\Delta p}}$ in terms of condenser performance data, which consist of the condenser heat load, the circulating water inlet temperature, and flow rate.

$$\dot{I}_{\Delta T} = \left(1 - \frac{T_0}{T_{\text{sat}}}\right)\dot{Q} - \dot{m}_w c_v \left[(T_2 - T_1) - T_0 \ln\frac{T_2}{T_1}\right]_w \tag{11-24}$$

and

$$\dot{I}_{w_{\Delta p}} = \dot{m}_w v_1(P_1 - P_2)_w \tag{11-25}$$

It should be noted that while T_2 is not directly available in the condenser performance data, it can be easily determined by the energy equation such as Eq. (11-13). The second law efficiency for the condenser may have two alternative forms. They are

$$\varepsilon_{\text{I}} = \frac{\dot{m}_w(a_{f_2} - a_{f_1})_w}{\dot{m}_s(a_{f_i} - a_{f_e})_s} \tag{11-26}$$

and

$$\varepsilon_{\text{II}} = \frac{\dot{m}_w(a_{f_2} - a_{f_1})_{w_{\Delta T}}}{\dot{m}_s(a_i - a_{f_e})_s} \tag{11-27}$$

The first form is simply the ratio of the process output to the process input and is identical to that defined in Chapter 4. In this form, the increase of the availability in cooling water is regarded as the process output, and the decrease of availability in condensing steam is regarded as the process input. However, when the hydraulic performance is separated from the heat transfer perfor-

Table 11-5 Performance data for an example condenser.

Condenser heat load (10^6 Btu/hr)	Condenser Pressure (in Hb abs.)
250	1.355
500	1.520
750	1.695
1000	1.880
1250	2.090

mance in the condenser, the second form is quite convenient and will truly indicate the condenser efficiency from a heat transfer (or thermal availability transfer) viewpoint. The second form may have advantages. In terms of condenser performance data, Eq. (11-27) can be further expressed as

$$\varepsilon_{\text{II}} = \frac{\dot{m}_w c_v \left[(T_2 - T_1) - T_0 \ln \dfrac{T_2}{T_1} \right]_w}{\left(1 - \dfrac{T_0}{T_{\text{sat}}} \right) \dot{Q}} \tag{11-28}$$

For illustration, we utilize the condenser performance data given in Table 11-5. These data were collected at the time when the water inlet temperature was 80°F and the flow rate was approximately 140,342 gallons per minute. The availability calculations are straightforward. Table 11-6 is the summary of condenser irreversibility and second law efficiency.

11-4 EVAPORATIVE COOLING TOWER PERFORMANCE

Evaporative cooling towers are devices that lower the temperature of water by bringing water into contact with air. The water and air flow are directed in order to provide maximum heat transfer. In the cooling towers the mass transfer takes place between the water and the air streams. In fact, the water evaporation and

Table 11-6 Condenser irreversibility and second law efficiency for the condenser in Table 11-5.

$\dot{Q}$ (10^6 Btu/hr)	$\dot{I}_{\Delta T}$ (10^6 Btu/hr)	ε_{II} (%)
250	2.99	64.55
500	7.58	62.19
750	13.59	60.81
1000	20.78	60.05
1250	29.52	59.26

vapor transfer constitute more than two-thirds of the total heat transfer. Like other equipment, evaporative cooling towers are operated at design as well as off-design points. The Cooling Tower Institute recommends that the tower performance curves are plotted with the tower cold-water temperature against the ambient wet-bulb temperature for different tower cooling ranges. Figure 11-5 presents typical performance curves for a mechanical-draft cooling tower at a given water flow rate. Generally the performance curves are available from the equipment vendor. They are prepared with design and field data. These curves would very from one tower model to another.

The availability analysis involving the tower performance data can be carried out in the same manner as the analysis for the condenser. We take the cooling tower in Fig. 11-6 as a control volume and have the availability equation as

$$\left(\dot{A}_1 - \dot{A}_2\right) + \dot{A}_3 + \dot{W}_f + \left(\dot{A}_4 - \dot{A}_5\right) - \dot{A}_{\text{dest}} = 0 \tag{11-29}$$

where $(\dot{A}_1 - \dot{A}_2)$ = availability decrease in cooling water
$\dot{A}_3$ = availability of tower makeup water
$\dot{W}_f$ = fan power
$(\dot{A}_5 - \dot{A}_4)$ = availability loss from cooling tower
$\dot{A}_{\text{dest}}$ = availability destruction in cooling tower operation

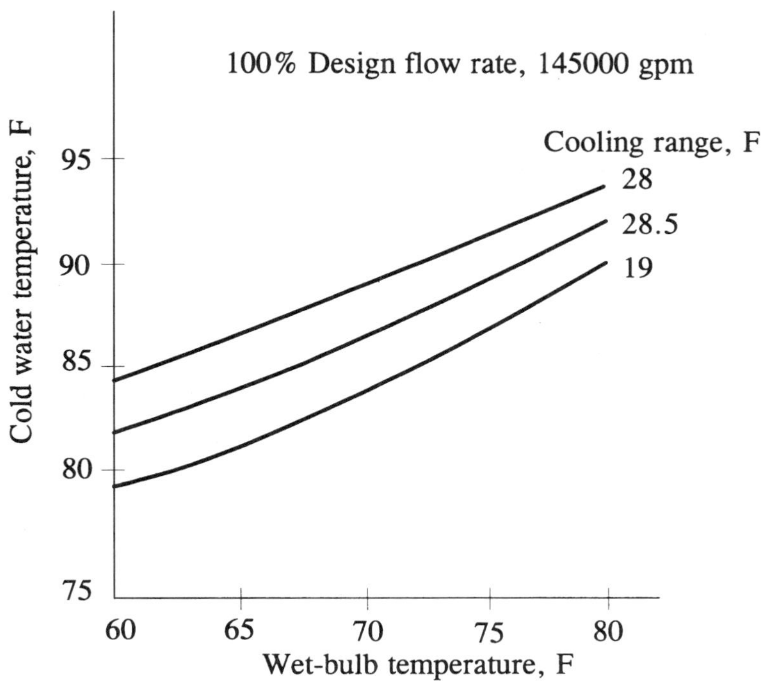

Figure 11-5 Typical cooling tower performance curves.

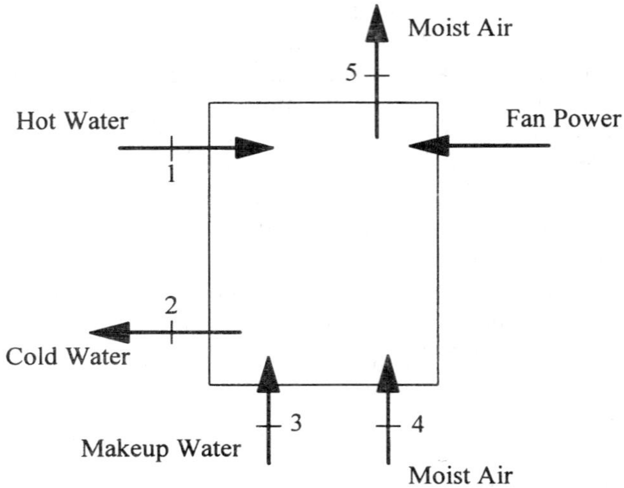

Figure 11-6 A schematic diagram for an evaporative cooling tower.

There should be some explanation for the availability destruction in cooling-tower operation. Generally, the availability destruction is caused by the finite difference in temperature and water vapor concentration in the tower heat-transfer process. The pressure drops in both the water and air sides also would affect the value of the availability destruction. As indicated in Fig. 11-6 usually a fan is required in the cooling tower. Its operation would contribute additional availability destruction. The term $(\dot{A}_5 - \dot{A}_4)$ is the availability loss from the cooling tower. It represents the amount of availability that cannot be utilized and so usually is combined with the availability destruction. The combined term is referred to as the tower availability destruction and loss or, for simplicity, as the tower irreversibility.

In a more convenient form, Eq. (11-29) can be expressed as

$$\dot{m}_w(a_{f_i} - a_{f_e})_w + \dot{m}_m a_{f_m} + \dot{W}_f - \dot{m}_a(a_{f_e} - a_{f_i})_a - \dot{A}_{\text{dest}} = 0 \quad (11\text{-}30)$$

In numerical calculations, we need equations for the flow availability for water as well as for moist air. These equations have been developed in Chapter 5:

$$a_{f_w} = \left[h_f(T) - h_g(T_0)\right] + v_f(T)\left[P - P_g(T)\right] - T_0\left[s_f(T) - s_g(T_0)\right] + R_v T_0 \ln\frac{P_g(T_0)}{x_v{}^0 P_0} \quad (11\text{-}31)$$

$$a_{f_a} = \hat{c}_p(T - T_0) - T_0\left[\hat{c}_p \ln\frac{T}{T_0} - \hat{R}\ln\frac{P}{P_0}\right] + T_0\left[\hat{R}\ln\frac{1 + c\omega^0}{1 + c\omega} + c\omega R_a \ln\frac{\omega}{\omega^0}\right] \quad (11\text{-}32)$$

When the cooling tower is treated as a heat exchanger, the second law efficiency for the evaporative cooling tower may be defined by

$$\varepsilon = \frac{\dot{m}_a(a_{f_e} - a_{f_i})_a}{\dot{m}_w(a_{f_i} - a_{f_e})_w + \dot{m}_w a_{f_m} + \dot{W}_f} \tag{11-33}$$

The following is an example showing the numerical calculation of the cooling tower irreversibility due to heat and mass transfer between the water and air streams. Also included is the second law efficiency for the cooling tower.

Consider the cooling tower with the performance data shown in Fig. 11-5. At one point, the cooling tower has the following performance characteristics

Ambient wet-bulb temperature	75°F (90°F and 50%RH)
Cooling range	23.5°F
Cold water temperature	89°F
Water flow rate	145,000 gpm

The tower vendor indicates that the cooling tower has been designed with a water-to-air ratio of 1.10. The make-up water has a temperature of 89°F and is supplied to the tower at the rate of 2920 gpm. The field measurement also provides the conditions of moist air leaving the tower at 99°F and 90 percent relative humidity. Figure 11-7 summarizes the conditions of water and moist air at various locations. It should be noted that in this calculation, we convert the water flow rate into English units by using 1 gpm = 500 lbm/hr and setting the dead-state conditions as P_0 = 14.7 psia, T_0 = 90°F, Φ_0 = 50% RH, and ω^0 = 0.151 lbm of vapor per lbm of dry air.

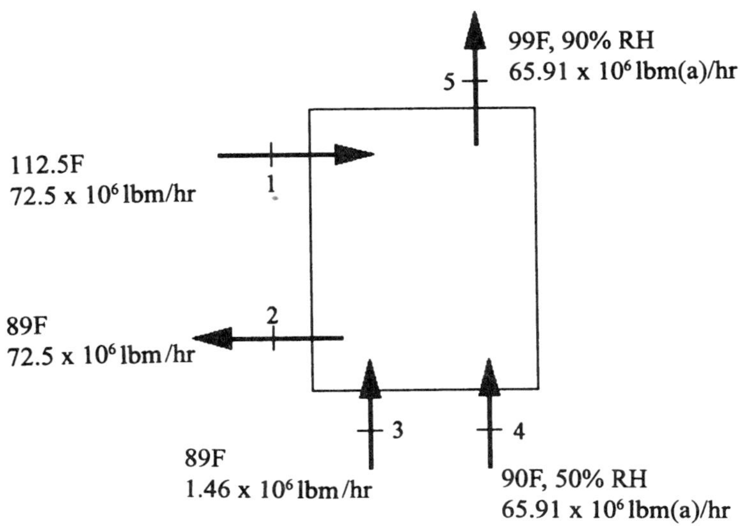

Figure 11-7 Summary of water and moist air conditions in a cooling tower.

Applying Eq. (11-31) we have the availability for the water at location 1 as

$$a_{f_1} = (80.52 - 1100.7) + \frac{0.01617 \times 144}{778}(14.7 - 1.372) - 550(0.1517 - 1.9512) + \frac{550 \times 85.76}{778}\ln\frac{0.6988}{0.024 \times 14.7}$$
$$= 11.04 \text{ Btu/lbm}$$

Similarly, we have $a_{f_2} = a_{f_3} = 10.61$ Btu/lbm. The moist air at location 4 has the conditions identical to those in the dead state. Thus, we have the flow availability $a_{f_4} = 0$.

For the moist air at location 5, we have by applying Eq. (11-32)

$$a_{f_5} = 0.2566(99 - 90) - 550 \times 0.2566\ln\frac{559}{550} + 550\left[0.0727\ln\frac{1 + 0.0243}{1 + 0.0598} + 1.6078 \times 0.0372 \times \frac{53.3}{778}\ln\frac{0.0372}{0.0151}\right]$$
$$= 0.6934 \text{ Btu/lbm}$$

Since the fan power is negligible, by using Eq. (11-30), the tower availability destruction is

$$\dot{A}_{\text{dest}} = 72.5 \times 10^6(11.04 - 10.61) + 1.46 \times 10^6(10.61) - 65.91 \times 10^6(0.6934 - 0)$$
$$= 0.97 \times 10^6 \text{ Btu/hr}$$

Finally, the tower second law efficiency is by Eq. (11-33)

$$\varepsilon = \frac{65.91 \times 10^6(0.6934 - 0)}{72.5 \times 10^6(11.04 - 10.61) + 1.46 \times 10^6(10.61)} = 0.979$$

11-5 PROBLEMS

11-1 A power plant condenser has the performance data shown in Table 11-7 at the cooling water flow rate 140,000 gpm. Determine the availability reduction of condensing steam in terms of the condenser heat load and circulating water inlet temperature. Present the results in a graphical form.

11-2 The second law efficiency for a condenser may be defined as

$$\varepsilon = \frac{\dot{m}_w(a_{f_e} - a_{f_i})_w}{\dot{m}_s(a_{f_i} - a_{f_e})_s}$$

Table 11-7 Performance data for the condenser in Problem 11-1.

Condenser pressure (in. Hg abs.)	Condenser heat load (MBtu/hr)	Circulating water Inlet temperature (°F)
0.980	250	70
1.100	500	70
1.230	750	70
1.380	1000	70
1.545	1250	70
1.355	250	80
1.520	500	80
1.695	750	80
1.880	1000	80
2.090	1250	80
1.845	250	90
2.055	500	90
2.280	750	90
2.530	1000	90
2.800	1250	90
1.965	250	92
2.180	500	92
2.425	750	92
2.680	1000	92
2.970	1250	92
2.145	250	95
2.380	500	95
2.640	750	95
2.925	1000	95
3.230	1250	95

where the subscripts s and w represent the condensing steam and the cooling water, respectively. For the condenser conditions specified in Problem 11-1, determine the second law efficiency and the availability destruction in terms of the condenser heat load and circulating water inlet temperature. Present the results in a graphical form.

11-3 The boiler feedwater pump has the following performance characteristics for 160°F water:

Flow Rate (gpm)	Developed Head (ft of water)	Efficiency (%)
0	7440	0
800	7420	27
1600	7330	46
2400	7200	58
3200	7010	67
4000	6730	74
5000 (design point)	6200	77
5600	5800	77

Calculate the availability destruction and the second law efficiency. Present the results in a graphical form.

11-4 A boiler forced-draft fan has the following performance characteristics for 100°F air:

Air Flow (cfm)	Static Pressure (in. of water)	Static Efficiency (%)
250,000	14.4	75
200,000	15.1	69
150,000	15.3	61
100,000	15.4	49
50,000	15.4	30
0	15.2	0

Calculate the availability destruction and the second law efficiency using the above performance data.

11-5 A pump has the following performance characteristics at the speed of 1750 rpm:

Capacity (gpm)	Head (ft of water)	Efficiency (%)
850	2005	74.8
800	2100	76.0
750	2180	76.8
700	2260	77.0
650	2325	76.3
600	2385	75.0
550	2425	72.8
500	2460	70.0
400	2510	62.0
300	2545	52.0
200	2565	39.0
100	2568	21.0
0	2570	0

The water temperature is assumed to be 160°F. Estimate the availability destruction and the second law efficiency for the pump under the operating conditions of 600 gpm.

11-6 A forced-draft fan is to be controlled by outlet damper and driven by two motors, one at 875 rpm and the other at 700 rpm. The fan performance at 875 rpm and 100°F air as follows:

Delivery (cfm)	Static Pressure (in. of water)	Static efficiency (%)
280,000	13.4	74.0
200,000	14.1	68.0
150,000	14.3	60.2
100,000	14.4	48.0
50,000	14.4	28.5
0	14.2	0

Estimate the second law efficiency under the following operating conditions: (a) 150,000 cfm at 875 rpm and (b) 200,000 cfm at 700 rpm.

11-7 Water enters a counterflow cooling tower at 100°F and leaves at 80°F. The air enters at 14.7 psia and 70°F and 40% RH and leaves at 95°F and 95% RH. If the water flow rate at the tower inlet is 25,000 lbm/min, what is the atmospheric air requirement in cfm? Also determine the availability loss and destruction in Btu/min.

11-8 A crossflow cooling tower operates under the following conditions:

Water side	
Flow rate	10×10^6 lbm/hr
Inlet temperature	110°F
Exit temperature	90°F
Make-up water temperature	90°F
Air side	
Flow rate	6.25×10^6 lbm/hr
Tower inlet	90°F, 50% RH
Tower exit	90% RH
Tower fan	
Power required	260 hp
Static efficiency	75%
Pressure developed	0.7 in H_2O

Determine the rate of flow availability in Btu/hr for various tower inlet and exit locations. Also calculate the second law efficiency for the cooling tower.

SELECTED REFERENCES

[A] SELECTED TEXTBOOKS

1. Ahern, J. E., The Energy Method of Energy Systems Analysis, John Wiley and Sons, 1980.
2. Bejan, A., Advanced Engineering Thermodynamics, John Wiley and Sons, 1988.
3. Bejan, A., Entropy Generation Through Heat and Fluid Flow, John Wiley and Sons, 1982.
4. Bruges, E. A., Available Energy and the Second-Law Analysis, Academic Press Inc., 1959.
5. Gyftopoulos, E. P., L. J. Lazaridis, and T. F. Widmer, Potential Fuel Effectiveness in Industry, Ballinger Publishing Company, 1974.
6. Horlock, J. H., Combined Power Plants, Pergamon Press, 1992.
7. Horlock, J. H., Cogeneration: Combined Heat and Power, Pergamon Press, 1987.
8. Kotas, T. J., The Energy Method of Thermal Plant Analysis, Butterworths, 1985.
9. Li, K. W., and A. P. Priddy, Power Plant System Design, John Wiley and Sons, 1985.
10. Moran, M. J., Availability Analysis, Prentice Hall, 1982.
11. Obert, E. F., Concepts of Thermodynamics, McGraw Hill, 1960.
12. Potter, P. J., Power Plant Theory and Design, Second Edition, Ronald Company, 1959.
13. Sussman, M. V., Availability (Exergy) Analysis, Mulliken House, 1980.
14. Van Wylen, G. J., and R. E. Sonntag, Fundamentals of Classical Thermodynamics, 3rd edition, John Wiley and Sons, 1986.

[B] SELECTED TECHNICAL PAPERS

1. Bidini, G. and C. N. Grimaldi, "Steam recompression in combined cycles with steam injection," International Gas Turbine Institute IGTI, Vol. 8, pp. 397–406, ASME Publication, 1993.
2. Bidini, G. and A. Bosie, "Second-law analysis of intercooled gas turbine combined cycles," International Gas Turbine Institute IGTI, Vol. 4, pp. 281–287, 1989.
3. Carrington, C. G. and E. Yasni, "Off-design exergy audit of a thermal power station," Journal of Engineering for Gas Turbine and Power, Vol. 110, No. 2, pp. 166–172, 1988.
4. El-Masri, M. A. and R. A. Gaggioli, "Exergy analysis and optimization of recuperated gas turbine cycles," AES Vol. 2-1, ASME Publication, 1986.

5. El-Masri, M. A., "Exergy balance analysis of the reheat gas turbine combined cycle," Proceeding of the ASME-JSME Thermal Engineering Joint Conference, Honolulu, HI, 1987.
6. El-Masri, M. A., "Energy analysis and optimization of recuperated gas turbine cycles," AES Vol. 1, pp. 125–134, ASME Publication, 1986.
7. Galletti, A., R. Gusso, and S. S. Stecco, "Energy-exergy analysis of an actual regenerative cogeneration gas turbine plant," AES Vol. 3, pp. 69–76, ASME Publication, 1986.
8. Habib, M. A. and S. M. Zubair, "Second-law based thermodynamic analysis of regenerative reheat Rankine-cycle power plants," Energy Vol. 17, No. 3, pp. 295–301, UK, March 1992.
9. Habib, M. A., "Thermodynamic analysis of the performance of cogeneration plants," Energy Vol. 17, No. 5, pp. 485–491, UK, 1992.
10. Habib, M. A., "First and second law analysis of cogeneration plants," AES Vol. 24, pp. 31–36, ASME Publication, 1991.
11. Habib, M. A. and S. M. Zubair, "Thermodynamics of regenerative Rankine cycle power plants," AES Vol. 19, pp. 29–34, ASME Publication, 1990.
12. Horlock, J. H. and R. W. Haywood, "Thermodynamic availability and its application to combined heat and power plant," Proceedings of Institute of Mechanical Engineers, Vol. 199, 1985.
13. Huang, F. F., "Performance evaluation of selected combustion gas turbine cogeneration system based on first and second-law analysis," Journal of Engineering for Gas Turbine and Power, Vol. 112, No. 1, pp. 117–121, January 1991.
14. Huang, F. F., "Methodology for overall performances evaluation of combined gas-steam power plants based on energy as well as exergy consideration," International Gas Turbine Institute IGTI, Vol. 4, pp. 447–454, 1989.
15. Huang, F. F., "Methodology of performance evaluation of combustion gas turbine cogeneration system based on first-law as well as second-law analysis," AES Vol. 6, pp. 35–40, ASME Publication, 1988.
16. Kotas, T. J., "Exergy criteria of performance for thermal plants," International Journal of Heat and Fluid Flow, Vol. 2, No. 4, pp. 147–163, Dec. 1980.
17. Li, K. W., "Thermodynamic analysis of combined-cycle cogeneration system," ASME Paper 87-JPG2-Pwr-44, 1987.
18. Manfrida, G. and S. S. Stecco, "Second-law analysis of large steam power plants at nominal and off-design loads," Proceedings of Intersociety Energy Conversion Engineering Conference, Vol. 2, pp. 919–924, August 1984.
19. Moran, M. J. and E. Sciubba, "Exergy analysis: Principle and practice," Journal of Engineering for Gas Turbines and Power, Vol. 116, No. 2, pp. 285–290, April 1994.
20. Murthy, T. C. and R. Natarajan, "First and second-law analysis of the Rankine steam power," International Journal of Mechanical Engineering Education, Vol. 16, No. 4, pp. 279–290, 1988.
21. Raghavan, P. and R. Natarajan, "First and second-law analysis of the reheat-steam power cycle," International Journal of Applied Engineering Education, Vol. 5, No. 5, pp. 607–619, UK, 1989.
22. Rodriguez, Luis S. J., "Calculation of available-energy quantities," in the ACS Symposium Series 122, American Chemical Society, 1980.
23. Sarabchi, K., "Parametric analysis of gas turbine cogeneration plant form first and second law viewpoints," International Gas Turbine Institute, IGTI. Vol. 7, pp. 485–491, ASME publication, 1992.
24. Sciubba, E. and T. M. Su, "Second law analysis of the steam turbine power cycle: A parametric study," AES Vol. 3, pp. 151–165, ASME publication, 1986.

APPENDIX

A

CONVERSION FACTORS

Length	1 cm = 0.3937 in. = 10^4 μm = 10^8 Å
	1 in. = 2.540 cm
	1 ft = 30.48 cm
Mass	1 lbm = 453.59 g_m = 7000 g
	1 kg_m = 2.205 lb_m
Force	1 lbf = 32.174 lbm-ft/s^2
	1 lbf = 444, 800 dyn = 4.448 N
	1 torr = 1 mmHg at 0°C = 10^3 μmHg
	= 1.933×10^{-2} psi
Volume	1 L = 0.0353 ft^3 = 0.2642 gal = 61.025 $in.^3$
	1 ft^3 = 28.316L = 7.4805 gal = 0.02832 m^3
	1 $in.^3$ = 16.387 cm^3
Density	1 lbm/ft^3 = 0.01602 g_m/cm^3
Energy	1 Btu = 778.16 ft-lbf = 1055 J
	1 Btu/lbm = 2.32 kJ/kg
Power	1 W = 1 J/s = 3.413 Btu/h
	1 hp = 746 W = 550 ft-lb_f/s = 2545 Btu/h
Velocity	1 mi/h = 0.447 m/s
Specific heat	1 Btu/lbm-R = 4.187 kJ/kg-K
Temperature	T(°R) = T(°F) + 459.67 = 1.8[T(°C) + 273.15] = 1.8T(°K)

APPENDIX
B
ESTIMATE OF COAL AVAILABILITY

In this illustration, the coal has the composition given in Table B-1. To determine the coal availability, we first prepare the combustion equation for 100 pounds of coal. That is,

$$\begin{aligned}&[6.07C + 2.40H_2 + 0.19O_2 + 0.05N_2 + 0.07S + 0.19H_2O(l)]\\&\quad + 7.15O_2 + 7.15 \times 3.76N_2\\&\quad \rightarrow 6.07CO_2 + 2.59H_2O(g) + 0.07SO_2 + 26.93N_2\end{aligned}$$

It should be pointed out that the coal and combustion air are assumed to be at the standard conditions (77°F and 14.7 psia) and that the combustion products are completely in thermal and chemical equilibrium with the environment.

The coal availability is simply the available work for the above combustion process. It is numerically given by the equation

$$A_{\text{coal}}^{\text{ch}} = -\Delta G(T_0, P_0) + \bar{R}T_0 \ln\left(\frac{\left(x_{O_2}^0\right)^{7.15}\left(x_{N_2}^0\right)^{7.15\times 3.76}}{\left(x_{CO_2}^0\right)^{6.07}\left(x_{H_2O}^0\right)^{2.59}\left(x_{SO_2}^0\right)^{0.07}\left(x_{N_2}^0\right)^{26.93}}\right) \tag{B-1}$$

where

$$\begin{aligned}-\Delta G(T_0, P_0) = {}& 6.07\bar{g}_C(T_0, P_0) + 2.40\bar{g}_{H_2}(T_0, P_0) + 0.19\bar{g}_{O_2}(T_0, P_0)\\&+ 0.05\bar{g}_{N_2}(T_0, P_0) + 0.07\bar{g}_S(T_0, P_0) + 0.19\bar{g}_{H_2O(l)}(T_0, P_0)\\&+ 7.15\bar{g}_{O_2}(T_0, P_0) + 26.88\bar{g}_{N_2}(T_0, P_0)\end{aligned}$$

Table B-1 Example composition of coal.

Element	Ultimate analysis (lbm/100 lbm of coal)	Molar analysis (Moles/100 lb of coal)
Carbon	72.8	6.07
Hydrogen	4.8	2.40
Oxygen	6.2	0.19
Nitrogen	1.5	0.05
Sulfur	2.2	0.07
Water	3.5	0.19
Ash	9.0	—

$$- 6.07\bar{g}_{CO_2}(T_0, P_0) - 2.59\bar{g}_{H_2O(g)}(T_0, P_0)$$
$$- 0.07\bar{g}_{SO_2}(T_0, P_0) - 26.93\bar{g}_{N_2}(T_0, P_0) \qquad \text{(B-2)}$$

Taking Gibbs function values for sulfur and sulfur dioxide

$$\bar{g}_S(T_0, P_0) = 0$$
$$\bar{g}_{SO_2}(T_0, P_0) = -129{,}150 \text{ Btu/lbmole}$$

and other values from Table C-6, we calculate the above change of the Gibbs function value as

$$\begin{aligned} -\Delta G(T_0, P_0) &= 0.19(-102{,}040) - 6.07(-169{,}680) - 2.59(-98{,}350) \\ &\quad - 0.07(-129{,}150) \\ &= 1{,}274{,}336 \text{ Btu/100 lbm of coal} \end{aligned}$$

Take the following model for the environment:

Pressure	14.7 psia
Temperature	77°F
Components	Mole fraction
Nitrogen	0.7567
Oxygen	0.2035
Water Vapor	0.0303
Carbon Dioxide	0.0003
Sulfur Dioxide	0.0003
Other gases	0.0089

We calculate the second term of Eq. (B-1) as

$$\bar{R}T_0 \ln\left[\frac{\left(x^0_{O_2}\right)^{7.15}\left(x^0_{N_2}\right)^{7.15\times 3.76}}{\left(x^0_{CO_2}\right)^{6.07}\left(x^0_{H_2O}\right)^{2.59}\left(x^0_{SO_2}\right)^{0.07}\left(x^0_{N_2}\right)^{26.93}}\right]$$

$$= \frac{1545 \times 537}{778}\ln\left[\frac{(0.2035)^{7.15}(0.7567)^{7.15\times 3.76}}{(0.0003)^{6.07}(0.0303)^{2.59}(0.0003)^{0.07}(0.7567)^{26.93}}\right]$$

$$= 56{,}135 \text{ Btu}/100 \text{ lbm of coal}$$

Finally the chemical availability per pound of coal at the standard conditions is given by

$$a_{\text{coal}}(T_0, P_0) = A^{\text{ch}}_{\text{coal}}/100 = \frac{1{,}274{,}336 + 56{,}135}{100} = 13{,}304 \text{ Btu/lbm of coal}$$

APPENDIX

C

TABLES AND CHARTS

Table C-1 Critical constants

Substance	Formula	Molecular weight	Temperature T_c (°R)	Pressure P_c (psia)	Specific volume v_c (ft^3/lbmole)
Ammonia	NH_3	17.03	729.8	1636	1.16
Carbon dioxide	CO_2	44.01	547.5	1071	1.51
Carbon monoxide	CO	28.01	240	507	1.49
Helium	He	4.003	9.5	33.2	0.926
Hydrogen (normal)	H_2	2.016	59.9	188.1	1.04
Nitrogen	N_2	28.016	227.1	492	1.44
Nitrous oxide	N_2O	44.02	557.4	1054	1.54
Oxygen	O_2	32.00	278.6	736	1.25
Sulfur dioxide	SO_2	64.06	775.2	1143	1.95
Water	H_2O	18.016	1165.3	3204	0.90
Dichlorodifluoromethane	CCl_2F_2	120.92	692.4	582	3.49
Methane	CH_4	16.042	343.9	673	1.59
Propane	C_3H_8	44.094	665.9	617	3.20

Data extracted from K. A. Kobe and R. E. Lynn, Jr., Chemical Review 52:117–236 (1953).

Table C-2 Thermodynamic properties of steam[a]

Table C-2.1 Saturated steam: temperature table

		Specific volume ft^3/lbm		Internal energy Btu/lbm			Enthalpy Btu/lbm			Entropy Btu/lbm-R		
Temp. °F T	Press. lbf/sq. in P	Sat. Liquid v_f	Sat. Vapor v_u	Sat. Liquid u_f	Evap. u_{fu}	Sat. Vapor u_u	Sat. Liquid h_f	Evap. h_{fu}	Sat. Vapor h_u	Sat. Liquid s_f	Evap. s_{fu}	Sat. Vapor s_u
32.018	0.088.66	0.016022	3302	0.00	1021.2	1021.2	0.01	1075.4	1075.4	0.00000	2.1869	2.1869
35	0.09992	0.016021	2948	2.99	1019.2	1022.2	3.00	1073.7	1076.7	0.00607	2.1704	2.1764
40	0.12166	0.016020	2445	8.02	1015.8	1023.9	8.02	1070.9	1078.9	0.01617	2.1430	2.1592
45	0.14748	0.016021	2037	13.04	1012.5	1025.5	13.04	1068.1	1081.1	0.02618	2.1162	2.1423
50	0.17803	0.016024	1704.2	18.06	1009.1	1027.2	18.06	1065.2	1083.3	0.03607	2.0899	2.1259
60	0.2563	0.016035	1206.9	28.08	1002.4	1030.4	28.08	1059.6	1087.7	0.05555	2.0388	2.0943
70	0.3632	0.016051	867.7	38.09	995.6	1033.7	38.09	1054.0	1092.0	0.07463	1.9896	2.0642
80	0.5073	0.016073	632.8	48.08	988.9	1037.0	48.09	1048.3	1096.4	0.09332	1.9423	2.0356
90	0.6988	0.016099	467.7	58.07	982.2	1040.2	58.07	1042.7	1100.7	0.11165	1.8966	2.0083
100	0.9503	0.016130	350.0	68.04	975.4	1043.5	68.05	1037.0	1105.0	0.12963	1.8526	1.9822
110	1.2763	0.016166	265.1	78.02	968.7	1046.7	78.02	1031.3	1109.3	0.14730	1.8101	1.9574
120	1.6945	0.016205	203.0	87.99	961.9	1049.9	88.00	1025.5	1113.5	0.16465	1.7690	1.9336
130	2.225	0.016247	157.17	97.97	955.1	1053.0	97.98	1019.8	1117.8	0.18172	1.7292	1.9109
140	2.892	0.016293	122.88	107.95	948.2	1056.2	107.96	1014.0	1121.9	0.19851	1.6907	1.8892
150	3.722	0.016343	96.99	117.95	941.3	1059.3	117.96	1008.1	1126.1	0.21503	1.6533	1.8684
160	4.745	0.016395	77.23	127.94	934.4	1062.3	127.96	1002.2	1130.1	0.23130	1.6171	1.8484
170	5.996	0.016450	62.02	137.95	927.4	1065.4	137.97	996.2	1134.2	0.24732	1.5819	1.8293
180	7.515	0.016509	50.20	147.97	920.4	1068.3	147.99	990.2	1138.2	0.26311	1.5478	1.8109
190	9.343	0.016570	40.95	158.00	913.3	1071.3	158.03	984.1	1142.1	0.27866	1.5146	1.7932
200	11.529	0.016634	33.63	168.04	906.2	1074.2	168.07	977.9	1145.9	0.29400	1.4822	1.7762
210	14.125	0.016702	27.82	178.10	898.9	1077.0	178.14	971.6	1149.7	0.30913	1.4508	1.7599
212	14.698	0.016716	26.80	180.11	897.5	1077.6	180.16	970.3	1150.5	0.31213	1.4446	1.7567
220	17.188	0.016772	23.15	188.17	891.7	1079.8	188.22	965.3	1153.5	0.32406	1.4201	1.7441
230	20.78	0.016845	19.386	198.26	884.3	1082.6	198.32	958.8	1157.1	0.33880	1.3901	1.7289
240	24.97	0.016922	16.327	208.36	876.9	1085.3	208.44	952.3	1160.7	0.35335	1.3609	1.7143

250	29.82	0.017001	13.826	218.49	869.4	1087.9	218.59	945.6	1164.2	0.36772	1.3324	1.7001
260	35.42	0.017084	11.768	228.64	861.8	1090.5	228.76	938.8	1167.6	0.38193	1.3044	1.6864
270	41.85	0.017170	10.066	238.82	854.1	1093.0	238.95	932.0	1170.9	0.39597	1.2771	1.6731
280	49.18	0.017259	8.650	249.02	846.3	1095.4	249.18	924.9	1174.1	0.40986	1.2504	1.6602
290	57.53	0.017352	7.467	259.25	838.5	1097.7	259.44	917.8	1177.2	0.42360	1.2241	1.6477
300	66.98	0.017448	6.472	269.52	830.5	1100.0	269.73	910.4	1180.2	0.43720	1.1984	1.6356
310	77.64	0.017548	5.632	279.81	822.3	1102.1	280.06	903.0	1183.0	0.45067	1.1731	1.6238
320	89.60	0.017652	4.919	290.14	814.1	1104.2	290.43	895.3	1185.8	0.46400	1.1483	1.6123
330	103.00	0.017760	4.312	300.51	805.7	1106.2	300.84	887.5	1188.4	0.47722	1.1238	1.6010
340	117.93	0.017872	3.792	310.91	797.1	1108.0	311.30	879.5	1190.8	0.49031	1.0997	1.5901
350	134.53	0.017988	3.346	321.35	788.4	1109.8	321.80	871.3	1193.1	0.50329	1.0760	1.5793
360	152.92	0.018108	2.961	331.84	779.6	1111.4	332.35	862.9	1195.2	0.51617	1.0526	1.5688
370	173.23	0.018233	2.628	342.37	770.6	1112.9	342.96	854.2	1197.2	0.52894	1.0295	1.5585
380	195.60	0.018363	2.339	352.95	761.4	1114.3	353.62	845.4	1199.0	0.54163	1.0067	1.5483
390	220.2	0.018498	2.087	363.58	752.0	1115.6	364.34	836.2	1200.6	0.55422	0.9841	1.5383
400	247.1	0.018638	1.8661	374.27	742.4	1116.6	375.12	826.8	1202.0	0.56672	0.9617	1.5284
410	276.5	0.018784	1.6726	385.01	732.6	1117.6	385.97	817.2	1203.1	0.57916	0.9395	1.5187
420	308.5	0.018936	1.5024	395.81	722.5	1118.3	396.89	807.2	1204.1	0.59152	0.9175	1.5091
430	343.5	0.019094	1.3521	406.68	712.2	1118.9	407.89	796.9	1204.8	0.60381	0.8957	1.4995
440	381.2	0.019260	1.2192	417.62	701.7	1119.3	418.98	786.3	1205.3	0.61605	0.8740	1.4900
450	422.1	0.019433	1.1011	428.6	690.9	1119.5	430.2	775.4	1205.6	0.6282	0.8523	1.4806
460	466.3	0.019614	0.9961	439.7	679.8	1119.6	441.4	764.1	1205.5	0.6404	0.8308	1.4712
470	514.1	0.019803	0.9025	450.9	668.4	1119.4	452.8	752.4	1205.2	0.6525	0.8093	1.4618
480	565.5	0.020002	0.8187	462.2	656.7	1118.9	464.3	740.3	1204.6	0.6646	0.7878	1.4524
490	620.7	0.020211	0.7436	473.6	644.7	1118.3	475.9	727.8	1203.7	0.6767	0.7663	1.4430
500	680.0	0.02043	0.6761	485.1	632.3	1117.4	487.7	714.8	1202.5	0.6888	0.7448	1.4335
520	811.4	0.02091	0.5605	508.5	606.2	1114.8	511.7	687.3	1198.9	0.7130	0.7015	1.4145
540	961.5	0.02145	0.4658	532.6	578.4	1111.0	536.4	657.5	1193.8	0.7374	0.6576	1.3950
560	1131.8	0.02207	0.3877	557.4	548.4	1105.8	562.0	625.0	1187.0	0.7620	0.6129	1.3749
580	1324.3	0.02278	0.3225	583.1	515.9	1098.9	588.6	589.3	1178.0	0.7872	0.5668	1.3540
600	1541.0	0.02363	0.2677	609.9	480.1	1090.0	616.7	549.7	1166.4	0.8130	0.5187	1.3317
620	1784.4	0.02465	0.2209	638.3	440.2	1078.5	646.4	505.0	1151.4	0.8398	0.4677	1.3075
640	2057.1	0.02593	0.1805	668.7	394.5	1063.2	678.6	453.4	1131.9	0.8681	0.4122	1.2803
660	2362	0.02767	0.14459	702.3	340.0	1042.3	714.4	391.1	1105.5	0.8990	0.3493	1.2483
680	2705	0.03032	0.11127	741.7	269.3	1011.0	756.9	309.8	1066.7	0.9350	0.2718	1.2068
700	3090	0.03666	0.07438	801.7	145.9	947.7	822.7	167.5	990.2	0.9902	0.1444	1.1346
705.44	3204	0.05053	0.05053	872.6	0	872.6	902.5	0	902.5	1.0580	0	1.0580

[a]Abridged from Joseph H. Keenan, Frederick G. Keyes, Philip G. Hill, and Joan G. Moore, *Steam Tables* (New York: John Wiley & Sons, Inc., 1969). Reprinted with permission of John Wiley & Sons, Inc.

Table C-2.2 Saturated steam: pressure table

		Specific volume		Internal energy			Enthalpy			Entropy		
Press lbf/sq. in P	Temp. Fahr. T	Sat. Liquid v_f	Sat. Vapor v_u	Sat. Liquid u_f	Evap. u_{fu}	Sat. Vapor u_u	Sat. Liquid h_f	Evap. h_{fu}	Sat. Vapor h_u	Sat. Liquid s_f	Evap. s_{fu}	Sat. Vapor s_u
1.0	101.70	0.016136	333.6	69.74	974.3	1044.0	69.74	1036.0	1105.8	0.13266	1.8453	1.9779
2.0	126.04	0.016230	173.75	94.02	957.8	1051.8	94.02	1022.1	1116.1	0.17499	1.7448	1.9198
3.0	141.43	0.016300	118.72	109.38	947.2	1056.6	109.39	1013.1	1122.5	0.20089	1.6852	1.8861
4.0	152.93	0.016358	90.64	120.88	939.3	1060.2	120.89	1006.4	1127.3	0.21983	1.6426	1.8624
5.0	162.21	0.016407	73.53	130.15	932.9	1063.0	130.17	1000.9	1131.0	0.23486	1.6093	1.8441
6.0	170.03	0.016451	61.98	137.98	927.4	1065.4	138.00	996.2	1134.2	0.24736	1.5819	1.8292
8.0	182.84	0.016526	47.35	150.81	918.4	1069.2	150.84	988.4	1139.3	0.26754	1.5383	1.8058
10	193.19	0.016590	38.42	161.20	911.0	1072.2	161.23	982.1	1143.3	0.28358	1.5041	1.7877
14.696	211.99	0.016715	26.80	180.10	897.5	1077.6	180.15	970.4	1150.5	0.31212	1.4446	1.7567
15	213.03	0.016723	26.29	181.14	896.8	1077.9	181.19	969.7	1150.9	0.31367	1.4414	1.7551
20	227.96	0.016830	20.09	196.19	885.8	1082.0	196.26	960.1	1156.4	0.33580	1.3962	1.7320
25	240.08	0.016922	16.306	208.44	876.9	1085.3	208.52	952.2	1160.7	0.35345	1.3607	1.7142
30	250.34	0.017004	13.748	218.84	869.2	1088.0	218.93	945.4	1164.3	0.36821	1.3314	1.6996
35	259.30	0.017073	11.900	227.93	862.4	1090.3	228.04	939.3	1167.4	0.38093	1.3064	1.6873
40	267.26	0.017146	10.501	236.03	856.2	1092.3	236.16	933.8	1170.0	0.39214	1.2845	1.6767
45	274.46	0.017209	9.403	243.37	850.7	1094.0	243.51	928.8	1172.3	0.40218	1.2651	1.6673
50	281.03	0.017269	8.518	250.08	845.5	1095.6	250.24	924.2	1174.4	0.41129	1.2476	1.6589
55	287.10	0.017325	7.789	256.28	840.8	1097.0	256.46	919.9	1176.3	0.41963	1.2317	1.6513
60	292.73	0.017378	7.177	262.06	836.3	1098.3	262.25	915.8	1178.0	0.42733	1.2170	1.6444
65	298.00	0.017429	6.657	267.46	832.1	1099.5	267.67	911.9	1179.6	0.43450	1.2035	1.6380
70	302.96	0.017478	6.209	272.56	828.1	1100.6	272.79	908.3	1181.0	0.44120	1.1909	1.6321
75	307.63	0.017524	5.818	277.37	824.3	1101.6	277.61	904.8	1182.4	0.44749	1.1790	1.6265
80	312.07	0.017570	5.474	281.95	820.6	1102.6	282.21	901.4	1183.6	0.45344	1.1679	1.6214
85	316.29	0.017613	5.170	286.30	817.1	1103.5	286.58	898.2	1184.8	0.45907	1.1574	1.6165
90	320.31	0.017655	4.898	290.46	813.8	1104.3	290.76	895.1	1185.9	0.46442	1.1475	1.6119
95	324.16	0.017696	4.654	294.45	810.6	1105.0	294.76	892.1	1186.9	0.46952	1.1380	1.6076
100	327.86	0.017736	4.434	298.28	807.5	1105.8	298.61	889.2	1187.8	0.47439	1.1290	1.6034
110	334.82	0.017813	4.051	305.52	801.6	1107.1	305.88	883.7	1189.6	0.48355	1.1122	1.5957

120	341.30	0.017886	3.730	312.27	796.0	1108.3	312.67	878.5	1191.1	0.49201	1.0966	1.5886
130	347.37	0.017957	3.457	318.61	790.7	1109.4	319.04	873.5	1192.5	0.49989	1.0822	1.5821
140	353.08	0.018024	3.221	324.58	785.7	1110.3	325.05	868.7	1193.8	0.50727	1.0688	1.5761
150	358.48	0.018089	3.016	330.24	781.0	1111.2	330.75	864.2	1194.9	0.51422	1.0562	1.5704
160	363.60	0.018152	2.836	335.63	776.4	1112.0	336.16	859.8	1196.0	0.52078	1.0443	1.5651
170	368.47	0.018214	2.676	340.76	772.0	1112.7	341.33	855.6	1196.9	0.52700	1.0330	1.5600
180	373.13	0.018273	2.533	345.68	767.7	1113.4	346.29	851.5	1197.8	0.53292	1.0223	1.5553
190	377.59	0.018331	2.405	350.39	763.6	1114.0	351.04	847.5	1198.6	0.53857	1.0122	1.5507
200	381.86	0.018387	2.289	354.9	759.6	1114.6	355.6	843.7	1199.3	0.5440	1.0025	1.5464
250	401.04	0.018653	1.8448	375.4	741.4	1116.7	376.2	825.8	1202.1	0.5680	0.9594	1.5274
300	417.43	0.018896	1.5442	393.0	725.1	1118.2	394.1	809.8	1203.9	0.5883	0.9232	1.5115
350	431.82	0.019124	1.3267	408.7	710.3	1119.0	409.9	795.0	1204.9	0.6060	0.8917	1.4978
400	444.70	0.019340	1.1620	422.8	696.7	1119.5	424.2	781.2	1205.5	0.6218	0.8638	1.4856
450	456.39	0.019547	1.0326	435.7	683.9	1119.6	437.4	768.2	1205.6	0.6360	0.8385	1.4746
500	467.13	0.019748	0.9283	447.7	671.7	1119.4	449.5	755.8	1205.3	0.6490	0.8154	1.4645
550	477.07	0.019943	0.8423	458.9	660.2	1119.1	460.9	743.9	1204.8	0.6611	0.7941	1.4551
600	486.33	0.02013	0.7702	469.4	649.1	1118.6	471.7	732.4	1204.1	0.6723	0.7742	1.4464
700	503.23	0.02051	0.6558	488.9	628.2	1117.0	491.5	710.5	1202.0	0.6927	0.7378	1.4305
800	518.36	0.02087	0.5691	506.6	608.4	1115.0	509.7	689.6	1199.3	0.7110	0.7050	1.4160
900	532.12	0.02123	0.5009	523.0	589.6	1112.6	526.6	669.5	1196.0	0.7277	0.6750	1.4027
1000	544.75	0.02159	0.4459	538.4	571.5	1109.9	542.4	650.0	1192.4	0.7432	0.6471	1.3903
1200	567.37	0.02232	0.3623	566.7	536.8	1103.5	571.7	612.3	1183.9	0.7712	0.5961	1.3673
1400	587.25	0.02307	0.3016	592.7	503.3	1096.0	598.6	575.5	1174.1	0.7964	0.5497	1.3461
1600	605.06	0.02386	0.2552	616.9	470.5	1087.4	624.0	538.9	1162.9	0.8196	0.5062	1.3258
1800	621.21	0.02472	0.2183	640.0	437.6	1077.7	648.3	502.1	1150.4	0.8414	0.4645	1.3060
2000	636.00	0.02565	0.18813	662.4	404.2	1066.6	671.9	464.4	1136.3	0.8623	0.4238	1.2861
2500	668.31	0.02860	0.13059	717.7	313.4	1031.0	730.9	360.5	1091.4	0.9131	0.3196	1.2327
3000	695.52	0.03431	0.08404	783.4	185.4	968.8	802.5	213.0	1015.5	0.9732	0.1843	1.1575
3203.6	705.44	0.05053	0.05053	872.6	0	872.6	902.5	0	902.5	1.0580	0	1.0580

Table C-2.3 Superheated vapor

°F	v	u	h	s	v	u	h	s	v	u	h	s
	P = 1.0(101.70)				P = 5.0(162.21)				P = 10.0(193.19)			
Sat	333.6	1044.0	1105.8	1.9779	73.53	1063.0	1131.0	1.8441	38.42	1072.2	1143.3	1.7877
200	392.5	1077.5	1150.1	2.0508	78.15	1076.3	1148.6	1.8715	38.85	1074.7	1146.6	1.7927
240	416.4	1091.2	1168.3	2.0775	83.00	1090.3	1167.1	1.8987	41.32	1089.0	1165.5	1.8205
280	440.3	1105.0	1186.5	2.1028	87.83	1104.3	1185.5	1.9244	43.77	1103.3	1184.3	1.8467
320	464.2	1118.9	1204.8	2.1269	92.64	1118.3	1204.0	1.9487	46.20	1117.6	1203.1	1.8714
360	488.1	1132.9	1223.2	2.1500	97.45	1132.4	1222.6	1.9719	48.62	1131.8	1221.8	1.8948
400	511.9	1147.0	1241.8	2.1720	102.24	1146.6	1241.2	1.9941	51.03	1146.1	1240.5	1.9171
440	535.8	1161.2	1260.4	2.1932	107.03	1160.9	1259.9	2.0154	53.44	1160.5	1259.3	1.9385
500	571.5	1182.8	1288.5	2.2235	114.20	1182.5	1288.2	2.0458	57.04	1182.2	1287.7	1.9690
600	631.1	1219.3	1336.1	2.2706	126.15	1219.1	1335.8	2.0930	63.03	1218.9	1335.5	2.0164
700	690.7	1256.7	1384.5	2.3142	138.08	1256.5	1384.3	2.1367	69.01	1256.3	1384.0	2.0601
800	750.3	1294.9	1433.7	2.3550	150.01	1294.7	1433.5	2.1775	74.98	1294.6	1433.3	2.1009
1000	869.5	1373.9	1534.8	2.4294	173.86	1373.9	1534.7	2.2520	86.91	1373.8	1534.6	2.1755
1200	988.6	1456.7	1639.6	2.4967	197.70	1456.6	1639.5	2.3192	98.84	1456.5	1639.4	2.2428
1400	1107.7	1543.1	1748.1	2.5584	221.54	1543.1	1748.1	2.3810	110.76	1543.0	1748.0	2.3045
	P = 14.696(211.99)				P = 20(227.96)				P = 40(267.26)			
Sat	26.80	1077.6	1150.5	1.7567	20.09	1082.0	1156.4	1.7320	10.501	1092.3	1170.0	1.6767
240	28.00	1087.9	1164.0	1.7764	20.47	1086.5	1162.3	1.7405				
280	29.69	1102.4	1183.1	1.8030	21.73	1101.4	1181.8	1.7676	10.711	1097.3	1176.6	1.6857
320	31.36	1116.8	1202.1	1.8280	22.98	1116.0	1201.0	1.7930	11.360	1112.8	1196.9	1.7124
360	33.02	1131.2	1221.0	1.8516	24.21	1130.6	1220.1	1.8168	11.996	1128.0	1216.8	1.7373
400	34.67	1145.6	1239.9	1.8741	25.43	1145.1	1239.2	1.8395	12.623	1143.0	1236.4	1.7606
440	36.31	1160.1	1258.8	1.8956	26.64	1159.6	1258.2	1.8611	13.243	1157.8	1255.8	1.7828
500	38.77	1181.8	1287.3	1.9263	28.46	1181.5	1286.8	1.8919	14.164	1180.1	1284.9	1.8140
600	42.86	1218.6	1335.2	1.9737	31.47	1218.4	1334.8	1.9395	15.685	1217.3	1333.4	1.8621
700	46.93	1256.1	1383.8	2.0175	34.47	1255.9	1383.5	1.9834	17.196	1255.1	1382.4	1.9063
800	51.00	1294.4	1433.1	2.0584	37.46	1294.3	1432.9	2.0243	18.701	1293.7	1432.1	1.9474
1000	59.13	1373.7	1534.5	2.1330	43.44	1373.5	1534.3	2.0989	21.70	1373.1	1533.8	2.0223
1200	67.25	1456.5	1639.3	2.2003	49.41	1456.4	1639.2	2.1663	24.69	1456.1	1638.9	2.0897
1400	75.36	1543.0	1747.9	2.2621	55.37	1542.9	1747.9	2.2281	27.68	1542.7	1747.6	2.1515
1600	83.47	1633.2	1860.2	2.3194	61.33	1633.2	1860.1	2.2854	30.66	1633.0	1859.9	2.2089

	P = 60(292.73)				P = 80(312.07)				P = 100(327.86)			
Sat	7.177	1098.3	1178.0	1.6444	5.474	1102.6	1183.6	1.6214	4.434	1105.8	1187.8	1.6034
320	7.485	1109.5	1192.6	1.6634	5.544	1106.0	1188.0	1.6271				
360	7.924	1125.3	1213.3	1.6893	5.886	1122.5	1209.7	1.6541	4.662	1119.7	1205.9	1.6259
400	8.353	1140.8	1233.5	1.7134	6.217	1138.5	1230.6	1.6790	4.934	1136.2	1227.5	1.6517
440	8.775	1156.0	1253.4	1.7360	6.541	1154.2	1251.0	1.7022	5.199	1152.3	1248.5	1.6755
500	9.399	1178.6	1283.0	1.7678	7.017	1177.2	1281.1	1.7346	5.587	1175.7	1279.1	1.7085
600	10.425	1216.3	1332.1	1.8165	7.794	1215.3	1330.7	1.7838	6.216	1214.2	1329.3	1.7582
700	11.440	1254.4	1381.4	1.8609	8.561	1253.6	1380.3	1.8285	6.834	1252.8	1379.2	1.8033
800	12.448	1293.0	1431.2	1.9022	9.321	1292.4	1430.4	1.8700	7.445	1291.8	1429.6	1.8449
1000	14.454	1372.7	1533.2	1.9773	10.831	1372.3	1532.6	1.9453	8.657	1371.9	1532.1	1.9204
1200	16.452	1455.8	1638.5	2.0448	12.333	1455.5	1638.1	2.0130	9.861	1455.2	1637.7	1.9882
1400	18.445	1542.5	1747.3	2.1067	13.830	1542.3	1747.0	2.0749	11.060	1542.0	1746.7	2.0502
1600	20.44	1632.8	1859.7	2.1641	15.324	1632.6	1859.5	2.1325	12.257	1632.4	1859.3	2.1076
1800	22.43	1726.7	1975.7	2.2179	16.818	1726.5	1975.5	2.1861	13.452	1726.4	1975.3	2.1614
2000	24.41	1824.0	2095.1	2.2685	18.310	1823.9	2094.9	2.2367	14.647	1823.7	2094.8	2.2121

	P = 120(341.30)				P = 140(353.08)				P = 160(363.60)			
Sat	3.730	1108.3	1191.1	1.5886	3.221	1110.3	1193.8	1.5761	2.836	1112.0	1196.0	1.5651
360	3.844	1116.7	1202.0	1.6021	3.259	1113.5	1198.0	1.5812				
400	4.079	1133.8	1224.4	1.6288	3.466	1131.4	1221.2	1.6088	3.007	1128.8	1217.8	1.5911
450	4.360	1154.3	1251.2	1.6590	3.713	1152.4	1248.6	1.6399	3.228	1150.5	1246.1	1.6230
500	4.633	1174.2	1277.1	1.6868	3.952	1172.7	1275.1	1.6682	3.440	1171.2	1273.0	1.6518
550	4.900	1193.8	1302.6	1.7127	4.184	1192.6	1300.9	1.6944	3.646	1191.3	1299.2	1.6784
600	5.164	1213.2	1327.8	1.7371	4.412	1212.1	1326.4	1.7191	3.848	1211.1	1325.0	1.7034
700	5.682	1252.0	1378.2	1.7825	4.860	1251.2	1377.1	1.7648	4.243	1250.4	1376.0	1.7494
800	6.195	1291.2	1428.7	1.8243	5.301	1290.5	1427.9	1.8068	4.631	1289.9	1427.0	1.7916
1000	7.208	1371.5	1531.5	1.9000	6.173	1371.0	1531.0	1.8827	5.397	1370.6	1530.4	1.8677
1200	8.213	1454.9	1637.3	1.9679	7.036	1454.6	1636.9	1.9507	6.154	1454.3	1636.5	1.9358
1400	9.214	1541.8	1746.4	2.0300	7.895	1541.6	1746.1	2.0129	6.906	1541.4	1745.9	1.9980
1600	10.212	1632.3	1859.0	2.0875	8.752	1632.1	1858.8	2.0704	7.656	1631.9	1858.6	2.0556
1800	11.209	1726.2	1975.1	2.1413	9.607	1726.1	1975.0	2.1242	8.405	1725.9	1974.8	2.1094
2000	12.205	1823.6	2094.6	2.1919	10.461	1823.5	2094.5	2.1749	9.153	1823.3	2094.3	2.1601

Table C-2.3 Superheated vapor (Continued)

°F	v	u	h	s	v	u	h	s	v	u	h	s
	P = 180(373.13)				P = 200(381.86)				P = 225(391.87)			
Sat	2.533	1113.4	1197.8	1.5553	2.289	1114.6	1199.3	1.5464	2.043	1115.8	1200.8	1.5365
400	2.648	1126.2	1214.4	1.5749	2.361	1123.5	1210.8	1.5600	2.073	1119.9	1206.2	1.5427
450	2.850	1148.5	1243.4	1.6078	2.548	1146.4	1240.7	1.5938	2.245	1143.8	1237.3	1.5779
500	3.042	1169.6	1270.9	1.6372	2.724	1168.0	1268.8	1.6239	2.405	1165.9	1266.1	1.6087
550	3.228	1190.0	1297.5	1.6642	2.893	1188.7	1295.7	1.6512	2.558	1187.0	1293.5	1.6366
600	3.409	1210.0	1323.5	1.6893	3.058	1208.9	1322.1	1.6767	2.707	1207.5	1320.2	1.6624
700	3.763	1249.6	1374.9	1.7357	3.379	1248.8	1373.8	1.7234	2.995	1247.7	1372.4	1.7095
800	4.110	1289.3	1426.2	1.7781	3.693	1288.6	1425.3	1.7660	3.276	1287.8	1424.2	1.7523
900	4.453	1329.4	1477.7	1.8175	4.003	1328.9	1477.1	1.8055	3.553	1328.3	1476.2	1.7920
1000	4.793	1370.2	1529.8	1.8545	4.310	1369.8	1529.3	1.8425	3.827	1369.3	1528.6	1.8292
1200	5.467	1454.0	1636.1	1.9227	4.918	1453.7	1635.7	1.9109	4.369	1453.4	1635.3	1.8977
1400	6.137	1541.2	1745.6	1.9849	5.521	1540.9	1745.3	1.9732	4.906	1540.7	1744.9	1.9600
1600	6.804	1631.7	1858.4	2.0425	6.123	1631.6	1858.2	2.0308	5.441	1631.3	1857.9	2.0177
1800	7.470	1725.8	1974.6	2.0964	6.722	1725.6	1974.4	2.0847	5.975	1725.4	1974.2	2.0716
2000	8.135	1823.2	2094.2	2.1470	7.321	1823.0	2094.0	2.1354	6.507	1822.9	2093.8	2.1223
	P = 250(401.04)				P = 275(409.52)				P = 300(417.43)			
Sat	1.8448	1116.7	1202.1	1.5274	1.6813	1117.5	1203.1	1.5192	1.5442	1118.2	1203.9	1.5115
450	2.002	1141.1	1233.7	1.5632	1.8026	1138.3	1230.0	1.5495	1.6361	1135.4	1226.2	1.5365
500	2.150	1163.8	1263.3	1.5948	1.9407	1161.7	1260.4	1.5820	1.7662	1159.5	1257.5	1.5701
550	2.290	1185.3	1291.3	1.6233	2.071	1183.6	1289.0	1.6110	1.8878	1181.9	1286.7	1.5997
600	2.426	1206.1	1318.3	1.6494	2.196	1204.7	1316.4	1.6376	2.004	1203.2	1314.5	1.6266
650	2.558	1226.5	1344.9	1.6739	2.317	1225.3	1343.2	1.6623	2.117	1224.1	1341.6	1.6516
700	2.688	1246.7	1371.1	1.6970	2.436	1245.7	1369.7	1.6856	2.227	1244.6	1368.3	1.6751
800	2.943	1287.0	1423.2	1.7401	2.670	1286.2	1422.1	1.7289	2.442	1285.4	1421.0	1.7187
900	3.193	1327.6	1475.3	1.7799	2.898	1327.0	1474.5	1.7689	2.653	1326.3	1473.6	1.7589
1000	3.440	1368.7	1527.9	1.8172	3.124	1368.2	1527.2	1.8064	2.860	1367.7	1526.5	1.7964
1200	3.929	1453.0	1634.8	1.8858	3.570	1452.6	1634.3	1.8751	3.270	1452.2	1633.8	1.8653
1400	4.414	1540.4	1744.6	1.9483	4.011	1540.1	1744.2	1.9376	3.675	1539.8	1743.8	1.9279
1600	4.896	1631.1	1857.6	2.0060	4.450	1630.9	1857.3	1.9954	4.078	1630.7	1857.0	1.9857
1800	5.376	1725.2	1974.0	2.0599	4.887	1725.0	1973.7	2.0493	4.479	1724.9	1973.5	2.0396
2000	5.856	1822.7	2093.6	2.1106	5.323	1822.5	2093.4	2.1000	4.879	1822.3	2093.2	2.0904

	P = 350(431.82)				P = 400(444.70)				P = 450(456.39)			
Sat	1.3267	1119.0	1204.9	1.4978	1.1620	1119.5	1205.5	1.4856	1.0326	1119.6	1205.6	1.4746
450	1.3733	1129.2	1218.2	1.5125	1.1745	1122.6	1209.6	1.4901				
500	1.4913	1154.9	1251.5	1.5482	1.2843	1150.1	1245.2	1.5282	1.1226	1145.1	1238.5	1.5097
550	1.5998	1178.3	1281.9	1.5790	1.3833	1174.6	1277.0	1.5605	1.2146	1170.7	1271.9	1.5436
600	1.7025	1200.3	1310.6	1.6068	1.4760	1197.3	1306.6	1.5892	1.2996	1194.3	1302.5	1.5732
650	1.8013	1221.6	1338.3	1.6323	1.5645	1219.1	1334.9	1.6153	1.3803	1216.6	1331.5	1.6000
700	1.8975	1242.5	1365.4	1.6562	1.6503	1240.4	1362.5	1.6397	1.4580	1238.2	1359.6	1.6248
800	2.085	1283.8	1418.8	1.7004	1.8163	1282.1	1416.6	1.6844	1.6077	1280.5	1414.4	1.6701
900	2.267	1325.0	1471.8	1.7409	1.9776	1323.7	1470.1	1.7252	1.7524	1322.4	1468.3	1.7113
1000	2.446	1366.6	1525.0	1.7787	2.136	1365.5	1523.6	1.7632	1.8941	1364.4	1522.2	1.7495
1200	2.799	1451.5	1632.8	1.8478	2.446	1450.7	1631.8	1.8327	2.172	1450.0	1630.8	1.8192
1400	3.148	1539.3	1743.1	1.9106	2.752	1538.7	1742.4	1.8956	2.444	1538.1	1741.7	1.8823
1600	3.494	1630.2	1856.5	1.9685	3.055	1629.8	1855.9	1.9535	2.715	1629.3	1855.4	1.9403
1800	3.838	1724.5	1973.1	2.0225	3.357	1724.1	1972.6	2.0076	2.983	1723.7	1972.1	1.9944
2000	4.182	1822.0	2092.8	2.0733	3.658	1821.6	2092.4	2.0584	3.251	1821.3	2092.0	2.0453

	P = 500(467.13)				P = 600(486.33)				P = 700(503.23)			
Sat	0.9283	1119.4	1205.3	1.4645	0.7702	1118.6	1204.1	1.4464	0.6558	1117.0	1202.0	1.4305
500	0.9924	1139.7	1231.5	1.4923	0.7947	1128.0	1216.2	1.4592				
550	1.0792	1166.7	1266.6	1.5279	0.8749	1158.2	1255.4	1.4990	0.7275	1149.0	1243.2	1.4723
600	1.1583	1191.1	1298.3	1.5585	0.9456	1184.5	1289.5	1.5320	0.7929	1177.5	1280.2	1.5081
650	1.2327	1214.0	1328.0	1.5860	1.0109	1206.6	1320.9	1.5609	0.8520	1203.1	1313.4	1.5387
700	1.3040	1236.0	1356.7	1.6112	1.0727	1231.5	1350.6	1.5872	0.9073	1226.9	1344.4	1.5661
800	1.4407	1278.8	1412.1	1.6571	1.1900	1275.4	1407.6	1.6343	1.0109	1272.0	1402.9	1.6145
900	1.5723	1321.0	1466.5	1.6987	1.3021	1318.4	1462.9	1.6766	1.1089	1315.6	1459.3	1.6576
1000	1.7008	1363.3	1520.7	1.7371	1.4108	1361.2	1517.8	1.7155	1.2036	1358.9	1514.9	1.6970
1100	1.8271	1406.0	1575.1	1.7731	1.5173	1404.2	1572.7	1.7519	1.2960	1402.4	1570.2	1.7337
1200	1.9518	1449.2	1629.8	1.8072	1.6222	1447.7	1627.8	1.7861	1.3868	1446.2	1625.8	1.7682
1400	2.198	1537.6	1741.0	1.8704	1.8289	1536.5	1739.5	1.8497	1.5652	1535.3	1738.1	1.8321
1600	2.442	1628.9	1854.8	1.9285	2.033	1628.0	1853.7	1.9080	1.7409	1627.1	1852.6	1.8906
1800	2.684	1723.3	1971.7	1.9827	2.236	1722.6	1970.8	1.9622	1.9152	1721.8	1969.9	1.9449
2000	2.926	1820.9	2091.6	2.0335	2.438	1820.2	2090.8	2.0131	2.0887	1819.5	2090.1	1.9958

Table C-2.3 Superheated vapor (Continued)

°F	v	u	h	s	v	u	h	s	v	u	h	s
	P = 800(518.36)				P = 1000(544.75)				P = 1250(572.56)			
Sat	0.5691	1115.0	1199.3	1.4160	0.4459	1109.9	1192.4	1.3903	0.3454	1101.7	1181.6	1.3619
550	0.6154	1138.8	1229.9	1.4469	0.4534	1114.8	1198.7	1.3966				
600	0.6776	1170.1	1270.4	1.4861	0.5140	1153.7	1248.8	1.4450	0.3786	1129.0	1216.6	1.3954
650	0.7324	1197.2	1305.6	1.5186	0.5637	1184.7	1289.1	1.4822	0.4267	1167.2	1266.0	1.4410
700	0.7829	1222.1	1338.0	1.5471	0.6080	1212.0	1324.6	1.5135	0.4670	1198.4	1306.4	1.4767
750	0.8306	1245.7	1368.6	1.5730	0.6490	1237.2	1357.3	1.5412	0.5030	1226.1	1342.4	1.5070
800	0.8764	1268.5	1398.2	1.5969	0.6878	1261.2	1388.5	1.5664	0.5364	1251.8	1375.8	1.5341
900	0.9640	1312.9	1455.6	1.6408	0.7610	1307.3	1448.1	1.6120	0.5984	1300.0	1438.4	1.5820
1000	1.0482	1356.7	1511.9	1.6807	0.8305	1352.2	1505.9	1.6530	0.6563	1346.4	1498.2	1.6244
1100	1.1300	1400.5	1567.8	1.7178	0.8976	1396.8	1562.9	1.6908	0.7116	1392.0	1556.6	1.6631
1200	1.2102	1444.6	1623.8	1.7526	0.9630	1441.5	1619.7	1.7261	0.7652	1437.5	1614.5	1.6991
1400	1.3674	1534.2	1736.6	1.8167	1.0905	1531.9	1733.7	1.7909	0.8689	1529.0	1730.0	1.7648
1600	1.5218	1626.2	1851.5	1.8754	1.2152	1624.4	1849.3	1.8499	0.9699	1622.2	1846.5	1.8243
1800	1.6749	1721.0	1969.0	1.9298	1.3384	1719.5	1967.2	1.9046	1.0693	1717.6	1965.0	1.8791
2000	1.8271	1818.8	2089.3	1.9808	1.4608	1817.4	2087.7	1.9557	1.1678	1815.7	2085.8	1.9304
	P = 1500(596.39)				P = 1750(617.31)				P = 2000(636.00)			
Sat	0.2769	1091.8	1168.7	1.3359	0.2268	1080.2	1153.7	1.3109	0.18813	1066.6	1136.3	1.2861
600	0.2816	1096.6	1174.8	1.3416								
650	0.3329	1147.0	1239.4	1.4012	0.2627	1122.5	1207.6	1.3603	0.2057	1091.1	1167.2	1.3141
700	0.3716	1183.4	1286.6	1.4429	0.3022	1166.7	1264.6	1.4106	0.2487	1147.7	1239.8	1.3782
750	0.4049	1214.1	1326.5	1.4767	0.3341	1201.3	1309.5	1.4485	0.2803	1187.3	1291.1	1.4216
800	0.4350	1241.8	1362.5	1.5058	0.3622	1231.3	1348.6	1.4802	0.3071	1220.1	1333.8	1.4562
850	0.4631	1267.7	1396.2	1.5320	0.3878	1258.8	1384.4	1.5081	0.3312	1249.5	1372.0	1.4860
900	0.4897	1292.5	1428.5	1.5562	0.4119	1284.8	1418.2	1.5334	0.3534	1276.8	1407.6	1.5126
1000	0.5400	1340.4	1490.3	1.6001	0.4569	1334.3	1482.3	1.5789	0.3945	1328.1	1474.1	1.5598
1100	0.5876	1387.2	1550.3	1.6399	0.4990	1382.2	1543.8	1.6197	0.4325	1377.2	1537.2	1.6017
1200	0.6334	1433.5	1609.3	1.6765	0.5392	1429.4	1604.0	1.6571	0.4685	1425.2	1598.6	1.6398
1400	0.7213	1526.1	1726.3	1.7431	0.6158	1523.1	1722.6	1.7245	0.5368	1520.2	1718.8	1.7082
1600	0.8064	1619.9	1843.7	1.8031	0.6896	1617.6	1841.0	1.7850	0.6020	1615.4	1838.2	1.7692
1800	0.8899	1715.7	1962.7	1.8582	0.7617	1713.9	1960.5	1.8404	0.6656	1712.0	1958.3	1.8249
2000	0.9725	1814.0	2083.9	1.9096	0.8330	1812.3	2082.0	1.8919	0.7284	1810.6	2080.2	1.8765

	$P = 2500(668.31)$				$P = 3000(695.52)$				$P = 3500$			
Sat	0.13059	1031.0	1091.4	1.2327	0.08404	968.8	1015.5	1.1575				
650									0.02491	663.5	679.7	0.8630
700	0.16839	1098.7	1176.6	1.3073	0.09771	1003.9	1058.1	1.1944	0.03058	759.5	779.3	0.9506
750	0.2030	1155.2	1249.1	1.3686	0.14831	1114.7	1197.1	1.3122	0.10460	1058.4	1126.1	1.2440
800	0.2291	1195.7	1301.7	1.4112	0.17572	1167.6	1265.2	1.3675	0.13626	1134.7	1223.0	1.3226
850	0.2513	1229.5	1345.8	1.4456	0.19731	1207.7	1317.2	1.4080	0.15818	1183.4	1285.9	1.3716
900	0.2712	1259.9	1385.4	1.4752	0.2160	1241.8	1361.7	1.4414	0.17625	1222.4	1336.5	1.4096
950	0.2896	1288.2	1422.2	1.5018	0.2328	1272.7	1402.0	1.4705	0.19214	1256.4	1380.8	1.4416
1000	0.3069	1315.2	1457.2	1.5262	0.2485	1301.7	1439.6	1.4967	0.2066	1287.6	1421.4	1.4699
1100	0.3393	1366.8	1523.8	1.5704	0.2772	1356.2	1510.1	1.5434	0.2328	1345.2	1496.0	1.5193
1200	0.3696	1416.7	1587.7	1.6101	0.3036	1408.0	1576.6	1.5848	0.2566	1399.2	1565.3	1.5624
1400	0.4261	1514.2	1711.3	1.6804	0.3524	1508.1	1703.7	1.6571	0.2997	1501.9	1696.1	1.6368
1600	0.4795	1610.8	1832.6	1.7424	0.3978	1606.3	1827.1	1.7201	0.3395	1601.7	1821.6	1.7010
1800	0.5312	1708.2	1954.0	1.7986	0.4416	1704.5	1949.6	1.7769	0.3776	1700.8	1945.4	1.7583
2000	0.5820	1807.2	2076.4	1.8506	0.4844	1803.9	2072.8	1.8291	0.4147	1800.6	2069.2	1.8108

	$P = 4000$				$P = 5000$				$P = 6000$			
650	0.02447	657.7	675.8	0.8574	0.02377	648.0	670.0	0.8482	0.02322	640.0	665.8	0.8405
700	0.02867	742.1	763.4	0.9345	0.02676	721.8	746.6	0.9156	0.02563	708.1	736.5	0.9028
750	0.06331	960.7	1007.5	1.1395	0.03364	821.4	852.6	1.0049	0.02978	788.6	821.7	0.9746
800	0.10522	1095.0	1172.9	1.2740	0.05932	987.2	1042.1	1.1583	0.03942	896.9	940.7	1.0708
850	0.12833	1156.5	1251.5	1.3352	0.08556	1092.7	1171.9	1.2596	0.05818	1018.8	1083.4	1.1820
900	0.14622	1201.5	1309.7	1.3789	0.10385	1155.1	1251.1	1.3190	0.07588	1102.9	1187.2	1.2599
950	0.16151	1239.2	1358.8	1.4144	0.11853	1202.2	1311.9	1.3629	0.09008	1162.0	1262.0	1.3140
1000	0.17520	1272.9	1402.6	1.4449	0.13120	1242.0	1363.4	1.3988	0.10207	1209.1	1322.4	1.3561
1100	0.19954	1333.9	1481.6	1.4973	0.15302	1310.6	1452.2	1.4577	0.12218	1286.4	1422.1	1.4222
1200	0.2213	1390.1	1553.9	1.5423	0.17199	1371.6	1530.8	1.5066	0.13927	1352.7	1507.3	1.4752
1300	0.2414	1443.7	1622.4	1.5823	0.18918	1428.6	1603.7	1.5493	0.15453	1413.3	1584.9	1.5206
1400	0.2603	1495.7	1688.4	1.6188	0.20517	1483.2	1673.0	1.5876	0.16854	1470.5	1657.6	1.5608
1600	0.2959	1597.1	1816.1	1.6841	0.2348	1587.9	1805.2	1.6551	0.19420	1578.7	1794.3	1.6307
1800	0.3296	1697.1	1941.1	1.7420	0.2626	1689.8	1932.7	1.7142	0.21801	1682.4	1924.5	1.6910
2000	0.3625	1797.3	2065.6	1.7948	0.2895	1790.8	2058.6	1.7676	0.24087	1784.3	2051.7	1.7450

Table C-2.4 Compressed liquid

T	v	u	h	s	v	u	h	s	v	u	h	s
	P = 500(467.13)				P = 1000(544.75)				P = 1500(596.39)			
Sat	0.019748	447.70	449.53	0.64904	0.021591	538.39	542.38	0.74320	0.023461	604.97	611.48	0.80824
32	0.015994	0.00	1.49	0.00000	0.015967	0.03	2.99	0.00005	0.015939	0.05	4.47	0.00007
50	0.015998	18.02	19.50	0.03599	0.015972	17.99	20.94	0.03592	0.015946	17.95	22.38	0.03584
100	0.016106	67.87	69.36	0.12932	0.016082	67.70	70.68	0.12901	0.016058	67.53	71.99	0.12870
150	0.016318	117.66	119.17	0.21457	0.016293	117.38	120.40	0.21410	0.016268	117.10	121.62	0.21364
200	0.016608	167.65	169.19	0.29341	0.016580	167.26	170.32	0.29281	0.016554	166.87	171.46	0.29221
250	0.016972	217.99	219.56	0.36702	0.016941	217.47	220.61	0.36628	0.016910	216.96	221.65	0.36554
300	0.017416	268.92	270.53	0.43641	0.017379	268.24	271.46	0.43552	0.017343	267.58	272.39	0.43463
350	0.017954	320.71	322.37	0.50249	0.017909	319.83	323.15	0.50140	0.017865	318.98	323.94	0.50034
400	0.018608	373.68	375.40	0.56604	0.018550	372.55	375.98	0.56472	0.018493	371.45	376.59	0.56343
450	0.019420	428.40	430.19	0.62798	0.019340	426.89	430.47	0.62632	0.019264	425.44	430.79	0.62470
500					0.02036	483.8	487.5	0.6874	0.02024	481.8	487.4	0.6853
550									0.02158	542.1	548.1	0.7469
	P = 2000(636.00)				P = 3000(695.52)				P = 5000			
Sat	0.025649	662.40	671.89	0.86227	0.034310	783.45	802.50	0.97320				
32	0.015912	0.06	5.95	0.00008	0.015859	0.09	8.90	0.00009	0.015755	0.11	14.70	0.00001
50	0.015920	17.91	23.81	0.03575	0.015870	17.84	26.65	0.03555	0.015773	17.67	32.26	0.03508
100	0.016034	67.37	73.30	0.12839	0.015987	67.04	75.91	0.12777	0.015897	66.40	81.11	0.12651
200	0.016527	166.49	172.60	0.29162	0.016476	165.74	174.89	0.29046	0.016376	164.32	179.47	0.28818
300	0.017308	266.93	273.33	0.43376	0.017240	265.66	275.23	0.43205	0.017110	263.25	279.08	0.42875
400	0.018439	370.38	377.21	0.56216	0.018334	368.32	378.50	0.55970	0.018141	364.47	381.25	0.55506
450	0.019191	424.04	431.14	0.62313	0.019053	421.36	431.93	0.62011	0.018803	416.44	433.84	0.61451
500	0.02014	479.8	487.3	0.6832	0.019944	476.2	487.3	0.6794	0.019603	469.8	487.9	0.6724
560	0.02172	551.8	559.8	0.7565	0.021382	546.2	558.0	0.7508	0.020835	536.7	556.0	0.7411
600	0.02330	605.4	614.0	0.8086	0.02274	597.0	609.6	0.8004	0.02191	584.0	604.2	0.7876
640					0.02475	654.3	668.0	0.8545	0.02334	634.6	656.2	0.8357
680					0.02879	728.4	744.3	0.9226	0.02535	690.6	714.1	0.8873
700									0.02676	721.8	746.6	0.9156

Table C-3 Thermodynamic properties of Freon-12 (Dichlorodifluoromethane)[a]

Table C-3.1 Saturated Freon-12

Temp °F T	Abs. Press. lbf/in^3 P	Specific Volume ft^3/lbm Sat. Liquid v_f	Evap. v_{fg}	Sat. Vapor v_g	Enthalpy Btu/lbm Sat. Liquid h_f	Evap. h_{fg}	Sat. Vapor h_g	Entropy Btu/lbm-R Sat. Liquid s_f	Evap. s_{fg}	Sat. Vapor s_g
−130	0.41224	0.009736	70.7203	70.730	−18.609	81.577	62.968	−0.04983	0.24743	0.19760
−120	0.64190	0.009816	46.7312	46.741	−16.565	80.617	64.052	−0.04372	0.23731	0.19359
−110	0.97034	0.009899	31.7671	31.777	−14.518	79.663	65.145	−0.03779	0.22780	0.19002
−100	1.4280	0.009985	21.1541	22.164	−12.466	78.714	66.248	−0.03200	0.21883	0.18683
−90	2.0509	0.010073	15.8109	15.821	−10.409	77.764	67.355	−0.02637	0.21034	0.18398
−80	2.8807	0.010164	11.5228	11.533	−8.3451	76.812	68.467	−0.02086	0.020229	0.18143
−70	3.9651	0.010259	8.5584	8.5687	−6.2730	75.853	69.580	−0.01548	0.19464	0.17916
−60	5.3575	0.010357	6.4670	6.4774	−4.1919	74.885	70.693	−0.01021	0.18716	0.17714
−50	7.1168	0.010459	4.9637	4.9742	−2.1011	73.906	71.805	−0.00506	0.18038	0.17533
−40	9.3076	0.010564	3.8644	3.8750	0	72.913	72.913	0	0.17373	0.17373
−30	11.999	0.010674	3.0478	3.0585	2.1120	71.903	74.015	0.00496	0.16733	0.17229
−20	15.267	0.010788	2.4321	2.4429	4.2357	70.874	75.110	0.00983	0.16119	0.17102
−10	19.189	0.010906	1.9628	1.9727	6.3716	69.824	76.196	0.01462	0.15527	0.16989
0	23.849	0.011030	1.5979	1.6089	8.5207	68.750	77.271	0.01932	0.14956	0.16888
10	29.335	0.011160	1.3129	1.3241	10.684	67.651	78.335	0.02395	0.14403	0.16798
20	35.736	0.011296	1.0875	1.0988	12.863	66.522	79.385	0.02852	0.13867	0.16719
30	43.148	0.011438	0.90736	0.91880	15.058	65.361	80.419	0.03301	0.13347	0.16648
40	51.667	0.011588	0.76198	0.77357	17.273	64.163	81.436	0.03745	0.12841	0.16586
50	61.394	0.011746	0.64362	0.65537	19.507	62.926	82.433	0.04184	0.12346	0.16530

Table C-3.1 Saturated Freon-12 (Continued)

Temp °F T	Abs. Press. lbf/in^3 P	Specific Volume ft^3/lbm Sat. Liquid v_f	Evap. v_{fg}	Sat. Vapor v_g	Enthalpy Btu/lbm Sat. Liquid h_f	Evap. h_{fg}	Sat. Vapor h_g	Entropy Btu/lbm-R Sat. Liquid s_f	Evap. s_{fg}	Sat. Vapor s_g
60	72.433	0.011913	0.54648	0.55839	21.766	61.643	83.409	0.04618	0.11861	0.16479
70	84.888	0.012089	0.46609	0.47818	24.050	60.309	84.359	0.05048	0.11386	0.16434
80	98.870	0.012277	0.39907	0.41135	26.365	58.917	85.282	0.05475	0.10917	0.16392
90	114.49	0.012478	0.34281	0.35529	28.713	57.461	86.174	0.05900	0.10453	0.16353
100	131.86	0.012693	0.29525	0.30794	31.100	55.929	87.029	0.06323	0.09992	0.16315
110	151.11	0.012924	0.25577	0.26769	33.531	54.313	87.844	0.06745	0.09534	0.16279
120	172.35	0.013174	0.22019	0.23326	36.013	52.597	88.610	0.07168	0.09073	0.16241
130	195.71	0.013447	0.19019	0.20364	38.553	50.768	89.321	0.07583	0.08609	0.16202
140	221.32	0.013746	0.16424	0.17799	41.162	48.805	89.967	0.08021	0.08138	0.16159
150	249.31	0.014078	0.14156	0.15564	43.850	46.684	90.534	0.08453	0.07657	0.16110
160	279.82	0.014449	0.12159	0.13604	46.633	44.373	91.006	0.08893	0.07260	0.16053
170	313.00	0.014871	0.10386	0.11873	49.529	41.830	91.359	0.09342	0.06643	0.15985
180	349.00	0.015360	0.08794	0.10330	52.562	38.999	91.561	0.09804	0.06096	0.15900
190	387.98	0.015942	0.073476	0.089418	55.769	35.792	91.561	0.10284	0.05511	0.15793
200	430.09	0.016659	0.060069	0.076728	59.203	32.075	91.278	0.10789	0.04862	0.15651
210	475.52	0.017601	0.047242	0.064843	62.959	27.599	90.558	0.11332	0.03921	0.15453
220	524.43	0.018986	0.035154	0.053140	67.246	21.790	89.036	0.11943	0.03206	0.15149
230	577.03	0.021854	0.017581	0.039435	72.893	12.229	85.122	0.12739	0.01773	0.14512
233.6	596.9	0.02870	0	0.02870	78.86	0	78.86	0.1359	0	0.1359

[a] Data from Freon Products Division, E.I. duPont de Nemours & Company, Inc. Reprinted by permission.

Table C-3.2 Superheated Freon-12

Temp °F	v	h	s	v	h	s	v	h	s
	5 lbf/in.^{2}			10 lbf/in.^{2}			15 lbf/in.^{2}		
0	8.0611	78.582	0.19663	3.9809	78.246	0.18471	2.6201	77.902	0.17751
20	8.4265	81.309	0.20244	4.1691	81.014	0.19061	2.7494	80.712	0.18349
40	8.7903	84.090	0.20812	4.3556	83.828	0.19635	2.8770	83.561	0.18931
60	9.1528	86.922	0.21367	4.5408	86.689	0.20197	3.0031	86.451	0.19498
80	9.5142	89.806	0.21912	4.7248	89.596	0.20746	3.1281	89.383	0.20051
100	9.8747	92.738	0.22445	4.9079	92.548	0.21283	3.2521	92.357	0.20593
120	10.234	95.717	0.22968	5.0903	95.546	0.21809	3.3754	95.373	0.21122
140	10.594	98.743	0.23481	5.2720	98.586	0.22325	3.4981	98.429	0.21640
160	10.952	101.812	0.23985	5.4533	101.669	0.22830	3.6202	101.525	0.22148
180	11.311	104.925	0.24479	5.6341	104.793	0.23326	3.7419	104.661	0.22646
200	11.668	108.079	0.24964	5.8145	107.957	0.23813	3.8632	107.835	0.23135
220	12.026	111.272	0.25441	5.9946	111.159	0.24291	3.9841	111.046	0.23614
	20 lbf/in.^{2}			25 lbf/in.^{2}			30 lbf/in.^{2}		
20	2.0391	80.403	0.17829	1.6125	80.088	0.17414	1.3278	79.765	0.17065
40	2.1373	83.289	0.18419	1.6932	83.012	0.18012	1.3969	82.730	0.17671
60	2.2340	86.210	0.18992	1.7723	85.965	0.18591	1.4644	85.716	0.18257
80	2.3295	89.168	0.19550	1.8502	88.950	0.19155	1.5306	88.729	0.18826
100	2.4241	92.164	0.20095	1.9271	91.968	0.19704	1.5957	91.770	0.19379
120	2.5179	95.198	0.20628	2.0032	95.021	0.20240	1.6600	94.843	0.19918
140	2.6110	98.270	0.21149	2.0786	98.110	0.20763	1.7237	97.948	0.20445
160	2.7036	101.380	0.21659	2.1535	101.234	0.21276	1.7868	101.086	0.20960
180	2.7957	104.528	0.22159	2.2279	104.393	0.21778	1.8494	104.258	0.21463
200	2.8874	107.712	0.22649	2.3019	107.588	0.22269	1.9116	107.464	0.21957
220	2.9789	110.932	0.23130	2.3756	110.817	0.22752	1.9735	110.702	0.22440
240	3.0700	114.186	0.23602	2.4491	114.080	0.23225	2.0351	113.973	0.22915
	35 lbf/in.^{2}			40 lbf/in.^{2}			50 lbf/in.^{2}		
40	1.1850	82.442	0.17375	1.0258	82.148	0.17112	0.80248	81.540	0.16655
60	1.2442	85.463	0.17968	1.0789	85.206	0.17712	0.84713	84.676	0.17271
80	1.3021	88.504	0.18542	1.1306	88.277	0.18292	0.89025	87.811	0.17862
100	1.3589	91.570	0.19100	1.1812	91.367	0.18854	0.93216	90.953	0.18434
120	1.4148	94.663	0.19643	1.2309	94.480	0.19401	0.97313	94.110	0.18988
140	1.4701	97.785	0.20172	1.2798	97.620	0.19933	1.0133	97.286	0.19527
160	1.5248	100.938	0.20689	1.3282	100.788	0.20453	1.0529	100.485	0.20051
180	1.5789	104.122	0.21195	1.3761	103.985	0.20961	1.0920	103.708	0.20563
200	1.6327	107.338	0.21690	1.4236	107.212	0.21457	1.1307	106.958	0.21064
220	1.6862	110.586	0.22175	1.4707	110.469	0.21944	1.1690	110.235	0.21553
240	1.7394	113.865	0.22651	1.5176	113.757	0.22420	1.2070	113.539	0.22032
260	1.7923	117.175	0.23117	1.5642	117.074	0.22888	1.2447	116.871	0.22502

Table C-3.2 Superheated Freon-12 (Continued)

Temp °F	*v*	*h*	*s*	*v*	*h*	*s*	*v*	*h*	*s*
	60 lbf/in.^{2}			70 lbf/in.^{2}			80 lbf/in.^{2}		
60	0.69210	84.126	0.16892	0.58088	83.552	0.16556	—	—	—
80	0.72964	87.330	0.17497	0.61458	86.832	0.17175	0.52795	86.316	0.16885
100	0.76588	90.528	0.18079	0.64685	90.091	0.17768	0.55734	89.640	0.17489
120	0.80110	93.731	0.18641	0.67803	93.343	0.18339	0.58556	92.945	0.18070
140	0.83551	96.945	0.19186	0.70836	96.597	0.18891	0.61286	96.242	0.18629
160	0.86928	100.776	0.19716	0.73800	99.862	0.19427	0.63943	99.542	0.19170
180	0.90252	103.427	0.20233	0.76708	103.141	0.19948	0.66543	102.851	0.19696
200	0.93531	106.700	0.20736	0.79571	106.439	0.20455	0.69095	106.174	0.20207
220	0.96775	109.997	0.21229	0.82397	109.756	0.20951	0.71609	109.513	0.20706
240	0.99988	113.319	0.21710	0.85191	113.096	0.21435	0.74090	112.872	0.21193
260	1.0318	116.666	0.22182	0.87959	116.459	0.21909	0.76544	116.251	0.21669
280	1.0634	120.039	0.22644	0.90705	119.846	0.22373	0.78975	119.652	0.22135
	90 lbf/in.^{2}			100 lbf/in.^{2}			125 lbf/in.^{2}		
100	0.48749	89.175	0.17234	0.43138	88.694	0.16996	0.32943	87.407	0.16455
120	0.51346	92.536	0.17824	0.45562	92.116	0.17597	0.35086	91.008	0.17087
140	0.53845	95.879	0.18391	0.47881	95.507	0.18172	0.37098	94.537	0.17686
160	0.56268	99.216	0.18938	0.50118	98.884	0.18726	0.39015	98.023	0.18258
180	0.58629	102.557	0.19469	0.52291	102.257	0.19262	0.40857	101.484	0.18807
200	0.60941	105.905	0.19984	0.54413	105.633	0.19782	0.42642	104.934	0.19338
220	0.63213	109.267	0.20486	0.56492	109.018	0.20287	0.44380	108.380	0.19853
240	0.65451	112.644	0.20976	0.58538	112.415	0.20780	0.46081	111.829	0.20353
260	0.67662	116.040	0.21455	0.60554	115.828	0.21261	0.47750	115.287	0.20840
280	0.69849	119.456	0.21923	0.62546	119.258	0.21731	0.49394	118.756	0.21316
300	0.72016	122.892	0.22381	0.64518	122.707	0.22191	0.51016	122.238	0.21780
320	0.74166	126.349	0.22830	0.66472	126.176	0.22641	0.52619	125.737	0.22235
	150 lbf/in.^{2}			175 lbf/in.^{2}			200 lbf/in.^{2}		
120	0.28007	89.800	0.16629	—	—	—	—	—	—
140	0.29845	93.498	0.17256	0.24595	92.373	0.16859	0.20579	91.137	0.16480
160	0.31566	97.112	0.17849	0.26198	96.142	0.17478	0.22121	95.100	0.17130
180	0.33200	100.675	0.18415	0.27697	99.823	0.18062	0.23535	98.921	0.17737
200	0.34769	104.206	0.18958	0.29120	103.447	0.18620	0.24860	102.652	0.18311
220	0.36285	107.720	0.19483	0.30485	107.036	0.19156	0.26117	106.325	0.18860
240	0.37761	111.226	0.19992	0.31804	110.605	0.19674	0.27323	109.962	0.19387
260	0.39203	114.732	0.20485	0.33087	114.162	0.20175	0.28489	113.576	0.19896
280	0.40617	118.242	0.20967	0.34339	117.717	0.20662	0.29623	117.178	0.20390
300	0.42008	121.761	0.21436	0.35567	121.273	0.21137	0.30730	120.775	0.20870
320	0.43379	125.290	0.21894	0.36773	124.835	0.21599	0.31815	124.373	0.21337
340	0.44733	128.833	0.22343	0.37963	128.407	0.22052	0.32881	127.974	0.21793

Table C-3.2 Superheated Freon-12 (Continued)

Temp °F	*v*	*h*	*s*	*v*	*h*	*s*	*v*	*h*	*s*
	250 lbf/in.^{2}			300 lbf/in.^{2}			400 lbf/in.^{2}		
160	0.16249	92.717	0.16462	—	—	—	—	—	—
180	0.17605	96.925	0.17130	0.13482	94.556	0.16537	—	—	—
200	0.18824	100.930	0.17747	0.14697	98.975	0.17217	0.091005	93.718	0.16092
220	0.19952	104.809	0.18326	0.15774	103.136	0.17838	0.10316	99.046	0.16888
240	0.21014	108.607	0.18877	0.16761	107.140	0.18419	0.11300	103.735	0.17568
260	0.22027	112.351	0.19404	0.17685	111.043	0.18969	0.12163	108.105	0.18183
280	0.23001	116.060	0.19913	0.18562	114.879	0.19495	0.12949	112.286	0.18756
300	0.23944	119.747	0.20405	0.19402	118.670	0.20000	0.13680	116.343	0.19298
320	0.24862	123.420	0.20882	0.20214	122.430	0.20489	0.14372	120.318	0.19814
340	0.25759	127.088	0.21346	0.21002	126.171	0.20963	0.15032	124.235	0.20310
360	0.26639	130.754	0.21799	0.21770	129.900	0.21423	0.15668	128.112	0.20789
380	0.27504	134.423	0.22241	0.22522	133.624	0.21872	0.16285	131.961	0.21258
	500 lbf/in.^{2}			600 lbf/in.^{2}					
220	0.064207	92.397	0.15683	—	—	—			
240	0.077620	99.218	0.16672	0.047488	91.024	0.15335			
260	0.087054	104.526	0.17421	0.061922	99.741	0.16566			
280	0.094923	109.277	0.18072	0.070859	105.637	0.17374			
300	0.10190	113.729	0.18666	0.078059	110.729	0.18053			
320	0.10829	117.997	0.19221	0.084333	115.420	0.18663			
340	0.11426	122.143	0.19746	0.090017	119.871	0.19227			
360	0.11992	126.205	0.20247	0.095289	124.167	0.19757			
380	0.12533	130.207	0.20730	0.10025	128.355	0.20262			
400	0.13054	134.166	0.21196	0.10498	132.466	0.20746			
420	0.13559	138.096	0.21648	0.10952	136.523	0.21213			
440	0.14051	142.004	0.22087	0.11391	140.539	0.21664			

Table C-4 Thermodynamic properties of ammonia[a]

Table C-4.1 Saturated ammonia

Temp. °F	Abs. Press. lbf/in^2 P	Specific Volume ft^2/lbm Sat. Liquid v_f	Evap. v_{fg}	Sat. Vapor v_g	Enthalpy Btu/lbm Sat. Liquid h_f	Evap. h_{fg}	Sat. Vapor h_g	Entropy Btu/lbm R Sat. Liquid s_f	Evap. s_{fg}	Sat. Vapor s_g
−60	5.55	0.0228	44.707	44.73	−21.2	610.8	589.6	−0.0517	1.5286	1.4769
−55	6.54	0.0229	38.357	38.38	−15.9	607.5	591.6	−0.0386	1.5017	1.4631
−50	7.67	0.0230	33.057	33.08	−10.6	604.3	593.7	−0.0256	1.4753	1.4497
−45	8.95	0.0231	28.597	28.62	−5.3	600.9	595.6	−0.0127	1.4495	1.4368
−40	10.41	0.02322	24.837	24.86	0	597.6	597.6	0.000	1.4242	1.4242
−35	12.05	0.02333	21.657	21.68	5.3	594.2	599.5	0.0126	1.3994	1.4120
−30	13.90	0.0235	18.947	18.97	10.7	590.7	601.4	0.0250	1.3751	1.4001
−25	15.98	0.0236	16.636	16.66	16.0	587.2	603.2	0.0374	1.3512	1.3886
−20	18.30	0.0237	14.656	14.68	21.4	583.6	605.0	0.0497	1.3277	1.3774
−15	20.88	0.02381	12.946	12.97	26.7	580.0	606.7	0.0618	1.3044	1.3664
−10	23.74	0.02393	11.476	11.50	32.1	576.4	608.5	0.0738	1.2820	1.3558
−5	26.92	0.02406	10.206	10.23	37.5	572.6	610.1	0.0857	1.2597	1.3454
0	30.42	0.02419	9.092	9.116	42.9	568.9	611.8	0.0975	1.2377	1.3352
5	34.27	0.02432	8.1257	8.150	48.3	565.0	613.3	0.1092	1.2161	1.3253
10	38.51	0.02446	7.2795	7.304	53.8	561.1	614.9	0.1208	1.1949	1.3157
15	43.14	0.02460	6.5374	6.562	59.2	557.1	616.3	0.1323	1.1739	1.3062
20	48.21	0.02474	5.8853	5.910	64.7	553.1	617.8	0.1437	1.1532	1.2969
25	53.73	0.02488	5.3091	5.334	70.2	548.9	619.1	0.1551	1.1328	1.2879
30	59.74	0.02503	4.8000	4.825	75.7	544.8	620.5	0.1663	1.1127	1.2790
35	66.26	0.02518	4.3478	4.373	81.2	540.5	621.7	0.1775	1.0929	1.2704
40	73.32	0.02533	3.9457	3.971	86.8	536.2	623.0	0.1885	1.0733	1.2618
45	80.96	0.02548	3.5885	3.614	92.3	531.8	624.1	0.1996	1.0539	1.2535
50	89.19	0.02564	3.2684	3.294	97.9	527.3	625.2	0.2105	1.0348	1.2453
55	98.06	0.02581	2.9822	3.008	103.5	522.8	626.3	0.2214	1.0159	1.2373
60	107.6	0.02597	2.7250	2.751	109.2	518.1	627.3	0.2322	0.9972	1.2294
65	117.8	0.02614	2.4939	2.520	114.8	513.4	628.2	0.2430	0.9786	1.2216
70	128.8	0.02632	2.2857	2.312	120.5	508.6	629.1	0.2537	0.9603	1.2140
75	140.5	0.02650	2.0985	2.125	126.2	503.7	629.9	0.2643	0.9422	1.2065
80	153.0	0.02668	1.9283	1.955	132.0	498.7	630.7	0.2749	0.9242	1.1991
85	166.4	0.02687	1.7741	1.801	137.8	493.6	631.4	0.2854	0.9064	1.1918
90	180.6	0.02707	1.6339	1.661	143.5	488.5	632.0	0.2958	0.8888	1.1846
95	195.8	0.02727	1.5067	1.534	149.4	483.2	632.6	0.3062	0.8713	1.1775
100	211.9	0.02747	1.3915	1.419	155.2	477.8	633.0	0.3166	0.8539	1.1705
105	228.9	0.02769	1.2853	1.313	161.1	472.3	633.4	0.3269	0.8366	1.1635
110	247.0	0.02790	1.1891	1.217	167.0	466.7	633.7	0.3372	0.8194	1.1566
115	266.2	0.02813	1.0999	1.128	173.0	460.9	633.9	0.3474	0.8023	1.1497
120	286.4	0.02836	1.0186	1.047	179.0	455.0	634.0	0.3576	0.7851	1.1427
125	307.8	0.02860	0.9444	0.973	185.1	448.9	634.0	0.3679	0.7679	1.1358

[a] Reprinted from National Bureau of Standards Circular No. 142, *Tables of Thermodynamics Properties of Ammonia*

Table C-4.2 Superheated ammonia

Abs. Press lbf/in² (Sat. Temp.)		Temperature, °F 0	20	40	60	80	100	120	140	160	180	200	220
	v	28.58	29.90	31.20	32.49	33.78	35.07	36.35	37.62	38.90	40.17	41.45	
10	h	618.9	629.1	639.3	649.5	659.7	670.0	680.3	690.6	701.1	711.6	722.2	
(−41.34)	s	1.477	1.499	1.520	1.540	1.559	1.578	1.596	1.614	1.631	1.647	1.664	
	v	18.92	19.82	20.70	21.58	22.44	23.31	24.17	25.03	25.88	26.74	27.59	
15	h	617.2	627.8	638.2	648.5	658.9	669.2	679.6	690.0	700.5	711.1	721.7	
(−27.29)	s	1.427	1.450	1.471	1.491	1.511	1.529	1.548	1.566	1.583	1.599	1.616	
	v	14.09	14.78	15.45	16.12	16.78	17.43	18.08	18.73	19.37	20.02	20.66	21.3
20	h	615.5	626.4	637.0	647.5	658.0	668.5	678.9	689.4	700.0	710.6	721.2	732.0
(−16.64)	s	1.391	1.414	1.436	1.456	1.476	1.495	1.513	1.531	1.549	1.565	1.582	1.598
	v	11.19	11.75	12.30	12.84	13.37	13.90	14.43	14.95	15.47	15.99	16.50	17.02
25	h	613.8	625.0	635.8	646.5	657.1	667.7	678.2	688.8	699.4	710.1	720.8	731.6
(−7.96)	s	1.362	1.386	1.408	1.429	1.449	1.468	1.486	1.504	1.522	1.539	1.555	1.571
	v	9.25	9.731	10.20	10.65	11.10	11.55	11.99	12.43	12.87	13.30	13.73	14.16
30	h	611.9	623.5	634.6	645.5	656.2	666.9	677.5	688.2	698.8	709.6	720.3	731.1
(−0.57)	s	1.337	1.362	1.385	1.406	1.426	1.446	1.464	1.482	1.500	1.517	1.533	1.550
	v		8.287	8.695	9.093	9.484	9.869	10.25	10.63	11.00	11.38	11.75	12.12
35	h		622.0	633.4	644.4	655.3	666.1	676.8	687.6	698.3	709.1	719.9	730.7
(5.89)	s		1.341	1.365	1.386	1.407	1.427	1.445	1.464	1.481	1.498	1.515	1.531
	v		7.203	7.568	7.922	8.268	8.609	8.945	9.278	9.609	9.938	10.27	10.59
40	h		620.4	632.1	643.4	654.4	665.3	676.1	686.9	697.7	708.5	719.4	730.3
(11.66)	s		1.323	1.347	1.369	1.390	1.410	1.429	1.447	1.465	1.482	1.499	1.515
	v		6.363	6.694	7.014	7.326	7.632	7.934	8.232	8.528	8.822	9.115	9.406
45	h		618.8	630.8	642.3	653.5	664.6	675.5	686.3	697.2	708.0	718.9	729.9
(16.87)	s		1.307	1.331	1.354	1.375	1.395	1.414	1.433	1.450	1.468	1.485	1.501
	v			5.988	6.280	6.564	6.843	7.117	7.387	7.655	7.921	8.185	8.448
50	h			629.5	641.2	652.6	663.7	674.7	685.7	696.6	707.5	718.5	729.4
(21.67)	s			1.317	1.340	1.361	1.382	1.401	1.420	1.437	1.455	1.472	1.488
	v			4.933	5.184	5.428	5.665	5.897	6.126	6.352	6.576	6.798	7.019
60	h			626.8	639.0	650.7	662.1	673.3	684.4	695.5	706.5	717.5	728.6
(30.21)	s			1.291	1.315	1.337	1.358	1.378	1.397	1.415	1.432	1.449	1.466

Table C-4.2 Superheated ammonia (Continued)

Abs. Press. lbf/in^2 (Sat. Temp.)		Temperature, °F 60	80	100	120	140	160	180	200	240	280	320	360
	v	4.401	4.615	4.822	5.025	5.224	5.420	5.615	5.807	6.187	6.563		
70	h	636.6	648.7	660.4	671.8	683.1	694.3	705.5	716.6	738.9	761.4		
(37.7)	s	1.294	1.317	1.338	1.358	1.377	1.395	1.413	1.430	1.463	1.494		
	v	3.812	4.005	4.190	4.371	4.548	4.722	4.893	5.063	5.398	5.73		
80	h	634.3	646.7	658.7	670.4	681.8	693.2	704.4	715.6	738.1	760.7		
(44.4)	s	1.275	1.298	1.320	1.340	1.360	1.378	1.396	1.414	1.447	1.478		
	v	3.353	3.529	3.698	3.862	4.021	4.178	4.332	4.484	4.785	5.081		
90	h	631.8	644.7	657.0	668.9	680.5	692.0	703.4	714.7	737.3	760.0		
(50.47)	s	1.257	1.281	1.304	1.325	1.344	1.363	1.381	1.400	1.432	1.464		
	v	2.985	3.149	3.304	3.454	3.600	3.743	3.883	4.021	4.294	4.562		
100	h	629.3	642.6	655.2	667.3	679.2	690.8	702.3	713.7	736.5	759.4		
(56.05)	s	1.241	1.266	1.289	1.310	1.331	1.349	1.368	1.385	1.419	1.451		
	v		2.166	2.288	2.404	2.515	2.622	2.727	2.830	3.030	3.227	3.420	
140	h		633.8	647.8	661.1	673.7	686.0	698.0	709.9	733.3	756.7	780.0	
(74.79)	s		1.214	1.240	1.263	1.284	1.305	1.324	1.342	1.376	1.409	1.440	
	v			1.720	1.818	1.910	1.999	2.084	2.167	2.328	2.484	2.637	
180	h			639.9	654.4	668.0	681.0	693.6	705.9	730.1	753.9	777.7	
(89.78)	s			1.199	1.225	1.248	1.269	1.289	1.308	1.344	1.377	1.408	
	v				1.443	1.525	1.601	1.675	1.745	1.881	2.012	2.140	2.265
220	h				647.3	662.0	675.8	689.1	701.9	726.8	751.1	775.3	799.5
(102.42)	s				1.192	1.217	1.239	1.260	1.280	1.317	1.351	1.383	1.413
	v				1.302	1.380	1.452	1.521	1.587	1.714	1.835	1.954	2.069
240	h				643.5	658.8	673.1	686.7	699.8	725.1	749.8	774.1	798.4
(108.09)	s				1.176	1.203	1.226	1.248	1.268	1.305	1.339	1.371	1.402
	v				1.182	1.257	1.326	1.391	1.453	1.572	1.686	1.796	1.904
260	h				639.5	655.6	670.4	684.4	697.7	723.4	748.4	772.9	797.4
(113.42)	s				1.162	1.189	1.213	1.235	1.256	1.294	1.329	1.361	1.391
	v				1.078	1.151	1.217	1.279	1.339	1.451	1.558	1.661	1.762
280	h				635.4	652.2	667.6	681.9	695.6	721.8	747.0	771.7	796.3
(118.45)	s				1.147	1.176	1.201	1.224	1.245	1.283	1.318	1.351	1.382

Table C-5 Ideal gas properties of selected gases[a]

Table C-5a Ideal gas properties of air h and u, Btu / lbm, s' Btu / lbm-R

T, °R	h	u	s'	T, °R	h	u	s'
300	71.61	51.04	0.46007	1400	342.90	246.93	0.83604
320	76.40	54.46	0.47550	1500	369.17	266.34	0.85416
340	81.18	57.87	0.49002	1600	395.74	286.06	0.87130
360	85.97	61.29	0.50369	1700	422.59	306.06	0.88758
380	90.75	64.70	0.51663	1800	449.71	326.32	0.90308
400	95.53	68.11	0.52890	1900	477.09	346.85	0.91788
420	100.32	71.52	0.54058	2000	504.71	367.61	0.93205
440	105.11	74.93	0.55172	2100	532.55	388.60	0.94564
460	109.90	78.36	0.56235	2200	560.59	409.78	0.95868
480	114.69	81.77	0.57255	2300	588.82	431.16	0.97123
500	119.48	85.20	0.58233	2400	617.22	452.70	0.98331
520	124.27	88.62	0.59173	2500	645.78	474.40	0.99497
537	128.34	91.53	0.59945	2600	674.49	496.26	1.00623
540	129.06	92.04	0.60078	2700	703.35	518.26	1.01712
560	133.86	95.47	0.60950	2800	732.33	540.40	1.02767
580	138.66	98.90	0.61793	2900	761.45	562.66	1.03788
600	143.47	102.34	0.62607	3000	790.68	585.04	1.04779
620	148.28	105.78	0.63395	3100	820.03	607.53	1.05741
640	153.09	109.21	0.64159	3200	849.48	630.12	1.06676
660	157.92	112.67	0.64902	3300	879.02	652.81	1.07585
680	162.73	116.12	0.65621	3400	908.66	675.60	1.08470
700	167.56	119.58	0.66321	3500	938.40	698.48	1.09332
720	172.39	123.04	0.67002	3600	968.21	721.44	1.10172
740	177.23	126.51	0.67665	3700	998.11	744.48	1.10991
760	182.08	129.99	0.68312	3800	1028.09	767.60	1.11791
780	186.94	133.47	0.68942	3900	1058.14	790.80	1.12571
800	191.81	136.97	0.69558	4000	1088.26	814.06	1.13334
820	196.69	140.47	0.70160	4100	1118.5	837.4	1.14079
840	201.56	143.98	0.70747	4200	1148.7	860.8	1.14809
860	206.46	147.50	0.71323	4300	1179.0	884.3	1.15522
880	211.35	151.02	0.71886	4400	1209.4	907.8	1.16221
900	216.26	154.57	0.72438	4500	1239.9	931.4	1.16905
920	221.18	158.12	0.72979	4600	1270.4	955.0	1.17575
940	226.11	161.68	0.73509	4700	1300.9	978.7	1.18232
960	231.06	165.26	0.74030	4800	1331.5	1002.5	1.18876
980	236.02	168.83	0.74540	4900	1362.2	1026.3	1.19508
1000	240.98	172.43	0.75042	5000	1392.9	1050.1	1.20129
1100	265.99	190.58	0.77426	5100	1423.6	1074.0	1.20738
1200	291.30	209.05	0.79628	5200	1454.4	1098.0	1.21336
1300	316.94	227.83	0.81680	5300	1485.3	1122.0	1.21923

[a] Data extracted from J. H. Keenan and J. Kaye, "Gas Tables," John Wiley and Sons, New York, 1945. Reprinted by permission of John Wiley & Sons, Inc.

Table C-5b Ideal gas enthalpy, internal energy, and absolute entropy of diatomic nitrogen, N_2 $\bar{h}$ and $\bar{u}$, Btu / lbmol; $\bar{s}'$, Btu / lbmol °R

T, °R	$\bar{h}$	$\bar{u}$	$\bar{s}'$	T, °R	$\bar{h}$	$\bar{u}$	$\bar{s}'$
300	2082.0	1486.2	41.695	1400	9897	7117	52.551
320	2221.0	1585.5	42.143	1500	10649	7670	53.071
340	2360.0	1684.8	42.564	1600	11410	8232	53.561
360	2498.9	1784.0	42.962	1700	12179	8803	54.028
380	2638.0	1883.4	43.337	1800	12956	9382	54.472
400	2777.0	1982.6	43.694	1900	13742	9968	54.896
420	2916.1	2082.0	44.034	2000	14534	10563	55.303
440	3055.1	2181.3	44.357	2100	15334	11164	55.694
460	3194.1	2280.6	44.665	2200	16140	11771	56.068
480	3333.1	2379.9	44.962	2300	16951	12384	56.429
500	3472.2	2479.3	45.246	2400	17768	13002	56.777
520	3611.3	2578.6	45.519	2500	18590	13625	57.112
537	3729.5	2663.1	45.743	2600	19416	14253	57.436
540	3750.3	2678.0	45.781	2700	20246	14885	57.750
560	3889.5	2777.4	46.034	2800	21081	15521	58.053
580	4028.7	2876.9	46.278	2900	21920	16161	58.348
600	4167.9	2976.4	46.514	3000	22762	16804	58.632
620	4307.1	3075.9	46.742	3100	23607	17451	58.910
640	4446.4	3175.5	46.964	3200	24455	18100	59.179
660	4585.8	3275.2	47.178	3300	25306	18753	59.442
680	4725.3	3374.9	47.386	3400	26160	19408	59.697
700	4864.9	3474.8	47.588	3500	27016	20065	59.944
720	5004.5	3574.7	47.785	3600	27874	20725	60.186
740	5144.3	3674.7	47.977	3700	28735	21387	60.422
760	5284.1	3774.9	48.164	3800	29598	22052	60.652
780	5424.2	3875.2	48.345	3900	30463	22718	60.877
800	5564.4	3975.7	48.522	4000	31329	23386	61.097
820	5704.7	4076.3	48.696	4100	32198	24056	61.310
840	5845.3	4177.1	48.865	4200	33068	24728	61.520
860	5985.9	4278.1	49.031	4300	33940	25401	67.726
880	6126.9	4379.4	49.193	4400	34813	26075	61.927
900	6268.1	4480.8	49.352	4500	35688	26751	62.123
920	6409.6	4582.6	49.507	4600	36564	27429	62.316
940	6551.2	4684.5	49.659	4700	37441	28108	62.504
960	6693.1	4786.7	49.808	4800	38320	28787	62.689
980	6835.4	4889.3	49.955	4900	39199	29468	62.870
1000	6978	4992	50.099	5000	40080	30151	63.049
1100	7695	5511	50.783	5100	40962	30834	63.223
1200	8420	6037	51.413	5200	41844	31518	63.395
1300	9154	6572	52.001	5300	42728	32203	63.563

Table C-5c Ideal gas enthalpy, internal energy, and absolute entropy of diatomic oxygen, O_2 $\bar{h}$ and $\bar{u}$, Btu / lbmol; $\bar{s}'$, Btu / lbmol °R

T, °R	$\bar{h}$	$\bar{u}$	$\bar{s}'$	T, °R	$\bar{h}$	$\bar{u}$	s'
300	2073.5	1477.8	44.927	1400	10210	7430	56.099
320	2212.6	1577.1	45.375	1500	11017	8038	56.656
340	2351.7	1676.5	45.797	1600	11833	8655	57.182
360	2490.8	1775.9	46.195	1700	12656	9280	57.680
380	2630.0	1875.3	46.571	1800	13486	9911	58.155
400	2769.1	1974.8	46.927	1900	14322	10549	58.607
420	2908.3	2074.3	47.267	2000	15164	11192	59.039
440	3047.5	2173.8	47.591	2100	16011	11841	59.451
460	3186.9	2273.4	47.900	2200	16863	12494	59.848
480	3326.5	2373.3	48.198	2300	17719	13151	60.228
500	3466.2	2473.2	48.483	2400	18579	13813	60.594
520	3606.1	2573.4	48.757	2500	19443	14479	60.946
537	3725.1	2658.7	48.982	2600	20311	15148	61.287
540	3746.2	2673.8	49.021	2700	21183	15821	61.616
560	3886.6	2774.5	49.276	2800	22058	16497	61.934
580	4027.3	2875.5	49.522	2900	22936	17177	62.242
600	4168.3	2976.8	49.762	3000	23818	17860	62.540
620	4309.7	3078.4	49.993	3100	24703	18546	62.831
640	4451.4	3180.4	50.218	3200	25591	19236	63.113
660	4593.5	3282.9	50.437	3300	26482	19928	63.386
680	4736.2	3385.8	50.650	3400	27376	20624	63.654
700	4879.3	3489.2	50.858	3500	28273	21323	63.914
720	5022.9	3593.1	51.059	3600	29174	22025	64.168
740	5167.0	3697.4	51.257	3700	30078	22730	64.415
760	5311.4	3802.2	51.450	3800	30984	23438	64.657
780	5456.4	3907.5	51.638	3900	31894	24149	64.893
800	5602.0	4013.3	51.821	4000	32806	24863	65.123
820	5748.1	4119.7	52.002	4100	33722	25580	65.350
840	5894.8	4226.6	52.179	4200	34640	26299	65.571
860	6041.9	4334.1	52.352	4300	35561	27022	65.788
880	6189.6	4442.0	52.522	4400	36485	27747	66.000
900	6337.9	4550.6	52.688	4500	37412	28475	66.208
920	6486.7	4659.7	52.852	4600	38341	29206	66.413
940	6636.1	4769.4	53.012	4700	39274	29940	66.613
960	6786.0	4879.5	53.170	4800	40209	30676	66.809
980	6936.4	4990.3	53.326	4900	41146	31415	67.003
1000	7088	5102	53.477	5000	42086	32157	67.193
1100	7850	5666	54.204	5100	43029	32901	67.380
1200	8626	6243	54.879	5200	43974	33648	67.562
1300	9413	6831	55.508	5300	44922	34397	67.743

Table C-5d Ideal gas enthalpy, internal energy, and absolute entropy of carbon monoxide, CO $\bar{h}$ and $\bar{u}$, Btu / lbmol; $\bar{s}'$, Btu / lbmol °R

T, °R	$\bar{h}$	$\bar{u}$	$\bar{s}'$	T, °R	$\bar{h}$	$\bar{u}$	$\bar{s}'$
300	2081.9	1486.1	43.223	1400	9948	7168	54.129
320	2220.9	1585.4	43.672	1500	10711	7732	54.655
340	2359.9	1684.7	44.093	1600	11483	8306	55.154
360	2498.8	1783.9	44.490	1700	12264	8888	55.628
380	2637.9	1883.3	44.866	1800	13053	9479	56.078
400	2776.9	1982.6	45.223	1900	13850	10077	56.509
420	2916.0	2081.9	45.563	2000	14653	10682	56.922
440	3055.0	2181.2	45.886	2100	15463	11293	57.317
460	3194.0	2280.5	46.194	2200	16279	11911	57.696
480	3333.0	2379.8	46.491	2300	17101	12534	58.062
500	3472.1	2479.2	46.775	2400	17927	13161	58.414
520	3611.2	2578.6	47.048	2500	18759	13794	58.754
537	3729.5	2663.1	47.272	2600	19594	14431	59.081
540	3750.3	2677.9	47.310	2700	20434	15072	59.398
560	3889.5	2777.4	47.563	2800	21277	15717	59.705
580	4028.7	2876.9	47.807	2900	22124	16365	60.002
600	4168.0	2976.5	48.044	3000	22973	17016	60.290
620	4307.4	3076.2	48.272	3100	23826	17670	60.569
640	4446.9	3175.9	48.494	3200	24681	18326	60.841
660	4586.5	3275.8	48.709	3300	25539	18986	61.105
680	4726.2	3375.8	48.917	3400	26399	29647	61.362
700	4866.0	3475.9	49.120	3500	27262	20311	61.612
720	5006.1	3576.3	49.317	3600	28127	20978	61.855
740	5146.4	3676.9	49.509	3700	28994	21646	62.093
760	5286.8	3777.5	49.697	3800	29862	22316	62.325
780	5427.4	3878.4	49.880	3900	30733	22988	62.511
800	5568.2	3979.5	50.058	4000	31605	23662	62.772
820	5709.4	4081.0	50.232	4100	32479	24337	62.988
840	5850.7	4182.6	50.402	4200	33354	25014	63.198
860	5992.3	4284.5	50.569	4300	34231	25692	63.405
880	6134.2	4386.6	50.732	4400	35109	26371	63.607
900	6276.4	4489.1	50.892	4500	35989	27052	63.805
920	6419.0	4592.0	51.048	4600	36869	27734	63.998
940	6561.7	4695.0	51.202	4700	37751	28417	64.188
960	6704.9	4798.5	51.353	4800	38634	29102	64.374
980	6848.4	4902.3	51.501	4900	39518	29787	64.556
1000	6992	5006	51.646	5000	40403	30473	64.735
1100	7717	5532	52.337	5100	41289	31161	64.910
1200	8451	6068	52.976	5200	42176	31849	65.082
1300	9195	6613	53.571	5300	43063	32538	65.252

Table C-5e Ideal gas enthalpy, internal energy, and absolute entropy of carbon dioxide, CO_2 $\bar{h}$ and $\bar{u}$, Btu / lbmol; $\bar{s}'$, Btu / lbmol °R

T, °R	$\bar{h}$	$\bar{u}$	$\bar{s}'$	T, °R	$\bar{h}$	$\bar{u}$	$\bar{s}'$
300	2108.2	1512.4	46.353	1400	13345	10565	61.124
320	2256.6	1621.1	46.832	1500	14576	11597	61.974
340	2407.3	1732.1	47.289	1600	15829	12652	62.783
360	2560.5	1845.6	47.728	1700	17101	13725	63.555
380	2716.4	1961.8	48.148	1800	18392	14817	64.292
400	2874.7	2080.4	48.555	1900	19698	15925	64.999
420	3035.7	2201.7	48.947	2000	21019	17047	65.676
440	3199.4	2325.6	49.329	2100	22353	18182	66.327
460	3365.7	2452.2	49.698	2200	23699	19330	66.953
480	3534.7	2581.5	50.058	2300	25056	20489	67.557
500	3706.2	2713.3	50.408	2400	26424	21658	68.139
520	3880.3	2847.7	50.750	2500	27801	22837	68.702
537	4030.2	2963.8	51.032	2600	29187	24024	69.245
540	4056.8	2984.4	51.082	2700	30581	25219	69.771
560	4235.8	3123.7	51.408	2800	31983	26422	70.282
580	4417.2	3265.4	51.726	2900	33392	27633	70.776
600	4600.9	3409.4	52.038	3000	34807	28849	71.255
620	4786.8	3555.6	52.343	3100	36228	30072	71.722
640	4974.9	3704.0	52.641	3200	37655	31300	72.175
660	5165.2	3854.6	52.934	3300	39087	32533	72.616
680	5357.6	4007.2	53.225	3400	40524	33772	73.045
700	5552.0	4161.9	53.503	3500	41965	35015	73.462
720	5748.4	4318.6	53.780	3600	43411	36262	73.870
740	5946.8	4477.3	54.051	3700	44861	37513	74.267
760	6147.0	4637.9	54.319	3800	46314	38768	74.655
780	6349.1	4800.1	54.582	3900	47771	40026	75.033
800	6552.9	4964.2	54.839	4000	49231	41288	75.404
820	6758.3	5129.9	55.093	4100	50695	42553	75.765
840	6965.7	5297.6	55.343	4200	52162	43821	76.119
860	7174.7	5466.9	55.589	4300	53632	45093	76.464
880	7385.3	5637.7	55.831	4400	55015	46367	76.803
900	7597.6	5810.3	56.070	4500	56581	47645	77.135
920	7811.4	5984.4	56.305	4600	58060	48925	77.460
940	8026.8	6160.1	56.536	4700	59541	50208	77.779
960	8243.8	6337.4	56.765	4800	61025	51493	78.091
980	8462.2	6516.1	56.990	4900	62511	52781	78.398
1000	8682	6696	57.212	5000	64000	54071	78.698
1100	9803	7618	58.281	5100	65491	55363	78.994
1200	10955	8572	59.283	5200	66984	56658	79.284
1300	12137	9555	60.229	5300	68479	57954	79.569

Table C-5f Ideal gas enthalpy, internal energy, and absolute entropy of water, H_2O $\bar{h}$ and $\bar{u}$, Btu / lbmol; $\bar{s}'$, Btu / lbmol °R

T, °R	$\bar{h}$	$\bar{u}$	$\bar{s}'$	T, °R	$\bar{h}$	$\bar{u}$	$\bar{s}'$
300	2367.6	1771.8	40.439	1400	11625	8845	53.168
320	2526.8	1891.3	40.952	1500	12551	9573	53.808
340	2686.0	2010.8	41.435	1600	13495	10318	54.418
360	2845.1	2130.2	41.889	1700	14455	11079	54.999
380	3004.4	2249.8	42.320	1800	15433	11858	55.559
400	3163.8	2369.4	42.728	1900	16428	12654	56.097
420	3323.2	2489.1	43.117	2000	17439	13467	56.617
440	3482.7	2608.9	43.487	2100	18467	14297	57.119
460	3642.3	2728.8	43.841	2200	19511	15142	57.605
480	3802.0	2848.8	44.182	2300	20571	16003	58.077
500	3962.0	2969.1	44.508	2400	21646	16880	58.535
520	4122.0	3089.4	44.821	2500	22735	17771	58.980
537	4258.3	3191.9	45.079	2600	23840	18676	59.414
540	4282.4	3210.0	45.124	2700	24957	19595	59.837
560	4442.8	3330.7	45.415	2800	26088	20528	60.248
580	4603.7	3451.9	45.696	2900	27231	21472	60.650
600	4764.7	3573.2	45.970	3000	28386	22429	61.043
620	4926.1	3694.9	46.235	3100	29553	23397	61.426
640	5087.8	3816.8	46.492	3200	30730	34375	61.801
660	5250.0	3939.3	46.741	3300	31918	25365	62.167
680	5412.5	4062.1	46.984	3400	33116	26364	62.526
700	5575.4	4185.3	47.219	3500	34324	27373	62.876
720	5738.8	4309.0	47.450	3600	35540	28391	63.221
740	5902.6	4433.1	47.673	3700	36765	29418	63.557
760	6066.9	4557.6	47.893	3800	37999	30453	63.887
780	6231.7	4682.7	48.106	3900	39240	31495	64.210
800	6396.9	4808.2	48.316	4000	40489	32546	64.528
820	6562.6	4934.2	48.520	4100	41745	33603	64.839
840	6728.9	5060.8	48.721	4200	43008	34668	65.144
860	6895.6	5187.8	48.916	4300	44278	35739	65.444
880	7062.9	5315.3	49.109	4400	45554	36816	65.738
900	7230.9	5443.6	49.298	4500	46836	37900	66.028
920	7399.4	5572.4	49.483	4600	48124	38989	66.312
940	7568.4	5701.7	49.665	4700	49417	40083	66.591
960	7738.0	5831.6	49.843	4800	50716	41183	66.866
980	7908.2	5962.0	50.019	4900	52019	42288	67.135
1000	8079	6093	50.191	5000	53327	43398	67.401
1100	8942	6758	51.013	5100	54640	44512	67.662
1200	9820	7437	51.777	5200	55957	45631	67.918
1300	10715	8133	52.494	5300	57279	46754	68.172

Table C-5g Ideal gas enthalpy, internal energy, and absolute entropy of diatomic hydrogen, H_2 $\bar{h}$ and $\bar{u}$, Btu / lbmol; $\bar{s}'$, Btu / lbmol °R

T, °R	$\bar{h}$	$\bar{u}$	$\bar{s}'$	T, °R	$\bar{h}$	$\bar{u}$	$\bar{s}'$
300	2063.5	1467.7	27.337	1400	9674	6894	37.883
320	2189.4	1553.9	27.742	1500	10382	7403	38.372
340	2317.2	1642.0	28.130	1600	11093	7915	38.830
360	2446.8	1731.9	28.501	1700	11807	8431	39.264
380	2577.8	1823.2	28.856	1800	12527	8952	39.675
400	2710.2	1915.8	29.195	1900	13251	9478	40.067
420	2843.7	2009.6	29.520	2000	13980	10008	40.441
440	2978.1	2104.3	29.833	2100	14715	10544	40.799
460	3113.5	2200.0	30.133	2200	15454	11086	41.143
480	3249.4	2296.2	30.424	2300	16200	11632	41.475
500	3386.1	2393.2	30.703	2400	16951	12185	41.794
520	3523.3	2490.6	30.972	2500	17707	12743	42.104
537	3640.3	2573.9	31.194	2600	18470	13306	42.403
540	3660.9	2588.5	31.232	2700	19238	13876	42.692
560	3798.8	2686.7	31.482	2800	20012	14451	42.973
580	3937.1	2785.3	31.724	2900	20792	15033	43.247
600	4075.6	2884.1	31.959	3000	21577	15619	43.514
620	4214.3	2983.1	32.187	3100	22368	16212	43.773
640	4353.1	3082.1	32.407	3200	23164	16809	44.026
660	4492.1	3181.4	32.621	3300	23966	17412	44.273
680	4631.1	3280.7	32.829	3400	24772	18020	44.513
700	4770.2	3380.1	33.031	3500	25583	18632	44.748
720	4909.5	3479.6	33.226	3600	26399	19249	44.978
740	5048.8	3579.2	33.417	3700	27219	19871	45.203
760	5188.1	3678.8	33.603	3800	28043	20497	45.423
780	5327.6	3778.6	33.784	3900	28871	21126	45.638
800	5467.1	3878.4	33.961	4000	29704	21760	45.849
820	5606.7	3978.3	34.134	4100	30540	22398	46.056
840	5746.3	4078.2	34.302	4200	31380	23039	46.257
860	5885.9	4178.0	34.466	4300	32224	23684	46.456
880	6025.6	4278.0	34.627	4400	33071	24333	46.651
900	6165.3	4378.0	34.784	4500	33922	24985	46.842
920	6305.1	4478.1	34.938	4600	34776	25641	47.030
940	6444.9	4578.1	35.087	4700	35633	26299	47.215
960	6584.7	4678.3	35.235	4800	36493	26961	47.396
980	6724.6	4778.4	35.379	4900	37357	27626	47.574
1000	6865	4879	35.520	5000	38223	28294	47.749
1100	7565	5380	36.188	5100	39093	28965	47.921
1200	8266	5883	36.798	5200	39965	29639	48.090
1300	8969	6387	37.360	5300	40840	30315	48.257

Table C-6 Values of the enthalpy of formation, $\bar{h}_F$, Gibbs function of formation, $\bar{g}_F$, and absolute entropy, at 25°C (77°F) at 1 atm[a]

Substance	Formula	$\bar{h}_F$ Btu/lbmole	$\bar{g}_F$ Btu/lbmole	Absolute entropy Btu/lbmole-R
Carbon	C(s)	0	0	1.36
Hydrogen	H_2(g)	0	0	31.21
Nitrogen	N_2(g)	0	0	45.77
Oxygen	O_2(g)	0	0	49.00
Carbon monoxide	CO(g)	−47,550	−59,010	47.21
Carbon dioxide	CO_2(g)	−169,300	−169,680	51.07
Water	H_2O(g)	−104,040	−98,350	45.11
Water	H_2O(l)	−122,970	−102.040	16.71
Methane	CH_4(g)	−32,210	−21,860	44.49
Acetylene	C_2H_2(g)	+97,540	+89,990	48.00
Ethylene	C_2H_4(g)	+22,490	+29,310	52.45
Ethane	C_2H_6(g)	−36,430	−14,150	54.85
Propane	C_3H_8(g)	−44,680	−10,110	64.51
n-Octane	C_8H_{18}(g)	−89,680	+7,110	111.55
n-Octane	C_8H_{18}(l)	−107,530	+2,840	86.23

[a] From the JANAF Thermochemical Tables, Dow Chemical Co., 1971; *Selected Values of Chemical Thermodynamic Properties*, NBS Tech Note 270-3, 1968; and *API Research Project* 44, Carnegie Press, 1953.

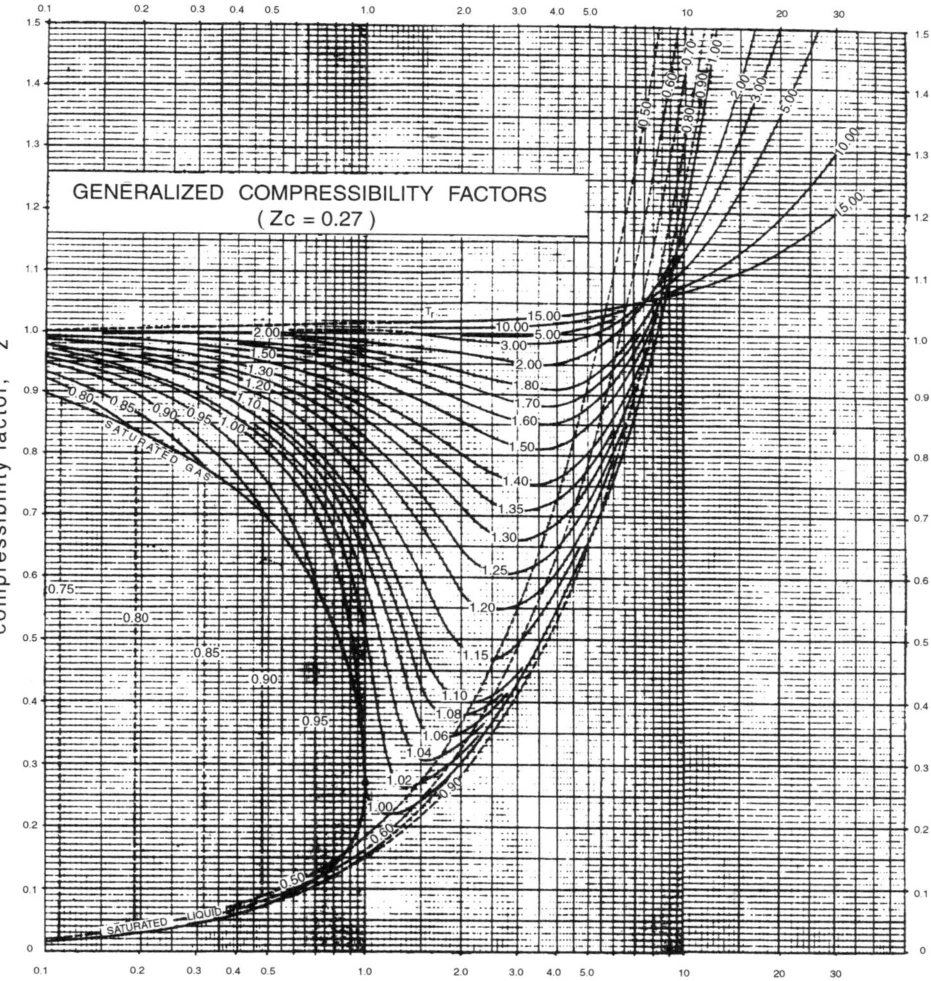

Figure C-1 Generalized Compressibility Chart. (From G. J. Van Wylen and R. E. Sonntag, *Fundamentals of Classical Thermodynamics*, 2nd ed., John Wiley & Sons, New York, 1973.)

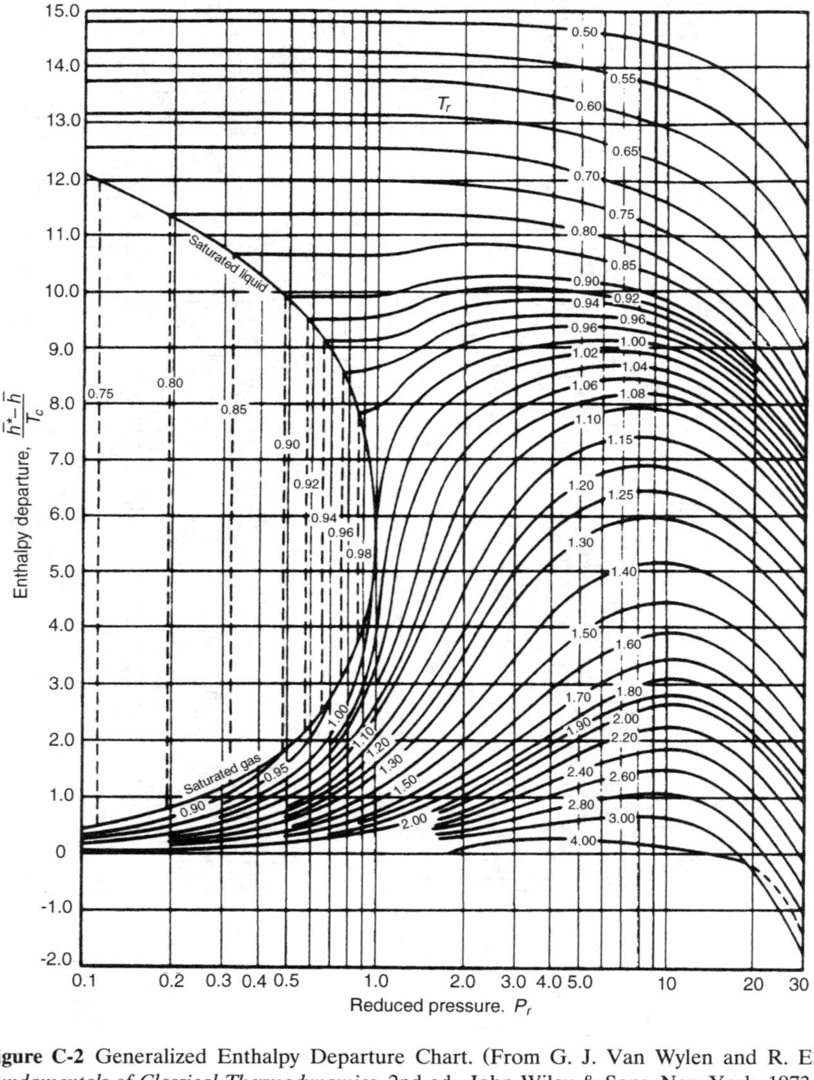

Figure C-2 Generalized Enthalpy Departure Chart. (From G. J. Van Wylen and R. E. Sonntag, *Fundamentals of Classical Thermodynamics*, 2nd ed., John Wiley & Sons, New York, 1973.)

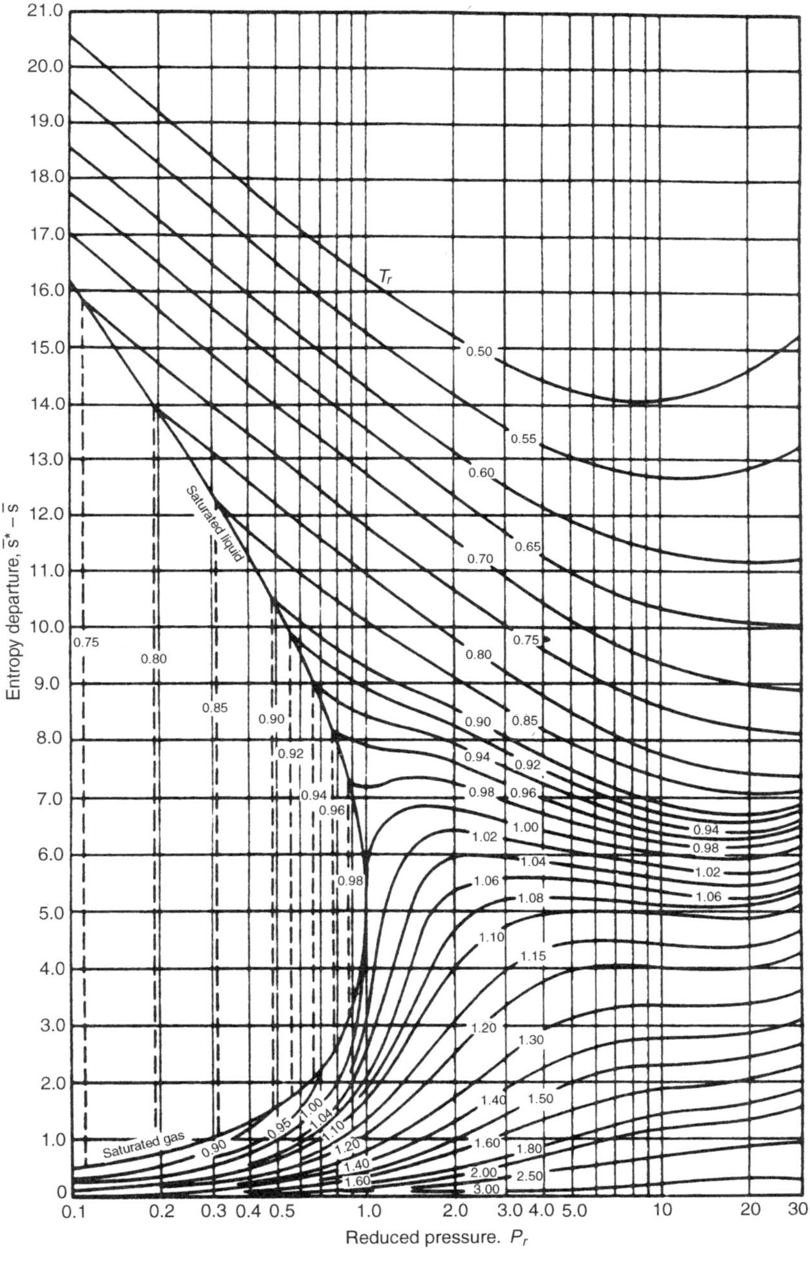

Figure C-3 Generalized Entropy Departure Chart. (From G. J. Van Wylen and R. E. Sonntag, Fundamentals of Classical Thermodynamics, 2nd ed., John Wiley & Sons, New York, 1973.)

APPENDIX

D

FLOW AND NONFLOW AVAILABILITY FOR AIR

	14.7 PSIA				50.0 PSIA			
T (°R)	h (Btu/lbm)	s (Btu/lbm-R)	a (Btu/lbm)	a_f (Btu/lbm)	h (Btu/lbm)	s (Btu/lbm-R)	a (Btu/lbm)	a_f (Btu/lbm)
300	71.61	0.46007	18.099	18.099	71.61	0.37617	48.637	63.156
320	76.39	0.47551	14.593	14.593	76.39	0.39160	44.163	59.650
340	81.17	0.49001	11.590	11.590	81.17	0.40610	40.192	56.647
360	85.96	0.50368	9.031	9.031	85.96	0.41978	36.666	54.088
380	90.74	0.51662	6.871	6.871	90.74	0.43271	33.537	51.927
400	95.53	0.52989	5.065	5.065	95.53	0.44498	30.764	50.122
420	100.31	0.54056	3.583	3.583	100.31	0.45665	28.313	48.640
440	105.10	0.55169	2.390	2.390	105.10	0.46779	26.153	47.447
460	109.88	0.56233	1.465	1.465	109.88	0.47842	24.260	46.522
480	114.67	0.57252	0.781	0.781	114.67	0.48861	22.608	45.838
500	119.46	0.58229	0.321	0.321	119.46	0.49839	21.180	45.377
520	124.25	0.59169	0.066	0.066	124.25	0.50778	19.958	45.123
540	129.05	0.60073	0.002	0.002	129.05	0.51683	18.926	45.059
560	133.84	0.60945	0.115	0.115	133.84	0.52555	18.071	45.172
580	138.64	0.61787	0.393	0.393	138.64	0.53397	17.380	45.450
600	143.44	0.62601	0.823	0.823	143.44	0.54211	16.843	45.880
620	148.25	0.63389	1.398	1.398	148.25	0.54999	16.450	46.455
640	153.06	0.64153	2.108	2.108	153.06	0.55762	16.192	47.165
660	157.87	0.64894	2.945	2.945	157.87	0.56503	16.060	48.002
680	162.69	0.65613	3.901	3.901	162.69	0.57223	16.049	48.958
700	167.52	0.66313	4.971	4.971	167.52	0.57922	16.151	50.028
720	172.35	0.66993	6.148	6.148	172.35	0.58602	16.360	51.205
740	177.19	0.67656	7.427	7.427	177.19	0.59265	16.671	52.484
760	182.03	0.68302	8.802	8.802	182.03	0.59911	17.078	53.859
780	186.88	0.68932	10.270	10.270	186.88	0.60541	17.579	55.327
800	191.74	0.69547	11.826	11.826	191.74	0.61156	18.166	56.883
820	196.61	0.70148	13.466	13.466	196.61	0.61757	18.839	58.523
840	201.49	0.70735	15.187	15.187	201.49	0.62345	19.592	60.244

860	206.37	0.71310	16.986	16.986	206.37	0.62920	20.423	62.043
880	211.27	0.71873	18.859	18.859	211.27	0.63482	21.328	63.916
900	216.17	0.72424	20.805	20.805	216.17	0.64033	22.306	65.862
920	221.08	0.72964	22.819	22.819	221.08	0.64573	23.352	67.876
940	226.01	0.73403	24.900	24.900	226.01	0.65103	24.465	69.957
960	230.95	0.74013	27.046	27.046	230.95	0.65623	25.644	72.103
980	235.89	0.74523	29.255	29.255	235.89	0.66133	26.884	74.312
1000	240.85	0.75024	31.524	31.524	240.85	0.66634	28.186	76.581

	100.0 PSIA				150.0 PSIA			
T (°R)	h (Btu/lbm)	s (Btu/lbm-R)	a (Btu/lbm)	a_f (Btu/lbm)	h (Btu/lbm)	s (Btu/lbm-R)	a (Btu/lbm)	a_f (Btu/lbm)
300	71.61	0.32866	71.126	88.668	71.61	0.30087	85.042	103.591
320	76.39	0.34409	66.451	85.162	76.39	0.31630	80.300	100.086
340	81.17	0.35860	62.278	82.159	81.17	0.33081	76.060	97.082
360	85.96	0.37227	58.550	79.600	85.96	0.34448	72.265	94.524
380	90.74	0.38520	55.220	77.439	90.74	0.35741	68.867	92.363
400	95.53	0.39747	52.245	75.634	95.53	0.36968	65.825	90.558
420	100.31	0.40914	49.593	74.151	100.31	0.38135	63.106	89.075
440	105.10	0.42028	47.231	72.959	105.10	0.39249	60.677	87.883
460	109.88	0.43091	45.136	72.034	109.88	0.40312	58.515	86.957
480	114.67	0.44110	43.283	71.350	114.67	0.41331	56.594	86.273
500	119.46	0.45088	41.653	70.889	119.46	0.42309	54.897	85.813
520	124.25	0.46027	40.230	70.635	124.25	0.43248	53.407	85.559
540	129.05	0.46932	38.996	70.571	129.05	0.44153	52.106	85.495
560	133.84	0.47804	37.940	70.684	133.84	0.45025	50.982	85.608
580	138.64	0.48646	37.048	70.962	138.64	0.45867	50.023	85.885
600	143.44	0.49460	36.309	71.392	143.44	0.46681	49.217	86.316
620	148.25	0.50248	35.714	71.967	148.25	0.47469	48.555	86.890
640	153.06	0.51012	35.255	72.677	153.06	0.48233	48.029	87.600
660	157.87	0.51752	34.922	73.514	157.87	0.48973	47.629	88.437
680	162.69	0.52472	34.709	74.470	162.69	0.49693	47.348	89.393
700	167.52	0.53171	34.609	75.540	167.52	0.50392	47.181	90.463
720	172.35	0.53852	34.617	76.717	172.35	0.51073	47.122	91.640
740	177.19	0.54514	34.726	77.996	177.19	0.51735	47.164	92.919
760	182.03	0.55160	34.932	79.371	182.03	0.52381	47.303	94.295
780	186.88	0.55790	35.231	80.839	186.88	0.53011	47.534	95.763
800	191.74	0.56406	35.617	82.395	191.74	0.53627	47.853	97.318
820	196.61	0.57006	36.088	84.035	196.61	0.54227	48.257	98.959
840	201.49	0.57594	36.639	85.756	201.49	0.54815	48.742	100.680
860	206.37	0.58169	37.269	87.555	206.37	0.55390	49.304	102.479
880	211.27	0.58731	37.972	89.428	211.27	0.55952	49.940	104.352
900	216.17	0.59282	38.749	91.374	216.17	0.56503	50.650	106.298
920	221.08	0.59822	39.594	93.388	221.08	0.57043	51.427	108.312
940	226.01	0.60352	40.505	95.469	226.01	0.57573	52.271	110.393
960	230.95	0.60872	41.482	97.615	230.95	0.58093	53.181	112.539
980	235.89	0.61382	42.521	99.824	235.89	0.58603	54.153	114.747
1000	240.85	0.61883	43.621	102.093	240.85	0.59104	55.185	117.017

	200.0 PSIA				250.0 PSIA			
T (°R)	h (Btu/lbm)	s (Btu/lbm-R)	a (Btu/lbm)	a_f (Btu/lbm)	h (Btu/lbm)	s (Btu/lbm-R)	a (Btu/lbm)	a_f (Btu/lbm)
300	71.61	0.28115	95.127	114.180	71.61	0.26585	103.037	122.393
320	76.39	0.29659	90.351	110.674	76.39	0.28129	98.241	118.887
340	81.17	0.31109	86.077	107.671	81.17	0.29579	93.947	115.884
360	85.96	0.32476	82.248	105.112	85.96	0.30947	90.099	113.325
380	90.74	0.33769	78.817	102.951	90.74	0.32240	86.648	111.164
400	95.53	0.34996	75.742	101.146	95.53	0.33467	83.552	109.359
420	100.31	0.36164	72.989	99.663	100.31	0.34634	80.779	107.876
440	105.10	0.37277	70.526	98.471	105.10	0.35747	78.296	106.684
460	109.88	0.38341	68.331	97.546	109.88	0.36811	76.080	105.759
480	114.67	0.39359	66.377	96.862	114.67	0.37830	74.106	105.075
500	119.46	0.40337	64.646	96.401	119.46	0.38808	72.355	104.614
520	124.25	0.41277	63.122	96.147	124.25	0.39747	70.811	104.360
540	129.05	0.42181	61.787	96.083	129.05	0.40652	69.456	104.296
560	133.84	0.43053	60.630	96.196	133.84	0.41524	68.279	104.409
580	138.64	0.43895	59.637	96.474	138.64	0.42366	67.266	104.687
600	143.44	0.44709	58.798	96.904	143.44	0.43180	66.406	105.117
620	148.25	0.45497	58.102	97.479	148.25	0.43968	65.691	105.692
640	153.06	0.46261	57.542	98.189	153.06	0.44731	65.110	106.402
660	157.87	0.47002	57.109	99.026	157.87	0.45472	64.656	107.239
680	162.69	0.47721	56.795	99.982	162.69	0.46192	64.322	108.195
700	167.52	0.48420	56.594	101.052	167.52	0.46891	64.102	109.265
720	172.35	0.49101	56.501	102.229	172.35	0.47571	63.989	110.442
740	177.19	0.49763	56.510	103.508	177.19	0.48234	63.977	111.721
760	182.03	0.50409	56.615	104.883	182.03	0.48880	64.062	113.096
780	186.88	0.51040	56.813	106.351	186.88	0.49510	64.240	114.564
800	191.74	0.51655	57.098	107.907	191.74	0.50125	64.505	116.120
820	196.61	0.52256	57.469	109.547	196.61	0.50726	64.855	117.760
840	201.49	0.52843	57.919	111.268	201.49	0.51314	65.286	119.481
860	206.37	0.53418	58.448	113.067	206.37	0.51888	65.794	121.280
880	211.27	0.53981	59.051	114.940	211.27	0.52451	66.377	123.153
900	216.17	0.54531	59.726	116.886	216.17	0.53002	67.033	125.099
920	221.08	0.55072	60.470	118.900	221.08	0.53542	67.756	127.113
940	226.01	0.55601	61.281	120.981	226.01	0.54072	68.547	129.104
960	230.95	0.56121	62.157	123.127	230.95	0.54591	69.403	131.340
980	235.89	0.56631	63.095	125.336	235.89	0.55102	70.321	133.549
1000	240.05	0.57132	64.094	127.605	240.85	0.55603	71.300	135.818

	300.0 PSIA				350.0 PSIA			
T (°R)	h (Btu/lbm)	s (Btu/lbm-R)	a (Btu/lbm)	a_f (Btu/lbm)	h (Btu/lbm)	s (Btu/lbm-R)	a (Btu/lbm)	a_f (Btu/lbm)
300	71.61	0.25336	109.546	129.103	71.61	0.24279	115.076	134.777
320	76.39	0.26880	104.737	125.598	76.39	0.25823	110.257	131.271
340	81.17	0.28330	100.429	122.594	81.17	0.27273	105.940	128.268
360	85.96	0.29697	96.567	120.036	85.96	0.28640	102.068	125.709
380	90.74	0.30990	93.103	117.875	90.74	0.29934	98.594	123.549
400	95.53	0.32217	89.994	116.070	95.53	0.31161	95.476	121.744
420	100.31	0.33385	87.207	114.587	100.31	0.32328	92.679	120.261
440	105.10	0.34498	84.711	113.395	105.10	0.33441	90.174	119.068
460	109.88	0.35562	82.482	112.469	109.88	0.34505	87.935	118.143
480	114.67	0.36580	80.494	111.785	114.67	0.35524	85.937	117.459

	300.0 PSIA				350.0 PSIA			
T (°R)	h (Btu/lbm)	s (Btu/lbm-R)	a (Btu/lbm)	a_f (Btu/lbm)	h (Btu/lbm)	s (Btu/lbm-R)	a (Btu/lbm)	a_f (Btu/lbm)
500	119.46	0.37558	78.730	111.325	119.46	0.36501	84.164	116.999
520	124.25	0.38498	77.172	111.071	124.25	0.37441	82.596	116.744
540	129.05	0.39402	75.804	111.007	129.05	0.38346	81.219	116.680
560	133.84	0.40274	74.613	111.120	133.84	0.39218	80.018	116.793
580	138.64	0.41116	73.587	111.397	138.64	0.40060	78.982	117.071
600	143.44	0.41930	72.714	111.828	143.44	0.40874	78.099	117.501
620	148.25	0.42718	71.985	112.402	148.25	0.41662	77.361	118.076
640	153.06	0.43482	71.391	113.112	153.06	0.42425	76.757	118.786
660	157.07	0.44223	70.924	113.949	157.87	0.43166	76.281	119.623
680	162.69	0.44942	70.576	114.905	162.69	0.43885	75.924	120.579
700	167.52	0.45641	70.342	115.975	167.52	0.44585	75.680	121.649
720	172.35	0.46322	70.215	117.152	172.35	0.45265	75.544	122.826
740	177.19	0.46984	70.191	118.431	177.19	0.45928	75.509	124.105
760	182.03	0.47630	70.262	119.807	182.03	0.46574	75.571	125.480
780	186.88	0.48260	70.426	121.275	186.88	0.47204	75.726	126.948
800	191.74	0.48876	70.678	122.830	191.74	0.47819	75.968	128.504
820	196.61	0.49477	71.015	124.471	196.61	0.48420	76.295	130.144
840	201.49	0.50064	71.432	126.192	201.49	0.49008	76.703	131.865
860	206.37	0.50639	71.927	127.990	206.37	0.49582	77.188	133.664
880	211.27	0.51202	72.496	129.864	211.27	0.50145	77.748	135.537
900	216.17	0.51752	73.138	131.810	216.17	0.50696	78.380	137.483
920	221.08	0.52293	73.849	133.824	221.08	0.51236	79.081	139.497
940	226.01	0.52822	74.626	135.905	226.01	0.51766	79.849	141.578
960	230.95	0.53342	75.468	138.051	230.95	0.52285	80.681	143.725
980	235.89	0.53852	76.373	140.259	235.89	0.52795	81.576	145.933
1000	240.85	0.54353	77.139	142.529	240.85	0.53296	82.532	148.202

	400.0 PSIA				450.0 PSIA			
T (°R)	h (Btu/lbm)	s (Btu/lbm-R)	a (Btu/lbm)	a_f (Btu/lbm)	h (Btu/lbm)	s (Btu/lbm-R)	a (Btu/lbm)	a_f (Btu/lbm)
300	71.61	0.23364	119.883	139.692	71.61	0.22557	124.134	144.027
320	76.39	0.24908	115.056	136.186	76.39	0.24101	119.302	140.521
340	81.17	0.26358	110.732	133.183	81.17	0.25551	114.972	137.518
360	85.96	0.27725	106.853	130.624	85.96	0.26918	111.088	134.959
380	90.74	0.29019	103.372	128.463	90.74	0.28211	107.601	132.799
400	95.53	0.30246	100.246	126.658	95.53	0.29438	104.470	130.993
420	100.31	0.31413	97.443	125.175	100.31	0.30606	101.661	129.511
440	105.10	0.32526	94.930	123.983	105.10	0.31719	99.142	128.318
460	109.88	0.33590	92.684	123.057	109.88	0.32782	96.890	127.393
480	114.67	0.34609	90.679	122.374	114.67	0.33801	94.880	126.709
500	119.46	0.35586	88.898	121.913	119.46	0.34779	93.094	126.248
520	124.25	0.36526	87.324	121.659	124.25	0.35719	91.513	125.994
540	129.05	0.37430	85.939	121.595	129.05	0.36623	90.123	125.930
560	133.04	0.38302	84.732	121.708	133.84	0.37495	88.910	126.043
580	138.64	0.39144	83.688	121.986	138.64	0.38337	87.861	126.321
600	143.44	0.39958	82.798	122.416	143.44	0.39151	86.965	126.751
620	148.25	0.40746	82.052	122.991	148.25	0.39939	86.214	127.326
640	153.06	0.41510	81.442	123.701	153.06	0.40703	85.598	128.036
660	157.87	0.42251	80.958	124.537	157.87	0.41443	85.108	128.873
680	162.69	0.42970	80.594	125.494	162.69	0.42163	84.738	129.829
700	167.52	0.43670	80.343	126.564	167.52	0.42862	84.482	130.899
720	172.35	0.44350	80.199	127.741	172.35	0.43543	84.333	132.076
740	177.19	0.45013	80.158	129.020	177.19	0.44205	84.286	133.355
760	182.03	0.45659	80.212	130.395	182.03	0.44851	84.335	134.730
780	186.88	0.46289	80.360	131.863	186.88	0.45481	84.477	136.198

	400.0 PSIA				450.0 PSIA			
T (°R)	h (Btu/lbm)	s (Btu/lbm-R)	a (Btu/lbm)	a_f (Btu/lbm)	h (Btu/lbm)	s (Btu/lbm-R)	a (Btu/lbm)	a_f (Btu/lbm)
800	191.74	0.46904	80.595	133.419	191.74	0.46097	84.706	137.754
820	196.61	0.47505	80.915	135.059	196.61	0.46698	85.020	139.394
840	201.49	0.48092	81.315	136.780	201.49	0.47285	85.415	141.115
860	206.37	0.48667	81.793	138.579	206.37	0.47860	85.888	142.914
880	211.27	0.49230	82.346	140.452	211.27	0.48422	86.415	144.787
900	216.17	0.49781	82.971	142.398	216.17	0.48973	87.054	146.733
920	221.08	0.50321	83.665	144.412	221.08	0.49514	87.742	148.747
940	226.01	0.50850	84.425	146.493	226.01	0.50043	88.497	150.828
960	230.95	0.51370	85.251	148.639	230.95	0.50563	89.317	152.974
980	235.89	0.51880	86.139	150.848	235.89	0.51073	90.199	155.183
1000	240.85	0.52381	87.087	153.117	240.85	0.51574	91.142	157.452

	500.0 PSIA				550.0 PSIA			
T (°R)	h (Btu/lbm)	s (Btu/lbm-R)	a (Btu/lbm)	a_f (Btu/lbm)	h (Btu/lbm)	s (Btu/lbm-R)	a (Btu/lbm)	a_f (Btu/lbm)
300	71.61	0.21835	127.945	147.905	71.61	0.21181	131.398	151.413
320	76.39	0.23378	123.108	144.399	76.39	0.22725	126.558	147.907
340	81.17	0.24829	118.774	141.396	81.17	0.24175	122.220	144.904
360	85.96	0.26196	114.885	138.837	85.96	0.25543	118.327	142.345
380	90.74	0.27489	111.394	136.676	90.74	0.26836	114.832	140.184
400	95.53	0.28716	108.258	134.871	95.53	0.28063	111.693	138.379
420	100.31	0.29883	105.444	133.388	100.31	0.29230	108.875	136.896
440	105.10	0.30997	102.921	132.196	105.10	0.30343	106.349	135.704
460	109.88	0.32060	100.665	131.271	109.88	0.31407	104.089	134.778
480	114.67	0.33079	98.651	130.587	114.67	0.32426	102.071	134.095
500	119.46	0.34057	96.860	130.126	119.46	0.33404	100.276	133.634
520	124.25	0.34996	95.275	129.872	124.25	0.34343	98.687	133.380
540	129.05	0.35901	93.880	129.808	129.05	0.35248	97.289	133.316
560	133.84	0.36773	92.662	129.921	133.84	0.36120	96.068	133.429
580	138.64	0.37615	91.609	130.199	138.64	0.36962	95.011	133.707
600	143.44	0.38429	90.709	130.629	143.44	0.37776	94.107	134.137
620	148.25	0.39217	89.953	131.204	148.25	0.38564	93.347	134.712
640	153.06	0.39980	89.332	131.914	153.06	0.39327	92.723	135.422
660	157.87	0.40721	88.838	132.751	157.87	0.40068	92.225	136.258
680	162.69	0.41441	88.464	133.707	162.69	0.40788	91.847	137.215
700	167.52	0.42140	88.203	134.777	167.52	0.41487	91.583	138.285
720	172.35	0.42821	88.049	135.954	172.35	0.42167	91.426	139.462
740	177.19	0.43483	87.998	137.233	177.19	0.42830	91.370	140.741
760	182.03	0.44129	88.042	138.608	182.03	0.43476	91.411	142.116
780	186.88	0.44759	88.180	140.076	186.88	0.44106	91.545	143.584
800	191.74	0.45374	88.405	141.632	191.74	0.44721	91.766	145.140
820	196.61	0.45975	88.715	143.272	196.61	0.45322	92.072	146.780
840	201.49	0.46563	89.105	144.993	201.49	0.45910	92.459	148.501
860	206.37	0.47138	89.573	146.792	206.37	0.46484	92.923	150.300
880	211.27	0.47700	90.115	148.665	211.27	0.47047	93.462	152.173
900	216.17	0.48251	90.731	150.611	216.17	0.47598	94.074	154.119
920	221.08	0.48791	91.414	152.625	221.08	0.48138	94.754	156.133
940	226.01	0.49321	92.164	154.706	226.01	0.48668	95.500	158.214
960	230.95	0.49841	92.980	156.852	230.95	0.49187	96.312	160.360
980	235.89	0.50351	93.858	159.061	235.89	0.49697	97.186	162.569
1000	240.85	0.50852	94.796	161.330	240.85	0.50198	98.121	164.838

	600.0 PSIA				650.0 PSIA			
T (°R)	h (Btu/lbm)	s (Btu/lbm-R)	a (Btu/lbm)	a_f (Btu/lbm)	h (Btu/lbm)	s (Btu/lbm-R)	a (Btu/lbm)	a_f (Btu/lbm)
300	71.61	0.20585	134.555	154.615	71.61	0.20036	137.462	157.561
320	76.39	0.22129	129.711	151.110	76.39	0.21580	132.616	154.056
340	81.17	0.23579	125.370	148.106	81.17	0.23030	128.273	151.052
360	85.96	0.24946	121.475	145.548	85.96	0.24398	124.374	148.494
380	90.74	0.26239	117.977	143.387	90.74	0.25691	120.874	146.333
400	95.53	0.27467	114.834	141.582	95.53	0.26918	117.728	144.528
420	100.31	0.28634	112.014	140.099	100.31	0.28085	114.906	143.045
440	105.10	0.29747	109.484	138.907	105.10	0.29198	112.373	141.853
460	109.88	0.30811	107.221	137.981	109.88	0.30262	110.108	140.927
480	114.67	0.31830	105.200	137.297	114.67	0.31281	108.084	140.243
500	119.46	0.32807	103.402	136.837	119.46	0.32259	106.284	139.783
520	124.25	0.33747	101.811	136.583	124.25	0.33198	104.689	139.529
540	129.05	0.34651	100.409	136.519	129.05	0.34103	103.285	139.465
560	133.84	0.35523	99.185	136.632	133.84	0.34975	102.059	139.578
580	138.64	0.36365	98.125	136.909	138.64	0.35817	100.996	139.855
600	143.44	0.37179	97.218	137.340	143.44	0.36631	100.086	140.286
620	148.25	0.37967	96.455	137.914	148.25	0.37419	99.321	140.860
640	153.06	0.38731	95.828	138.624	153.06	0.38182	98.691	141.570
660	157.87	0.39472	95.327	139.461	157.87	0.38923	98.188	142.407
680	162.69	0.40191	94.946	140.417	162.69	0.39643	97.804	143.363
700	167.52	0.40890	94.678	141.487	167.52	0.40342	97.534	144.433
720	172.35	0.41571	94.518	142.664	172.35	0.41022	97.371	145.610
740	177.19	0.42234	94.460	143.943	177.19	0.41685	97.310	146.889
760	182.03	0.42880	94.498	145.319	182.03	0.42331	97.346	148.265
780	186.88	0.43510	94.628	146.786	186.88	0.42961	97.474	149.733
800	191.74	0.44125	94.847	148.342	191.74	0.43576	97.689	151.288
820	196.61	0.44726	95.150	149.983	196.61	0.44177	97.990	152.929
840	201.49	0.45313	95.533	151.704	201.49	0.44765	98.371	154.650
860	206.37	0.45888	95.995	153.502	206.37	0.45339	98.830	156.449
880	211.27	0.46451	96.530	155.376	211.27	0.45902	99.363	158.322
900	216.17	0.47002	97.139	157.322	216.17	0.46453	99.969	160.268
920	221.08	0.47542	97.816	159.336	221.08	0.46993	100.643	162.282
940	226.01	0.48071	98.559	161.417	226.01	0.47523	101.384	164.363
960	230.95	0.48591	99.368	163.563	230.95	0.48042	102.190	166.509
980	235.89	0.49101	100.239	165.771	235.89	0.48553	103.059	168.717
1000	240.85	0.49602	101.171	168.041	240.85	0.49053	103.988	170.987

	700.0 PSIA				750.0 PSIA			
T (°R)	h (Btu/lbm)	s (Btu/lbm-R)	a (Btu/lbm)	a_f (Btu/lbm)	h (Btu/lbm)	s (Btu/lbm-R)	a (Btu/lbm)	a_f (Btu/lbm)
300	71.61	0.19528	140.156	160.289	71.61	0.19056	142.667	162.828
320	76.39	0.21072	135.308	156.783	76.39	0.20599	137.817	159.323
340	81.17	0.22522	130.963	153.780	81.17	0.22049	133.469	156.319
360	85.96	0.23890	127.062	151.221	85.96	0.23417	129.567	153.761
380	90.74	0.25183	123.559	149.061	90.74	0.24710	126.062	151.600
400	95.53	0.26410	120.412	147.256	95.53	0.25937	122.913	149.795
420	100.31	0.27577	117.587	145.773	100.31	0.27104	120.086	148.312
440	105.10	0.28690	115.052	144.580	105.10	0.28218	117.549	147.120
460	109.88	0.29754	112.784	143.655	109.88	0.29281	115.280	146.194
480	114.67	0.30773	110.758	142.971	114.67	0.30300	113.252	145.510
500	119.46	0.31751	108.956	142.511	119.46	0.31278	111.447	145.050
520	124.25	0.32690	107.359	142.256	124.25	0.32217	109.849	144.796

540	129.05	0.33595	105.953	142.192	129.05	0.33122	108.441	144.732
560	133.84	0.34467	104.724	142.305	133.84	0.33994	107.210	144.845
580	138.64	0.35309	103.659	142.583	138.64	0.34836	106.143	145.122
600	143.44	0.36123	102.748	143.013	143.44	0.35650	105.229	145.553
620	148.25	0.36911	101.980	143.588	148.25	0.36438	104.460	146.127
640	153.06	0.37674	101.348	144.298	153.06	0.37201	103.826	146.837
660	157.87	0.38415	100.842	145.135	157.87	0.37942	103.319	147.674
680	162.69	0.39135	100.457	146.091	162.69	0.38662	102.931	148.630
700	167.52	0.39834	100.184	147.161	167.52	0.39361	102.656	149.700
720	172.35	0.40514	100.019	148.338	172.35	0.40041	102.489	150.877
740	177.19	0.41177	99.956	149.617	177.19	0.40704	102.424	152.156
760	182.03	0.41823	99.989	150.992	182.03	0.41350	102.455	153.532
780	186.88	0.42453	100.115	152.460	186.88	0.41980	102.579	155.000
800	191.74	0.43068	100.328	154.016	191.74	0.42595	102.791	156.555
820	196.61	0.43669	100.627	155.656	196.61	0.43196	103.087	158.196
840	201.49	0.44257	101.005	157.377	201.49	0.43784	103.464	159.917
860	206.37	0.44831	101.462	159.176	206.37	0.44359	103.919	161.716
880	211.27	0.45394	101.993	161.049	211.27	0.44921	104.448	163.589
900	216.17	0.45945	102.597	162.995	216.17	0.45472	105.050	165.535
920	221.08	0.46485	103.269	165.009	221.08	0.46012	105.720	167.549
940	226.01	0.47015	104.007	167.090	226.01	0.46542	106.457	169.630
960	230.95	0.47534	104.811	169.236	230.95	0.47062	107.259	171.776
980	235.89	0.48045	105.678	171.445	235.89	0.47572	108.123	173.984
1000	240.85	0.48546	106.605	173.714	240.85	0.48073	109.048	176.254

	800.0 PSIA				850.0 PSIA			
T (°R)	h (Btu/lbm)	s (Btu/lbm-R)	a (Btu/lbm)	a_f (Btu/lbm)	h (Btu/lbm)	s (Btu/lbm-R)	a (Btu/lbm)	a_f (Btu/lbm)
300	71.61	0.18613	145.017	165.204	71.61	0.18198	147.226	167.435
320	76.39	0.20157	140.165	161.698	76.39	0.19741	142.373	163.929
340	81.17	0.21607	135.816	158.695	81.17	0.21192	138.022	160.926
360	85.96	0.22974	131.912	156.136	85.96	0.22559	134.117	158.368
380	90.74	0.24268	128.405	153.975	90.74	0.23852	130.609	156.207
400	95.53	0.25495	125.255	152.170	95.53	0.25079	127.456	154.402
420	100.31	0.26662	122.426	150.687	100.31	0.26246	124.626	152.919
440	105.10	0.27775	119.888	149.495	105.10	0.27360	122.087	151.726
460	109.88	0.28839	117.616	148.569	109.88	0.28423	119.814	150.801
480	114.67	0.29858	115.587	147.886	114.67	0.29442	117.783	150.117
500	119.46	0.30835	113.781	147.425	119.46	0.30420	115.975	149.657
520	124.25	0.31775	112.181	147.171	124.25	0.31359	114.374	149.402
540	129.05	0.32680	110.771	147.107	129.05	0.32264	112.962	149.338
560	133.04	0.33552	109.538	147.220	133.84	0.33136	111.728	149.452
580	138.64	0.34394	108.470	147.498	138.64	0.33978	110.658	149.729
600	143.44	0.35208	107.554	147.928	143.44	0.34792	109.741	150.159
620	148.25	0.35996	106.783	148.503	148.25	0.35580	108.969	150.734
640	153.06	0.36759	106.148	149.213	153.06	0.36344	108.332	151.444
660	157.87	0.37500	105.639	150.049	157.87	0.37084	107.821	152.281
680	162.69	0.38219	105.249	151.006	162.69	0.37804	107.430	153.237
700	167.52	0.38919	104.973	152.076	167.52	0.38503	107.152	154.307
720	172.35	0.39599	104.804	153.253	172.35	0.39184	106.982	155.484
740	177.19	0.40262	104.738	154.532	177.19	0.39846	106.914	156.763
760	182.03	0.40908	104.767	155.907	182.03	0.40492	106.942	158.138
780	186.88	0.41538	104.889	157.375	186.88	0.41122	107.063	159.606
800	191.74	0.42153	105.099	158.931	191.74	0.41738	107.271	161.162

820	196.61	0.42754	105.394	160.571	196.61	0.42338	107.565	162.803
840	201.49	0.43342	105.769	162.292	201.49	0.42926	107.938	164.523
860	206.37	0.43916	106.222	164.091	206.37	0.43501	108.390	166.322
880	211.27	0.44479	106.749	165.964	211.27	0.44063	108.915	168.195
900	216.17	0.45030	107.350	167.910	216.17	0.44614	109.514	170.141
920	221.08	0.45570	108.018	169.924	221.08	0.45154	110.181	172.155
940	226.01	0.46100	108.753	172.005	226.01	0.45684	110.915	174.236
960	230.95	0.46619	109.553	174.151	230.95	0.46204	111.714	176.383
980	235.89	0.47129	110.416	176.360	235.89	0.46714	112.575	178.591
1000	240.05	0.47630	111.340	178.629	240.85	0.47215	113.497	180.860

	900.0 PSIA				950.0 PSIA			
T (°R)	h (Btu/lbm)	s (Btu/lbm-R)	a (Btu/lbm)	a_f (Btu/lbm)	h (Btu/lbm)	s (Btu/lbm-R)	a (Btu/lbm)	a_f (Btu/lbm)
300	71.61	0.17806	149.310	169.539	71.61	0.17435	151.282	171.529
320	76.39	0.19350	144.456	166.033	76.39	0.18979	146.427	168.023
340	81.17	0.20800	140.104	163.030	81.17	0.20429	142.074	165.020
360	85.96	0.22167	136.197	160.471	85.96	0.21797	138.165	162.461
380	90.74	0.23460	132.687	158.311	90.74	0.23090	134.655	160.300
400	95.53	0.24687	129.534	156.505	95.53	0.24317	131.500	158.495
420	100.31	0.25855	126.702	155.023	100.31	0.25484	128.667	157.013
440	105.10	0.26968	124.161	153.830	105.10	0.26597	126.125	155.820
460	109.88	0.28032	121.887	152.905	109.88	0.27661	123.850	154.895
480	114.67	0.29051	119.855	152.221	114.67	0.28680	121.817	154.211
500	119.46	0.30028	118.046	151.760	119.46	0.29658	120.006	153.750
520	124.25	0.30968	116.443	151.506	124.25	0.30597	118.402	153.496
540	129.05	0.31872	115.030	151.442	129.05	0.31502	116.989	153.432
560	133.84	0.32744	113.795	151.555	133.84	0.32374	115.752	153.545
580	138.64	0.33586	112.724	151.833	138.64	0.33216	114.679	153.823
600	143.44	0.34400	111.806	152.263	143.44	0.34030	113.760	154.253
620	148.25	0.35188	111.032	152.838	148.25	0.34818	112.985	154.828
640	153.06	0.35952	110.393	153.548	153.06	0.35581	112.345	155.538
660	157.07	0.36693	109.881	154.385	157.87	0.36322	111.832	156.375
680	162.69	0.37412	109.489	155.341	162.69	0.37041	111.439	157.331
700	167.52	0.38111	109.210	156.411	167.52	0.37741	111.159	158.401
720	172.35	0.38792	109.039	157.588	172.35	0.38421	110.986	159.578
740	177.19	0.39454	108.969	158.867	177.19	0.39084	110.916	160.857
760	182.03	0.40100	108.996	160.242	182.03	0.39730	110.941	162.232
780	186.88	0.40731	109.115	161.710	186.88	0.40360	111.059	163.700
800	191.74	0.41346	109.322	163.266	191.74	0.40975	111.265	165.256
820	196.61	0.41947	109.614	164.906	196.61	0.41576	111.556	166.896
840	201.49	0.42534	109.987	166.627	201.49	0.42164	111.927	168.617
860	206.37	0.43109	110.437	168.426	206.37	0.42738	112.376	170.416
880	211.27	0.43672	110.961	170.299	211.27	0.43301	112.899	172.289
900	216.17	0.44223	111.559	172.245	216.17	0.43852	113.496	174.235
920	221.08	0.44763	112.224	174.259	221.08	0.44392	114.160	176.249
940	226.01	0.45292	112.957	176.340	226.01	0.44922	114.891	178.330
960	230.95	0.45812	113.754	178.486	230.95	0.45441	115.688	180.476
980	235.89	0.46322	114.614	180.695	235.89	0.45951	116.546	182.685
1000	240.85	0.46823	115.535	182.964	240.85	0.46452	117.466	184.954

	1000.0 PSIA				1050.0 PSIA			
T (°R)	h (Btu/lbm)	s (Btu/lbm-R)	a (Btu/lbm)	a_f (Btu/lbm)	h (Btu/lbm)	s (Btu/lbm-R)	a (Btu/lbm)	a_f (Btu/lbm)
300	71.61	0.17084	153.154	173.417	71.61	0.16749	154.936	175.213
320	76.39	0.18628	148.298	169.911	76.39	0.18293	150.078	171.707
340	81.17	0.20078	143.943	166.908	81.17	0.19743	145.723	168.703
360	85.96	0.21445	140.034	164.349	85.96	0.21111	141.813	166.145
380	90.74	0.22738	136.523	162.188	90.74	0.22404	138.300	163.984
400	95.53	0.23965	133.367	160.383	95.53	0.23631	135.143	162.179
420	100.31	0.25133	130.533	158.900	100.31	0.24798	132.309	160.696
440	105.10	0.26246	127.990	157.708	105.10	0.25911	129.765	159.504
460	109.88	0.27309	125.713	156.782	109.88	0.26975	127.487	158.578
480	114.67	0.28328	123.679	156.099	114.67	0.27994	125.452	157.895
500	119.46	0.29306	121.868	155.638	119.46	0.28972	123.640	157.434
520	124.25	0.30246	120.263	155.384	124.25	0.29911	122.034	157.180
540	129.05	0.31150	118.848	155.320	129.05	0.30816	120.618	157.116
560	133.84	0.32022	117.610	155.433	133.84	0.31688	119.379	157.229
580	138.64	0.32864	116.537	155.711	138.64	0.32530	118.305	157.506
600	143.44	0.33678	115.616	156.141	143.44	0.33344	117.383	157.937
620	148.25	0.34466	114.840	156.716	148.25	0.34132	116.606	158.512
640	153.06	0.35230	114.199	157.426	153.06	0.34895	115.965	159.222
660	157.87	0.35970	113.685	158.262	157.87	0.35636	115.449	160.058
680	162.69	0.36690	113.291	159.219	162.69	0.36356	115.054	161.015
700	167.52	0.37389	113.010	160.289	167.52	0.37055	114.772	162.084
720	172.35	0.38070	112.836	161.466	172.35	0.37735	114.597	163.261
740	177.19	0.38732	112.764	162.745	177.19	0.38398	114.525	164.541
760	182.03	0.39378	112.789	164.120	182.03	0.39044	114.548	165.916
780	186.88	0.40008	112.906	165.588	186.88	0.39674	114.664	167.384
800	191.74	0.40624	113.111	167.144	191.74	0.40289	114.868	168.939
820	196.61	0.41225	113.400	168.784	196.61	0.40890	115.157	170.580
840	201.49	0.41812	113.770	170.505	201.49	0.41478	115.526	172.301
860	206.37	0.42387	114.218	172.304	206.37	0.42052	115.973	174.100
880	211.27	0.42949	114.741	174.177	211.27	0.42615	116.494	175.973
900	216.17	0.43500	115.336	176.123	216.17	0.43166	117.088	177.919
920	221.08	0.44041	115.999	178.137	221.08	0.43706	117.751	179.933
940	226.01	0.44570	116.729	180.218	226.01	0.44236	118.480	182.014
960	230.95	0.45090	117.525	182.364	230.95	0.44755	119.274	184.160
980	235.89	0.45600	118.382	184.573	235.89	0.45266	120.131	186.368
1000	240.85	0.46101	119.301	186.842	240.85	0.45766	121.049	188.638

The reference temperature is 537.0°R
The reference pressure is 14.7 psia

INDEX